SYSTEM IDENTIFICATION

This is Volume 80 in
MATHEMATICS IN SCIENCE AND ENGINEERING
A series of monographs and textbooks
Edited by RICHARD BELLMAN, *University of Southern California*

A complete list of the books in this series appears at the end of this volume.

SYSTEM IDENTIFICATION

ANDREW P. SAGE

and JAMES L. MELSA

Information and Control Sciences Center
Institute of Technology
Southern Methodist University
Dallas, Texas

1971

ACADEMIC PRESS New York and London

ACADEMIC PRESS, INC.
111 Fifth Avenue, New York, New York 10003

United Kingdom Edition published by
ACADEMIC PRESS, INC. (LONDON) LTD.
Berkeley Square House, London W1X 6BA

LIBRARY OF CONGRESS CATALOG CARD NUMBER: 76-137606

AMS (MOS) 1970 Subject Classification 93B30

PRINTED IN THE UNITED STATES OF AMERICA

To LaVerne

APS

To Kathy

JLM

CONTENTS

PREFACE

In the last several years, much interest has been devoted to a study of modern system theory. An essential ingredient of many modern system theory problems is the requirement to accomplish system identification or modeling. As defined here, system identification or modeling is the process of determining a difference or differential equation (or the coefficient parameters of such an equation) such that it describes a physical process in accordance with some predetermined criterion. There are certainly broader definitions of system identification although they are not considered here.

The purpose of this text is to survey many, but certainly not all, of the existing techniques for system identification and to give reasonably in-depth treatment to modern digital computer oriented methods for system identification. These methods are based on optimization and estimation theory as developed in the original works of Bellman, Bryson, Bucy, Ho, Kalman, Middleton, Pontryagin, and others.

In order to keep the size of the text reasonable it has been necessary to assume that the reader is familiar with some of the fundamental concepts of optimization and estimation theory such as presented in Chapters 2–4, 6, and 8–10 of "Optimum Systems Control" (Sage, 1968) or Chapters 5–7 and 9 of "Estimation Theory with Applications to Communications and Control" (Sage and Melsa, 1971) or equivalent material. The text is presented in such a way that prior study of computational techniques for optimization, estimation, and identification is unnecessary.

Hopefully the text will be of interest to systems engineers in control, operations research, mathematical modeling, biomedical modeling, and other disciplines in which the determination of system models from observed data is of interest.

ACKNOWLEDGMENTS

The inspiring leadership of Dean Thomas L. Martin, Jr., of the SMU Institute of Technology has provided the atmosphere which made possible this effort and for this the authors express their heartfelt gratitude.

The original research of the authors upon which this text is based has been supported in part by the Air Force Office of Scientific Research, Office of Aerospace Research, under Contract F44620-68-0023. The authors wish to especially thank Project Scientist Dr. Joshua H. Rosenbloom for encouragement of the authors' research efforts and that of their students.

The authors also wish to thank Mary Lou Caruthers, Tommie Dawson, Patsy Wuensche, and Carolyn Hughes who patiently typed the manuscript.

SYSTEM IDENTIFICATION

1

INTRODUCTION

Many problems in systems engineering, particularly those associated with systems control, can be subdivided into four interrelated parts:

1. goal identification;
2. determination of position with respect to the goal;
3. determination of environmental factors affecting the past, present, and future, and consequent establishment of a system model;
4. determination of a policy in accordance with the goal definition (1), knowledge of the current state (2), and system model and environment (3).

Often the policy in Step (4) is determined in an optimum fashion, in which case the subject of optimum systems control arises.

In this text we will be concerned with Step (3) of the aforementioned four steps in a systems problem. In particular, we will concern ourselves with the determination of system models from records of system operation. Our problem in general terms is this: We observe a noise corrupted version of a system state vector $\mathbf{x}(t)$, input signal $\mathbf{u}(t)$, and input disturbance $\mathbf{w}(t)$

$$\mathbf{z}(t) = \mathbf{h}[\mathbf{x}(t), \mathbf{u}(t), \mathbf{w}(t), \mathbf{p}(t), \mathbf{v}(t), t]$$

In this observation model, $\mathbf{p}(t)$ represents unknown parameters of the system and $\mathbf{v}(t)$ represents observation noise. The state vector $\mathbf{x}(t)$ is assumed to evolve from the stochastic differential equation

$$d\mathbf{x}(t)/dt = \mathbf{f}[\mathbf{x}(t), \mathbf{u}(t), \mathbf{w}(t), \mathbf{p}(t), t]$$

In general, the order of the differential equation is not known, although for most of the identification schemes we propose, an assumed model of known order will be chosen. This general identification problem is illustrated schematically in Fig. 1–1. Solution of the system

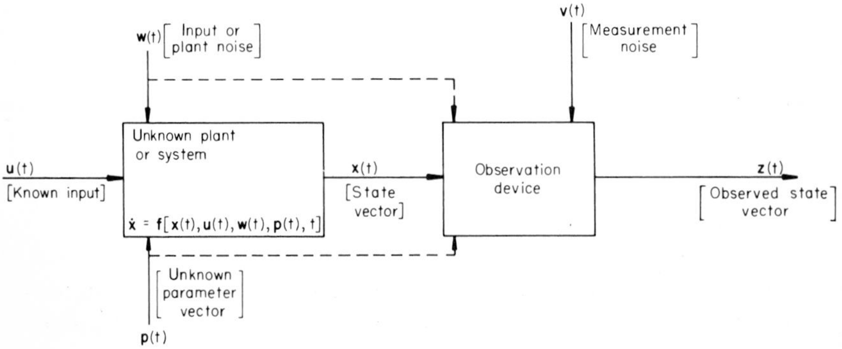

FIG. 1-1. General system identification problem.

identification problem will consist of determination of estimates of the unknown parameter vector $\mathbf{p}(t)$ and the order of $\mathbf{f}$ if it is unknown. The parameter vector $\mathbf{p}(t)$ may consist of coefficients of the system differential equation as well as mean and variance coefficients of the system noise $\mathbf{w}(t)$ and observation noise $\mathbf{v}(t)$.

Several subclasses of this general system identification problem may easily be discerned. Some of these are:

1. Noise free identification in which $\mathbf{w}(t)$ and $\mathbf{v}(t)$ are not present. This is the simplest type of system identification problem in which there is a known input $\mathbf{u}(t)$ and perfect observation of a function of the state vector.
2. The system and observation models are linear.
3. No observation of the input noise $\mathbf{w}(t)$ is possible.

We will consider some of these subclasses in the discussions which follow.

In general, it is neither possible nor desirable to determine all of the characteristics of a process for which we desire a system model. The characteristics which we must know are determined by the goal we have set for the system, the allowable degree of complexity of the overall system, and its associated computational requirements.

In classical systems design, modeling was accomplished once during the design stage, and this is still an important part of system identification and modeling. In many systems today, on line or recursive identification is desirable so as to make optimum adaptation to the system goal possible in the face of environmental uncertainty and changing environmental conditions. We will, therefore, develop methods of system identification which are useful both for on line as well as for off line use. Structurally, we may divide our efforts to follow in three classes:

1. "classical" methods of system identification,
2. cost functions for system identification,
3. computational techniques for system identification.

Chapter 2 will concern itself with a survey of various classical methods of system identification. We will discuss the simple time-honored problem of sinusoidal response testing both for the elementary case of constant coefficient linear systems as well as for the considerably more complex case where the system is linear but time varying.

A discussion of the identification of the system impulse response by inserting a white noise input to the system and crosscorrelating the system output with a delayed version of the system input will be followed by an application of the Kalman filter equations to the determination of the impulse response of a time-invariant linear system. A brief discussion of the use of the Wiener analytical theory of nonlinear systems will be presented, as well as a critical examination of the learning model approach to identification, which has proven to be very popular in identification for the model reference type of adaptive control.

Chapter 3 will be devoted entirely to the formulation of cost functions for system identification. Of particular concern will be maximum a posteriori (or Bayesian maximum likelihood) identification and classical maximum likelihood estimation. In maximum likelihood estimation, we desire to maximize the probability density of the observation sequence, conditioned upon certain constant but unknown parameters, by optimum choice of these parameters. In maximum a posteriori identification, we desire to maximize the probability density function of the unknown random parameters conditioned upon the output observation sequence. The also important subject of minimum variance or conditional mean identification will not be discussed as such in this chapter since not much is known at present regarding computational solutions for conditional mean estimates. It will be shown, however, that conditional mean state estimates result from the maximum likelihood cost function. Chapter 7 will also show the close relation between approximate sequential conditional mean identification algorithms and sequential maximum a posteriori identification algorithms for a fairly general class of problems.

Chapters 4–7 will discuss various methods for the computational solution of system identification problems. Chapter 4 will present direct computational methods based on the first and second variation. First- and second-order gradient methods, as well as the conjugate gradient method, will be developed for single-stage, multistage, and continuous systems.

Chapter 5 will present the first-order stochastic gradient or stochastic approximation method. An in depth study of the identification of linear system dynamics using stochastic approximation will be presented.

Chapters 6 and 7 will discuss two direct computational methods for system identification. Discrete and continuous Newton–Raphson methods or quasilinearization will be developed and applied to various system identification problems. Chapter 7 will exploit the discrete and continuous invariant imbedding methods which yield approximate sequential solutions for the two-point boundary value problems and cost functions of Chapter 3.

An extensive bibliography of contemporary research works which concern system identification will be given as the concluding item in our efforts. In order to improve usability, the bibliography will be keyed to the pertinent section of the text to which it refers.

2

CLASSICAL METHODS OF SYSTEM IDENTIFICATION

2.1. Introduction

In this chapter, we consider a number of the so-called "classical" methods of system identification. The terminology "classical" perhaps carries a connotation of obsolescence; this is *not* the intended meaning here. Rather, the term "classical" is used to represent "tried and proven" methods of long standing and to differentiate from the more recently developed "modern" methods which comprise the remainder of this book.

The methods of this chapter have been applied to a wide variety of practical problems with good results. The basic difficulty of the methods, with the exception of the learning model approach, is that they attack the problem in an indirect manner. This statement perhaps needs some clarification. The classical methods generally identify the impulse response or the transfer function of the system. However, what is usually desired is the basic differential equations describing the system; it is not necessarily an easy task to determine the system model from the impulse response or transfer function, particularly when they are given in graphical form, which is normally the case.

2.2. Deconvolution Methods for Impulse Response Identification

In this section, we wish to consider some simple methods for determining the impulse response of a linear, time-invariant system. For simplicity, we shall restrict our attention to single-input, single-output systems. The nature of our problem is represented in Fig. 2.2–1;

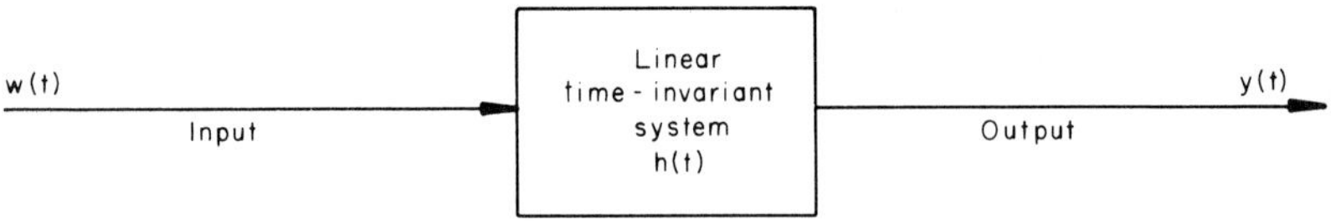

Fig. 2.2-1. Model for impulse response determination.

by the observation of the inputs and outputs of a linear, time-invariant system for a finite time $0 \leqslant t \leqslant T$, we wish to determine the impulse response of the system $h(t)$.

The above problem statement is a bit ambitious, and we will find it necessary to simplify the problem somewhat in order to make it mathematically tractable. We will only observe the input and output at N periodically sampled time values, say Δ seconds apart, in the time interval $[0, T]$, where $N\Delta = T$. On the basis of this data, we will approximately determine the impulse response at data points over the time interval $[0, T]$.

The output of the system due to an input $w(t)$ and zero initial conditions is given by the well-known convolution integral

$$y(t) = \int_0^t h(t - \tau)\, w(\tau)\, d\tau \tag{2.2-1}$$

Here we have also assumed that the input $w(\tau)$ is zero for $\tau < 0$. In addition, we require that $w(0) \neq 0$; if this restriction is not satisfied, then we may always make a finite translation of the time to achieve $w(0) \neq 0$.

Now we approximate the input time function $w(t)$ by assuming that it is a constant equal to the value at the lower sample point between each two adjacent sample points. Hence we have

$$w(t) \simeq w(n\Delta), \qquad \text{for} \quad n\Delta < t < (n+1)\Delta \tag{2.2-2}$$

We will similarly assume that $h(t)$ is constant between sample points, but in this case, we let the value be equal to the value at the midpoint of the interval so that

$$h(t) \simeq h\left(\frac{2n+1}{2}\Delta\right), \qquad \text{for} \quad n\Delta \leqslant t < (n+1)\Delta \tag{2.2-3}$$

We could also let $w(t) = 0.5\{w(n\Delta) + w[(n+1)\Delta]\}$, for $n\Delta \leqslant t < (n+1)\Delta$ or any number of other appoximations. However, if Δ is small enough, all of these approximations should lead to essentially the same answer. One of the best ways to determine if Δ is small enough is to let $\Delta' = \Delta/2$ and repeat the computations. If no appreciable change is noted, then Δ is small enough; if appreciable changes are noted, then we may let $\Delta'' = \Delta'/2$ and repeat them.

In terms of the above approximations for $w(t)$ and $h(t)$, the integral of Eq. (2.2–1) if evaluated for $t = n\Delta$ becomes

$$y(n\Delta) = \Delta \sum_{i=0}^{n-1} h\left(\frac{2n-1}{2}\Delta - i\Delta\right) w(i\Delta) \tag{2.2-4}$$

If we define the N vector of output observations as

$$\mathbf{y}^{\mathrm{T}}(T) = [y(\Delta)\, y(2\Delta) \cdots y(N\Delta)] \tag{2.2-5}$$

and the N vector of impulse response points as

$$\mathbf{h}^{\mathrm{T}}(T) = \left[h\left[\frac{\Delta}{2}\right] h\left[\frac{3\Delta}{2}\right] \cdots h\left[\frac{2N-1}{2}\Delta\right]\right] \tag{2.2-6}$$

then the expression of Eq. (2.2–4) may be written as

$$\mathbf{y}(T) = \Delta \mathbf{W}\mathbf{h}(T) \tag{2.2-7}$$

Here the matrix $\mathbf{W}$ is defined as

$$\mathbf{W} = \begin{bmatrix} w(0) & 0 & 0 & 0 & \cdots & 0 \\ w(\Delta) & w(0) & 0 & 0 & \cdots & 0 \\ w(2\Delta) & w(\Delta) & w(0) & 0 & \cdots & 0 \\ \cdots & \cdots & \cdots & \cdots & \cdots & \cdots \\ w[(N-1)\Delta] & w[(N-2)\Delta] & & \cdots & & w(0) \end{bmatrix}$$

Note the lower triangular form of $\mathbf{W}$.

The problem is now to solve Eq. (2.2-7) for $\mathbf{h}$, the vector of impulse response points. Because we required that $w(0) \neq 0$, we easily see that det $\mathbf{W} = [w(0)]^N \neq 0$ and $\mathbf{W}$ is nonsingular. Therefore we may formally write the solution of Eq. (2.3-7) as

$$\mathbf{h} = \mathbf{W}^{-1}\mathbf{y} \tag{2.2-8}$$

Due to the lower triangular form of $\mathbf{W}$, the solution for $\mathbf{h}$ can be easily written in a recursive form as

$$h_n = \frac{1}{w(0)}\left[\frac{y(n\Delta)}{\Delta} - \sum_{i=1}^{n-1} h_{n-i}w(i\Delta)\right] \tag{2.2-9}$$

where

$$h_n \triangleq h\left(\frac{2n-1}{2}\Delta\right) \tag{2.2-10}$$

and

$$h_1 = \frac{y(\Delta)}{\Delta w(0)} \tag{2.2-11}$$

Note that in Eq. (2.2-9) it is necessary to process an ever increasing amount of data in the summation. Approximately n multiplications and n summations are required to obtain h_n. This fact makes the sequential or on-line use of the algorithm impossible unless the time interval of interest is quite short. In addition, it means that numerical round-off errors will usually make the technique inaccurate if n becomes large. The procedure is nevertheless very simple and can be quite effective for many system identification problems. The fast Fourier transform may also be used to reduce the computational requirements of the numerical deconvolution technique significantly. Another advantage of this approach is that any input may be used. Hence it is not necessary to introduce special test inputs; normal operating records may be employed.

If a unit step input is used, then the algorithm of Eq. (2.2-9) can be considerably simplified. In this case $w(i\Delta) = 1$ for all i, and Eq. (2.2-9) is therefore

$$h_n = \frac{y(n\Delta)}{\Delta} - \sum_{i=1}^{n-1} h_{n-i} \tag{2.2-12}$$

We can put this expression into a particularly convenient form by defining

$$H_n = \sum_{i=1}^{n-1} h_{n-i} \tag{2.2-13}$$

so that

$$h_n = \frac{y(n\Delta)}{\Delta} - H_n \tag{2.2-14}$$

Here H_n is determined by the simple recursion relation

$$H_n = H_{n-1} + h_{n-1} \tag{2.2-15}$$

The algorithm of Eqs. (2.2-14) and (2.2-15) is now in a simple sequential form requiring only two additions and one division (by Δ) at each stage.

Example 2.2-1. Let us illustrate the use of the above results by considering a simple example. The problem is shown in Fig. 2.2-2;

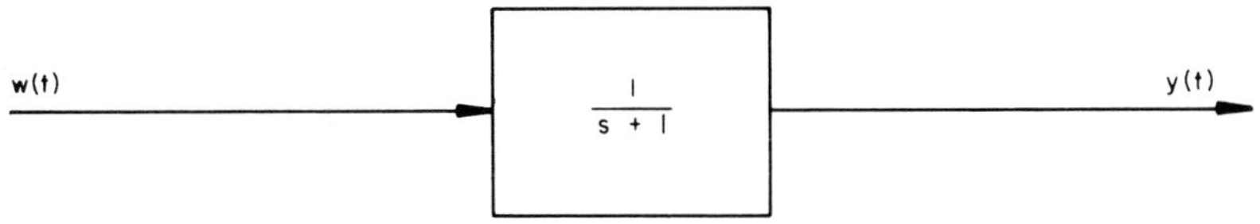

FIG. 2.2-2. Example 2.2-1.

the actual impulse response for this system is:

$$h(t) = e^{-t}$$

A unit step input is used for identification, and the output can be easily determined to be

$$y(t) = 1 - e^{-t}$$

Using $\Delta = 0.1$, the impulse response was determined over the interval $0 \leqslant t < 1$. The exact and approximate values of $h(t)$ are shown in Table 2.2-1 and indicate excellent correlation. Of course, this problem is very simple and one would hence expect good results.

In addition to the numerical difficulties noted above, one also encounters problems with this algorithm if the output measurement contain an appreciable amount of noise. Because of the one-to-one correspondence of data points and points determined on $h(t)$, there is no opportunity for noise averaging. One approach is to make a number of runs with identical inputs and use the average response to

TABLE 2.2-1

EXACT AND APPROXIMATE VALUES OF THE IMPULSE RESPONSE

	$h(t)$	
t	Exact	Approximate
0.05	0.951229	0.951625
0.15	0.860708	0.861068
0.25	0.778801	0.779125
0.35	0.704688	0.704981
0.45	0.637628	0.637894
0.55	0.576950	0.577189
0.65	0.522046	0.522264
0.75	0.472367	0.472561
0.85	0.427415	0.427594
0.95	0.386741	0.386903

compute the impulse response. If it is not possible to use the same input, then the impulse for each input–noisy-output set can be computed and averaged. This approach can require a very large amount of computation, however.

There is still another method for handling noisy output observations. However, it requires that we truncate the impulse response at some finite time. If the system is asymptotically stable so that $h(t) \to 0$ as $t \to \infty$, then the impulse response can probably be truncated without serious error. Let us suppose that $h(t) = 0$, for $t > T$, and that we take measurements of the input and output signal for $0 < t < t_f = m\Delta$, where $t_f \gg T$. Now, in terms of the above algorithm, we have a great deal of redundant information, since we need only the input and output observations for $0 \leqslant t \leqslant T$. We may use this data to improve the estimates of the impulse response. In order to illustrate how this can be done, let us consider the form of Eq. (2.2-7) given, if we make the assumption that $h(t)$ is truncated, by

$$\mathbf{y}(t_f) = \Delta \mathbf{W}_*(T)\,\mathbf{h}(T) + \mathbf{v}(t_f) \tag{2.2-16}$$

Here $\mathbf{y}(t_f)$ is

$$\mathbf{y}^T(t_f) = [y(\Delta)\, y(2\Delta) \cdots y(m\Delta)] \tag{2.2-17}$$

and

$$\mathbf{W}_*(T) = \begin{bmatrix} w(0) & 0 & \cdots & 0 \\ w(\Delta) & w(0) & \cdots & 0 \\ \cdots & \cdots & \cdots & \cdots \\ w[(N-1)\Delta] & w[(N-2)\Delta] & \cdots & w(0) \\ w(N\Delta) & w[(N-1)\Delta] & \cdots & w(\Delta) \\ \cdots & \cdots & \cdots & \cdots \\ w[(m-1)\Delta] & w[(m-2)\Delta] & \cdots & w[(m-N)\Delta] \end{bmatrix} \tag{2.2-18}$$

and $\mathbf{h}(T)$ is still defined by Eq. (2.2-6). The noise in the output measurement is assumed to be additive and zero-mean and is represented by the vector $\mathbf{v}(t_f)$ defined by

$$\mathbf{v}^T(t_f) = [v(\Delta)\ v(2\Delta) \cdots v(m\Delta)] \tag{2.2-19}$$

The variance matrix of $\mathbf{v}(t_f)$ is

$$\text{var}\{\mathbf{v}(t_f)\} = \mathbf{V}_v(t_f)$$

The estimation of $\mathbf{h}(T)$ based on the m-dimensional observation vector $\mathbf{y}(t_f)$ is a classical problem in point estimation [Sage and Melsa, 1971]. The minimum variance estimate of $\mathbf{h}(T)$ is given by

$$\hat{\mathbf{h}}(T) = [\mathbf{W}_*^T(T)\mathbf{V}_v^{-1}\mathbf{W}_*(T)]^{-1}\mathbf{W}_*^T(T)\mathbf{V}_v^{-1}\mathbf{y}(t_f) \tag{2.2-20}$$

Note that this result requires the inversion of a $N \times N$ matrix. Also, the result is nonsequential, since all of the data must be received before a computations can be made.

If the noise terms $v(i\Delta)$ are independent with variance $V_v(i\Delta)$, then a simple sequential form for $\hat{\mathbf{h}}(T)$ may be determined. Let us use the notation $\hat{\mathbf{h}}(T \mid n)$ to indicate the estimate based on $\mathbf{y}(n\Delta)$. Then we can easily show that (Sage and Melsa, 1971)

$$\hat{\mathbf{h}}(T \mid n) = \hat{\mathbf{h}}(T \mid n-1) + \mathbf{k}(n\Delta)[y(n\Delta) - \mathbf{c}(n\Delta)\hat{\mathbf{h}}(T \mid n-1)] \tag{2.2-21}$$

Here the N-vector $\mathbf{k}(n\Delta)$ is given by

$$\mathbf{k}(n\Delta) = \mathbf{V}[(n-1)\Delta]\,\mathbf{c}^T(n\Delta)\{\mathbf{c}(n\Delta)\,\mathbf{V}[(n-1)\Delta]\,\mathbf{c}^T(n\Delta) + V_v(n\Delta)\}^{-1} \tag{2.2-22}$$

and the $N \times N$ matrix $\mathbf{V}(n\Delta)$ is defined by the difference equation

$$\mathbf{V}(n\Delta) = [\mathbf{I} - \mathbf{k}(n\Delta)\,\mathbf{c}(n\Delta)]\,\mathbf{V}[(n-1)\Delta] \tag{2.2-23}$$

and the N-dimensional row vector $\mathbf{c}(n\Delta)$ is

$$\mathbf{c}(n\Delta) = \{w[(n-1)\Delta]\ w[(n-2)\Delta] \cdots w[(n-N)\Delta]\} \qquad (2.2\text{-}24)$$

We could initialize this algorithm by using the first N data points in Eq. (2.2-9) to obtain $\hat{\mathbf{h}}(t \mid N)$; then we would apply the recursive relation of Eq. (2.2-21) as each new observation point is obtained. Note that this sequential approach only requires that a scalar quantity be inversed.

Example 2.2-2. Let us consider the application of this sequential algorithm to the identification problem of Example 2.2-1. Here we might assume that $h(t) = 0$, for $t > 4$, with little error. Once again we will use a unit step input; in this case, the $\mathbf{c}(n\Delta)$ becomes, for $n > 40$, a constant 40-vector given by

$$\mathbf{c} = [1\ 1\ 1 \cdots 1]$$

The estimation algorithm of Eq. (2.2-21) now becomes

$$\hat{h}_i(4 \mid n) = \hat{h}_i(4 \mid n-1) + k_i(n\Delta)\Big[y(n\Delta) - \sum_{j=1}^{40} \hat{h}_j(4 \mid n-1)\Big],$$
$$i = 1, 2, \ldots, 40, \quad n > 40$$

while the components of the gain vector are given by

$$k_i(n\Delta) = \sum_{j=1}^{40} V_{ij}[(n-1)\Delta] \Big(\sum_{l=1}^{40} \sum_{p=1}^{40} V_{lp}[(n-1)\Delta] + V_v\Big)^{-1}$$

and

$$V_{ij}(n\Delta) = V_{ij}[(n-1)\Delta] - k_j(n\Delta) \sum_{l=1}^{40} V_{il}[(n-1)\Delta]$$

Note that the computation required by this method is not trivial since there are approximately $40 \times 40 = 1600$ additions at each step. We would normally initialize this algorithm by finding $\hat{\mathbf{h}}(4 \mid 40)$ by the use of Eqs. (2.2-14) and (2.2-15).

In order to obtain algorithms for numerical deconvolution for higher-order systems, it would normally be necessary to consider problems of numerical analysis which are beyond the scope of this brief treatment. The interested reader is directed to the numerous references in the bibliography which discuss these problems in detail. Much of the computational problem associated with this approach can be removed by the application of fast Fourier transform techniques.

2.3. System Identification Using Correlation Techniques

An often suggested method for identification of linear system dynamics involves the use of a white noise test signal and cross-correlation techniques. There are several advantages inherent in this technique:

1. The system identification is not critically dependent upon normal operating records.
2. By correlating over a sufficiently long period of time, the amplitude of the test signal can be set very low such that the plant is essentially undisturbed by the white noise test signal.
3. No a priori knowledge of the system to be identified is required.

Unfortunately, there are also serious disadvantages which restrict the use of the method. Among these are:

1. The identification time is often quite long.
2. The necessary white noise source represents additional system hardware or software.
3. The method is restricted to linear systems and, in fact, linear systems that are at most slowly time varying.

The basic identification problem which we will consider is illustrated in Fig. 2.3-1. In order to discern the high frequency components of

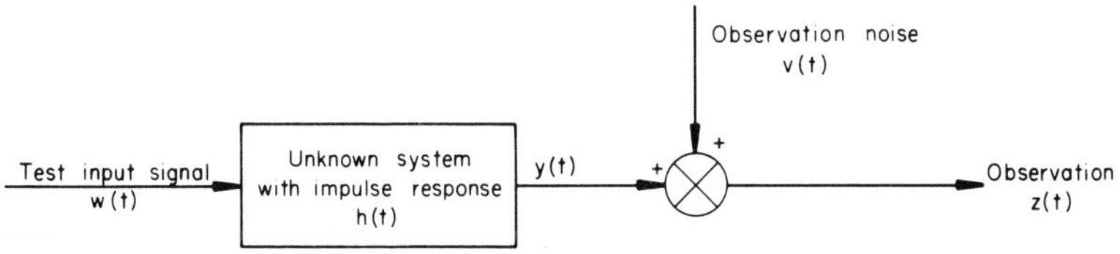

FIG. 2.3-1. Linear system identification problem with external test signal.

$h(t)$, it is necessary that $w(t)$ be a wide-bandwidth signal, and identification of $h(t)$ with zero error would, in fact, require $w(t)$ to have infinite bandwidth. In practice, it is almost always possible to provide a test signal with a bandwidth much wider than the system bandwidth.

Thus we will not consider errors due to a nonwhite input noise source although it is not at all difficult to do this. Rather than observe the output signal $y(t)$, only a noise corrupted version of $y(t)$, which we will denote by $z(t)$, is available.

A block diagram of the cross-correlation identification structure is shown in Fig. 2.3-2. It will be assumed that the system has been in

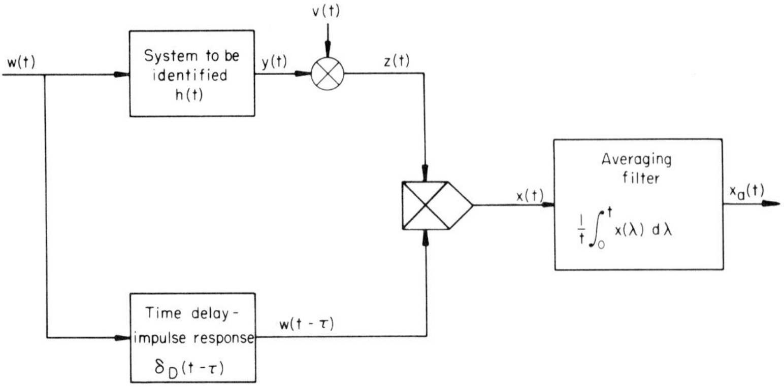

FIG. 2.3-2. Identification correlator.

operation for a sufficiently long period of time that steady state conditions have been reached. Also, the effect of normal operating records upon the identification will not be considered. The noise terms $w(t)$ and $v(t)$ will be assumed to be ergodic and Gaussian with zero mean. As will become apparent, the zero mean assumption is a very critical one and great care should be exercised to insure that the mean of the noises is zero (or known such that the effects of nonzero means can be subtracted out). The averaged output of the correlator is

$$x_a(t) = \frac{1}{t}\int_0^t x(\lambda)\,d\lambda \tag{2.3-1}$$

where the relations

$$x(t) = z(t)\,w(t-\tau) \tag{2.3-2}$$

$$z(t) = y(t) + v(t) \tag{2.3-3}$$

$$y(t) = \int_0^\infty h(\eta)\,w(t-\eta)\,d\eta \tag{2.3-4}$$

follow directly from Figs. 2.3–1 and 2.3–2.

The expected value of the correlator output is, with the ergodic assumptions and the definition $R_{wz}(\tau) = \mathscr{E}\{w(t)z(t+\tau)\}$,

$$\mathscr{E}\{x_a(t)\} = \frac{1}{t}\int_0^t \mathscr{E}\{x(\lambda)\}\,d\lambda = \mathscr{E}\{x\} = R_{wz}(\tau) \tag{2.3-5}$$

as we see from Eqs. (2.3-1) and (2.3-2). From Eqs. (2.3-3) and (2.3-4) and the zero mean assumption on $v(t)$, we have with $R_w(\tau) = \mathscr{E}\{w(t)w(t+\tau)\}$,

$$\mathscr{E}\{x_a(t)\} = R_{wz}(\tau) = \int_0^\infty h(\eta)R_w(\tau-\eta)\,d\eta \tag{2.3-6}$$

Now if we take Fourier transforms, we have the cross-spectral density relations

$$R_{wz}(s) = h(s)\,R_w(s) \tag{2.3-7}$$

If the bandwidth of $R_w(s)$ is considerably greater than that of $h(s)$, then approximately,

$$R_{wz}(s) = kh(s), \qquad R_{wz}(\tau) = kh(\tau) \tag{2.3-8}$$

This relation is exact if $w(t)$ is white such that, where δ_D is the Dirac delta,

$$R_w(\tau) = R_w\delta_D(\tau), \qquad R_w(s) = R_w \tag{2.3-9}$$

and then, for $R_w = 1$,

$$R_{wz}(t) = \mathscr{E}\{x_a(t)\} = h(\tau) \tag{2.3-10}$$

Complete system identification is then obtained by using N correlators in parallel, such that we measure

$$R_{wz}(\tau_i) = h(\tau_i), \qquad i = 1, 2,\ldots, N \tag{2.3-11}$$

We note that no use has been made of the Gaussian assumption thus far and therefore realize that sources other than Gaussian could be used for the test signals. The error analysis to follow is, however, dependent upon the Gaussian assumption. In order to determine the statistical errors associated with the identification, it is convenient to determine the correlation function associated with $x(t)$, the multiplier output.

We have

$$R_x(\gamma) = \mathscr{E}\{x(t)\,x(t+\gamma)\} = \mathscr{E}\{z(t)\,z(t+\gamma)\,w(t-\tau)\,w(t-\tau+\gamma)$$
$$= \mathscr{E}\{v(t)\,v(t+\gamma)\}\,\mathscr{E}\{w(t-\gamma)\,w(t-\tau+\gamma)\}$$
$$+ \mathscr{E}\left\{\int_0^\infty \int_0^\infty h(\lambda_1)\,h(\lambda_2)\,\mathscr{E}\{w(t-\tau)\,w(t+\gamma-\tau)\right.$$
$$\times\, w(t-\lambda_1)\,w(t+\gamma-\lambda_2)\,d\lambda_1\,d\lambda_2\} \qquad (2.3\text{-}12)$$

Now the fourth product moment for a Gaussian random variable (Sage and Melsa, 1971) is employed. This states that if a_1, a_2, a_3, and a_4 are jointly Gaussian,

$$\mathscr{E}\{a_1a_2a_3a_4\} = \mathscr{E}\{a_1a_2\}\,\mathscr{E}\{a_3a_4\} + \mathscr{E}\{a_1a_3\}\,\mathscr{E}\{a_2a_4\} + \mathscr{E}\{a_1a_4\}\,\mathscr{E}\{a_2a_3\} \quad (2.3\text{-}13)$$

We use this relation in Eq. (2.3-12) and obtain

$$R_x(\gamma) = R_v(\gamma)\,R_w(\gamma) + \int_0^\infty \int_0^\infty h(\lambda_1)\,h(\lambda_2)[R_w(\gamma)\,R_w(\gamma+\lambda_1-\lambda_2)$$
$$+ R_w(\tau-\lambda_1)\,R_w(\tau-\lambda_2)$$
$$+ R_w(\tau+\gamma-\lambda_2)\,R_w(\tau-\gamma-\lambda_1)]\,d\lambda_1\,d\lambda_2 \qquad (2.3\text{-}14)$$

Part of the foregoing expression is recognized as the signal, Eq. (2.3-6). The other term represents an error in the measurement.

We use the white noise assumption, $R_w(\gamma) = R_w\delta_D(\gamma)$, to obtain from Eq. (2.3-14)

$$R_x(\gamma) = R_v(\gamma)\,R_w\delta_D(\gamma) + h^2(\tau)\,R_w{}^2 + R_w{}^2h(\tau+\gamma)\,h(\tau-\gamma)$$
$$+ R_w{}^2\delta_D(\gamma)\int_0^\infty h(\lambda_1)\,h(\lambda_1+\gamma)\,d\lambda_1 \qquad (2.3\text{-}15)$$

If we define

$$R_x(\gamma) = R_{wz}^2(\tau) + R_e(\gamma) = h^2(\tau)\,R_w{}^2 + R_e(\gamma) \qquad (2.3\text{-}16)$$

we have

$$R_e(\gamma) = R_v(\gamma)\,R_w\delta_D(\gamma) + R_w{}^2h(\tau+\gamma)\,h(\tau-\gamma)$$
$$+ R_w{}^2\delta_D(\gamma)\int_0^\infty h(\lambda_1)\,h(\lambda_1+\gamma)\,d\lambda_1 \qquad (2.3\text{-}17)$$

This is the correlation function of the noise term in $x(t)$. The first term in the foregoing results from the external noise $v(t)$. The other two terms are a result of the system to be identified and the test signal only. Lindenlaub and Cooper (1963) have shown that by proper choice of a pseudorandom test signal, it is possible to eliminate the effects of the last two terms in the foregoing. This is also essentially the case in many practical identification problems in which the observation noise has a considerably larger variance than the test signal. The correlation of the noise term in $x(t)$ is, under these circumstances,

$$R_e(\gamma) = V_x(\gamma) = R_v(\gamma)\, R_w \delta_{\mathrm{D}}(\gamma)$$

which is the variance of $x(t)$ under the assumption that the last two terms in Eq. (2.3-17) are negligible. The problem we are faced with then is just that of estimation of a constant signal

$$R_{wz}(\tau) = R_w h(\tau)$$

imbedded in white Gaussian noise with zero mean and variance

$$V_x(\gamma) = R_v(\gamma)\, R_w \delta_{\mathrm{D}}(\gamma)$$

The optimum estimator under these conditions is just

$$x_{\mathrm{a}}(t) = \frac{1}{t} \int_0^t x(t)\, dt$$

as can easily be demonstrated (Sage and Melsa, 1971). The expected value of the filter output is just $R_w h(\tau)$. If the averaging filter output is divided by R_w, the expected averaging filter output is $h(\tau)$, the impulse response we are attempting to identify. The error variance in the identification is just

$$\operatorname{var}\{\tilde{h}(\tau)\} = (1/R_w{}^2) \operatorname{var}\{x_{\mathrm{a}}(t)\} = R_v(0)/tR_w$$

For a specified error variance, the minimum identification time

$$t_{\min} \geqslant R_v(0)/R_w \operatorname{var}\{\tilde{h}(\tau)\}$$

can be determined.

Example 2.3-1. In order to illustrate the various terms involved in an identification, we consider the case where $h(\tau) = e^{-a\tau}$. The exact expression for the correlation function of the multiplier output noise, Eq. (2.3-17), becomes

$$R_e(\gamma) = R_v(\gamma)\, R_w \delta_D(\gamma) + R_w{}^2 \Xi(\tau, \gamma) + R_w{}^2 \delta_D(\gamma)\, e^{-a\gamma}/2a$$

where

$$\Xi(\tau, \gamma) = \begin{cases} e^{-2a\tau}, & -\tau < \gamma < \tau \\ 0, & \text{otherwise} \end{cases}$$

Clearly, the second term in the foregoing is less than the first and third, so that a very close approximation

$$\operatorname{var}\{\tilde{h}(\tau)\} \simeq \frac{R_v(\gamma)}{tR_w} + \frac{1}{2at}$$

The error variance in the identification decreases with t, the observation interval. Consideration of the "self noise" introduced by the test signal itself does not materially complicate the problem. Also we may use an observation time sufficiently large and R_w sufficiently small, such that system normal operation is unaffected.

As we have already noted, a periodic pseudorandom test signal will often be preferable to Gaussian noise because of the ease of generation, ease of time delay with simple digital delay circuits, multiplication with simple digital logic switching circuits, and the fact that binary signals have a most favorable ratio of mean square input and maximum input amplitude.

Numerous investigations of identification using crosscorrelation technique have appeared in the literature. Many of these are cited in the bibliography. Among the most informative are works by Eykhoff (1963, 1968), Levin (1964), Litchtenberger (1961), Lindenlaub and Cooper (1963), and Turin (1957).

2.4. Identification of Time-Varying Linear Processes by Sinusoidal Response Measurement

One of the simplest methods of identifying linear time-*invariant* systems is by means of sinusoidal response measurements. If a linear time-invariant system with a transfer function $H(s)$ is excited by an input of the form $A \sin \omega t$, then the steady-state output is $AR(\omega) \sin[\omega t + \phi(\omega)]$. Here $R(\omega)$ is the ratio of the magnitude of the sinusoidal component of the output to the input, and $\phi(\omega)$ is the phase difference between the input and output. It is easy to show that $R(\omega)$ and $\phi(\omega)$ are related to $H(s)$ by the relations

$$R(\omega) = |H(s)|\,|_{s=j\omega} \tag{2.4-1}$$

$$\phi(\omega) = \arg H(s)\,|_{s=j\omega} \tag{2.4-2}$$

Hence by measuring the sinusoidal response, namely $R(\omega)$ and $\phi(\omega)$, for a number of values of ω, one is able to obtain a graphical plot of the magnitude and phase of the transfer function (Bode plot). The plot may be sufficient identification for some purposes such as stability analysis or compensation. If an analytic expression is needed for the transfer function, it is possible to use the straight-line approximations (Melsa and Schultz, 1969) to fit the experimental plot. In many cases, the form of the transfer function can be determined from the general nature of the system. When this is the case, it is usually fairly easy to determine the parameters by the use of this method. For a detailed discussion of this technique, see Melsa and Schultz (1969).

The sinusoidal response method can be extended to linear time-varying systems, although the approach becomes considerably more complicated than the simple procedure discussed above. In order to develop this algorithm we need to discuss a "transform" method for linear time-varying systems. The Laplace transform of a constant coefficient linear system is

$$\mathbf{\Phi}(s) = \int_{-\infty}^{\infty} \mathbf{\Phi}(t-\tau)\, e^{-s(t-\tau)}\, dt = (s\mathbf{I} - \mathbf{A})^{-1}$$

A fundamental characteristic of linear constant systems is that their impulse response matrix is dependent only on the "age variable" $(t - \tau)$. No such property is possessed by nonstationary systems, and the impulse response matrix must remain in the general form $\mathbf{\Phi}(t, \tau)$. A "transform" can still be defined by analogy to the above expression as

$$\mathbf{\Phi}(t, j\omega) = \int_{-\infty}^{\infty} \mathbf{\Phi}(t, \tau)\, e^{-j\omega(t-\tau)}\, d\tau \tag{2.4-3}$$

The upper limit can be replaced by t, since

$$\mathbf{\Phi}(t, \tau) = \mathbf{0}, \qquad \text{for} \quad \tau > t$$

It is convenient to define the vector $\mathbf{r}(t) = \mathbf{B}(t)\,\mathbf{u}(t)$ for the system

$$\dot{\mathbf{x}}(t) = \mathbf{A}(t)\,\mathbf{x}(t) + \mathbf{B}(t)\,\mathbf{u}(t)$$

It is then possible to show that a valid expression for the output state vector is given by

$$\mathbf{x}(t) = \frac{1}{2\pi} \int_{-\infty}^{\infty} \mathbf{\Phi}(t, j\omega)\,\mathbf{R}(j\omega)\,e^{j\omega t}\,d\omega \qquad (2.4\text{-}4)$$

which is just a generalization of the familiar matrix transformation for stationary systems.

An important property of $\mathbf{\Phi}(t, j\omega)$ is that it satisfies a linear differential equation which is easier to solve than the transform equation for the system transfer function. In addition, it yields a promising approach to the identification of time-varying linear plants. It is possible to show that the fundamental transfer function matrix must satisfy the linear differential equation

$$\frac{\partial}{\partial t}\,\mathbf{\Phi}(t, j\omega) + [j\omega\mathbf{I} - \mathbf{A}(t)]\,\mathbf{\Phi}(t, j\omega) = \mathbf{I} \qquad (2.4\text{-}5)$$

where $j\omega$ is regarded as a fixed parameter. This last equation is a state variable interpretation of Zadeh's equation (Zadeh, 1950) for the time-variable transform. If the system is stationary,

$$\frac{\partial}{\partial t}\,\mathbf{\Phi}(t, j\omega) = \mathbf{0} \qquad (2.4\text{-}6)$$

and

$$\mathbf{\Phi}(t, j\omega) = [j\omega\mathbf{I} - \mathbf{A}]^{-1}$$

which has been previously stated.

A scheme for the identification of time-varying linear systems which is based on this time-varying frequency transform of Zadeh will now be presented. Use is made of the property that this transform satisfies a linear ordinary differential equation closely related to the differential equation which describes the dynamics of the physical system. Provided the transform can be measured, it can be substituted into this differential equation and the coefficients of the equation determined. Through algebraic manipulation of these coefficients, the characteristic matrix of the differential equation describing the unknown system may be obtained. Periodic repetition of this procedure permits the time variation of the system to be discerned (Sage and Choate, 1965).

To formulate the problem in mathematical terms, the single-input linear system, described by

$$\dot{\mathbf{x}} = \mathbf{A}(t)\mathbf{x} + \mathbf{B}(t)\mathbf{u} \tag{2.4-7}$$

$$z = \mathbf{C}\mathbf{x} + \mathbf{D}\mathbf{u} \tag{2.4-8}$$

is considered. In these equation, $\mathbf{x}$ is the n-dimensional state vector and z is the 1-dimensional output vector defined by Eq. (2.4-8) in terms of the known matrices $\mathbf{C}$ and $\mathbf{D}$. We assume that at least the first component x_1 of $\mathbf{x}$ can be obtained instantaneously from z. $\mathbf{u}$ is the m-dimensional input vector, related to the single (scalar) input u, according to the equation

$$\mathbf{u}^{\mathrm{T}} = [u, pu, \ldots, p^{m-1}u] \tag{2.4-9}$$

where p indicates d/dt. Matrices $\mathbf{A}(t)$ and $\mathbf{B}(t)$ are not known completely, but are assumed to have the forms

$$\mathbf{A}(t) = [a_{ik}(t)] = \begin{bmatrix} 0 & 1 & 0 & \cdots & 0 \\ 0 & 0 & 1 & \cdots & 0 \\ \cdot & \cdot & \cdot & \cdots & \cdot \\ 0 & 0 & 0 & \cdots & 1 \\ a_{n1}(t) & a_{n2}(t) & a_{n3}(t) & \cdots & a_{nn}(t) \end{bmatrix} \tag{2.4-10}$$

$$\mathbf{B}(t) = [b_{ik}(t)] = \begin{bmatrix} 0 & 0 & 0 & \cdots & 0 \\ 0 & 0 & 0 & \cdots & 0 \\ \cdot & \cdot & \cdot & \cdots & \cdot \\ 0 & 0 & 0 & \cdots & 0 \\ b_{n1}(t) & b_{n2}(t) & b_{n3}(t) & \cdots & b_{nm}(t) \end{bmatrix} \tag{2.4-11}$$

or are assumed to be convertible to these forms by a linear nonsingular transformation of variable. Thus, phrased in terms of Eq. (2.4-7), the identification problem becomes that of determining the time functions

$$a_{ni}(t), \qquad i = 1, \ldots, n \tag{2.4-12}$$

and

$$b_{nk}(t), \qquad k = 1, \ldots, m \tag{2.4-13}$$

of matrices $\mathbf{A}(t)$ and $\mathbf{B}(t)$. In the general case, some of the $n + m$ functions may be known a priori. This situation, of course, simplifies the identification problem.

The time-varying frequency transform $\mathbf{h}(j\omega, t)$ is defined as

$$\mathbf{h}(j\omega, t) = \mathbf{T}(j\omega)\, \boldsymbol{\chi}(j\omega, t) \tag{2.4-14}$$

where

$$\mathbf{T}(j\omega) = [t_{ik}(j\omega)] \tag{2.4-15}$$

$$t_{ik}(j\omega) = \frac{i!(-j\omega)^{i-k}}{k!(i-k)!} = \left[\frac{i}{k}\right] (-j\omega)^{i-k}, \qquad i \geqslant k \tag{2.4-16}$$

$$t_{ik}(j\omega) = 0, \qquad i < k \tag{2.4-17}$$

$$|\mathbf{T}(j\omega)| = 1 \tag{2.4-18}$$

$$\boldsymbol{\chi}(j\omega, t) = \int_{-\infty}^{\infty} \mathbf{w}(t, \xi)\, e^{-j\omega(t-\xi)}\, d\xi \tag{2.4-19}$$

$\mathbf{w}(t, \xi)$ is the response of the state variable when the input is a unit impulse occurring at time $t = \xi$. We may readily verify that the definition of Eq. (2.4–14) gives the following property to the elements of $\mathbf{h}(j\omega, t)$:

$$h_i(j\omega, t) = ph_{i-1}(j\omega, t), \qquad i = 2,..., n \tag{2.4-20}$$

While $\mathbf{h}(j\omega, t)$ is defined in terms of the impulse response of Eq. (2.4-7) it is possible to express it in terms of the response of Eq. (2.4-7) to sinusoidal excitation. In fact, an alternate definition of $\mathbf{h}(j\omega, t)$ is

$$\mathbf{h}(j\omega, t) = e^{-j\omega t}\mathbf{T}(j\omega)\, \boldsymbol{\zeta}(j\omega, t) \tag{2.4-21}$$

where $\boldsymbol{\zeta}(j\omega, t)$ is the system response to $u = e^{j\omega t}$. The identification scheme which we develop makes use of Eq. (2.4-21) as a means for determining $\mathbf{h}(j\omega, t)$.

The elements of $\mathbf{h}(j\omega, t)$ belong to the field of complex numbers. Since measurements necessarily involve real numbers, it is convenient to break this vector into real and imaginary components. Using the abbreviation $\mathbf{h} \triangleq \mathbf{h}(j\omega, t)$ and the notation subscript "R" for "real part of ..." and subscript "I" for "imaginary part of ...," and regarding ω as a fixed parameter, it can be shown that

$$p\begin{bmatrix}\mathbf{h}_R\\ \mathbf{h}_I\end{bmatrix} = \begin{bmatrix}\mathbf{E}_R(j\omega, t) & -\mathbf{E}_I(j\omega, t)\\ \mathbf{E}_I(j\omega, t) & \mathbf{E}_R(j\omega, t)\end{bmatrix}\begin{bmatrix}\mathbf{h}_R\\ \mathbf{h}_I\end{bmatrix} + \begin{bmatrix}\mathbf{k}_R(j\omega, t)\\ \mathbf{k}_I(j\omega, t)\end{bmatrix} \tag{2.4-22}$$

where the $n \times n$ matrix $\mathbf{E}(j\omega, t)$ is of the general form of $\mathbf{A}(t)$ with elements $\{e_{nr}(j\omega, t)\}$ related to the elements $\{a_{ni}(t)\}$ of $\mathbf{A}(t)$ by the equations

$$e_{nr}(j\omega, t) = \sum_{k=r}^{n-1} \frac{(k-1)!(j\omega)^{k-r}}{(r-1)!(k-r)!} a_{nk}(t) \tag{2.4-23}$$

and $\mathbf{E}_R$ and $\mathbf{E}_I$ are the real and imaginary components of matrix $\mathbf{E}$. The additional element $a_{n,n+1}(t)$ is taken to be -1. The n-vector $\mathbf{k}(j\omega, t)$ is related to the matrix $\mathbf{B}(t)$ by

$$\mathbf{k}(j\omega, t) = \mathbf{B}(t) \begin{bmatrix} 1 \\ j\omega \\ \vdots \\ (j\omega)^{m-1} \end{bmatrix} \tag{2.4-24}$$

If $\mathbf{h}$ and $p\mathbf{h}$ are available, substitution in Eq. (2.4-22) gives us, with the aid of Eqs. (2.4-23) and (2.4-24), two algebraic equations in the functions $a_{ni}(t)$ and $b_{nk}(t)$. Measurements of $\mathbf{h}$ at q frequencies $\omega_1, \omega_2, \ldots, \omega_q$, gives $2q$ such equations. Assuming that s of the $n + m$ functions $a_{ni}(t)$ and $b_{nk}(t)$ are known a priori, it follows that the matrices $\mathbf{A}(t)$ and $\mathbf{B}(t)$ can be identified if $q = (n + m - s)/2$ or $q = (n + m + 1 - s)/2$, depending upon which expression results in an integer.

Thus if we define

$$\boldsymbol{\lambda}^T(t) = [a_{n1}(t), a_{n2}(t), \ldots, a_{nn}(t), b_{n1}(t), \ldots, b_{nm}(t)] \tag{2.4-25}$$

the identification problem may be more concisely stated as that of determining $\boldsymbol{\lambda}(t)$. To develop an expression for $\boldsymbol{\lambda}$, it is convenient to define

$$\mathbf{f}(j\omega, t) = (p + j\omega)\,\mathbf{h}(j\omega, t) \tag{2.4-26}$$

This definition may be justified on grounds other than purely notational convenience: We will show that this quantity can be measured directly under certain conditions. Substitution of Eq. (2.4-26) into Eq. (2.4-22) yields a vector equation whose nth row may be written

$$f_n(j\omega, t) = \mathbf{a}_n^T(t)\,\mathbf{h}(j\omega, t) + \mathbf{b}_n^T(t)\boldsymbol{\omega} \tag{2.4-27}$$

where f_n is the nth element of $\mathbf{f}$, and $\mathbf{a}_n^T$ and $\mathbf{b}_n^T$ are the nth rows of

$\mathbf{A}(t)$ and $\mathbf{B}(t)$, respectively. Equation (2.4-27) is equivalent to

$$f_n(j\omega, t) = [\mathbf{h}^{\mathrm{T}}(j\omega, t)\ \boldsymbol{\omega}^{\mathrm{T}}]\ \boldsymbol{\lambda}(t) \tag{2.4-28}$$

Because f_n and the elements of $\mathbf{h}$ and $\boldsymbol{\omega}$ belong to the field of complex numbers, and since measurements necessarily involve real numbers, it is desirable to break this relation into real and imaginary parts. We may rewrite Eq. (2.4-28) as the 2-vector equation

$$\begin{bmatrix} f_{n\mathrm{R}} \\ f_{n\mathrm{I}} \end{bmatrix} = \begin{bmatrix} \mathbf{h}_{\mathrm{R}}^{\mathrm{T}} & \boldsymbol{\omega}_{\mathrm{R}}^{\mathrm{T}} \\ \mathbf{h}_{\mathrm{I}}^{\mathrm{T}} & \boldsymbol{\omega}_{\mathrm{I}}^{\mathrm{T}} \end{bmatrix} \boldsymbol{\lambda} \tag{2.4-29}$$

Now letting a superscript placed to the left of a quantity designate the frequency parameter (e.g., ${}^{i}f_n = f_n(j\omega_i, t)$, ${}^{i}h^{\mathrm{T}} = h^{\mathrm{T}}(j\omega_i, t)$, Eq. (2.4-29) can be extended to yield

$$\begin{bmatrix} {}^{1}f_{n\mathrm{R}} \\ {}^{1}f_{n\mathrm{I}} \\ {}^{2}f_{n\mathrm{R}} \\ {}^{2}f_{n\mathrm{I}} \\ \vdots \end{bmatrix} = \begin{bmatrix} {}^{1}\mathbf{h}_{\mathrm{R}}^{\mathrm{T}} & {}^{1}\boldsymbol{\omega}_{\mathrm{R}}^{\mathrm{T}} \\ {}^{1}\mathbf{h}_{\mathrm{I}}^{\mathrm{T}} & {}^{1}\boldsymbol{\omega}_{\mathrm{I}}^{\mathrm{T}} \\ {}^{2}\mathbf{h}_{\mathrm{R}}^{\mathrm{T}} & {}^{2}\boldsymbol{\omega}_{\mathrm{R}}^{\mathrm{T}} \\ {}^{2}\mathbf{h}_{\mathrm{I}}^{\mathrm{T}} & {}^{2}\boldsymbol{\omega}_{\mathrm{I}}^{\mathrm{T}} \\ \vdots & \vdots \end{bmatrix} \boldsymbol{\lambda} \tag{2.4-30}$$

Denoting the vector on the left-hand side by $\boldsymbol{\nu}$ and the matrix on the right-hand side by $\mathbf{G}$, it is evident that $\mathbf{G}$ will be square if $\boldsymbol{\nu}$ is given the dimension of $\boldsymbol{\lambda}$. This requirement is met if the last element of $\boldsymbol{\lambda}$ is ${}^{q}f_{\mathrm{I}}$ $(q = n + m/2)$ for $(n + m)$ even, or ${}^{q}f_{\mathrm{R}}$ $(q = n + m\ 1/2)$ for $(n + m)$ odd. In general, $\mathbf{G}$ will be nonsingular; therefore our solution of Eq. (2.4-30) may be written

$$\boldsymbol{\lambda}(t) = \mathbf{G}^{-1}(j\omega_i, t)\ \boldsymbol{\gamma}(j\omega_i, t) \tag{2.4-31}$$

This equation provides us with a means for identifying the system provided that $\mathbf{G}$ and $\boldsymbol{\nu}$ can be determined from measurements made on the system. The matter of measurements will now be considered.

If Eq. (2.4-31) is to be used to identify the system, a means must be available for determining $\mathbf{h}(j\omega_i, t)$ and $f_n(j\omega_i, t)$, $i = 1, 2, \ldots, q$. In essence, the method to be used is to perturb the system with a test signal and to extract the needed information from the system's response. The choice as to what test signal to employ is suggested by Eq. (2.4-20), which displays a simple relation between $\mathbf{h}(j\omega, t)$ and the response of the system to $u_1 = e^{j\omega t}$. Since the latter is not physically realizable, the closely related and physically realizable input $u_1 = \cos \omega t$

is selected. Calling the response to this input $\boldsymbol{\psi}(j\omega, t)$, and noting that a real system contains only real-valued coefficients, it follows that

$$\boldsymbol{\psi}(j\omega, t) = \text{Re}\{\boldsymbol{\zeta}(j\omega, t)\} = \mathbf{h}_R(j\omega, t)\cos\omega t - \mathbf{h}_I(j\omega, t)\sin\omega t \qquad (2.4\text{-}32)$$

Using this result and definition (2.4-26), we see that

$$p\boldsymbol{\psi}(j\omega, t) = \mathbf{f}_R(j\omega, t)\cos\omega t - \mathbf{f}_I(j\omega, t)\sin\omega t \qquad (2.4\text{-}33)$$

These two relations show that for the measurement problem is that of demodulating $\boldsymbol{\psi}$ and $p\boldsymbol{\psi}$. Fortunately, it is not necessary to consider these equations separately, they may be combined into a single entry. To accomplish this, the $(n + 1)$-dimensional vector transform $\boldsymbol{\gamma}(j\omega, t)$ is defined:

$$\boldsymbol{\gamma}^T(j\omega, t) = [\mathbf{h}^T(j\omega, t), f_n(j\omega, t)] \qquad (2.4\text{-}34)$$

We note that $\boldsymbol{\gamma}$ consists of the quantities to be measured. Letting $\boldsymbol{\psi}^1(j\omega, t)$ be the response of the augmented state vector to $u_1 = \cos\omega t$, it follows that

$$\boldsymbol{\psi}^1(j\omega, t) = \boldsymbol{\gamma}_R(j\omega, t)\cos\omega t - \boldsymbol{\gamma}_I(j\omega, t)\sin\omega t \qquad (2.4\text{-}35)$$

and the task is to demodulate the single function $\boldsymbol{\psi}^1(j\omega, t)$. The method by which this is to be accomplished will now be discussed in detail for the determination of $\boldsymbol{\gamma}_R$. We will then mention those minor modifications needed to determine $\boldsymbol{\gamma}_I$.

The first step in the determination of $\boldsymbol{\gamma}_R$ is to multiply $\boldsymbol{\gamma}$ by $\cos\omega t$. This yields an $(n + 1)$-dimensional correlation product vector, which will be called $\boldsymbol{\rho}(j\omega, t)$. It follows from Eq. (2.4-35) that

$$\boldsymbol{\rho} = \boldsymbol{\psi}^1 \cos\omega t = \tfrac{1}{2}\{\boldsymbol{\gamma}_R + \boldsymbol{\gamma}_R\cos 2\omega t - \boldsymbol{\gamma}_I\sin 2\omega t\} \qquad (2.4\text{-}36)$$

Since an analysis of $\boldsymbol{\rho}(j\omega, t)$ in the frequency domain is desired, it is convenient to assign the symbol $\mathbf{P}(j\omega, j\mu)$ to designate its Fourier transform, where μ is the frequency parameter of the transform. But as the vector-valued function $\mathbf{P}$ is not conveniently represented graphically, a scalar function of $\mathbf{P}$ is sought. The function chosen should have the property of being proportional to the magnitude, or "length," of $\mathbf{P}$. This immediately suggests the choice of the norm of $\mathbf{P}$, defined

$$\|\mathbf{P}\| = (\mathbf{P}^\dagger\mathbf{P})^{1/2}$$

where the symbol "†" denotes Hermitian transpose.

First undertaking the analysis of the case in which the system is stationary, we see that γ will be time independent, i.e., $\gamma(i\omega, t) = \gamma(i\omega)$. This causes the spectrum of $\rho(j\omega, t)$ to consist entirely of impulses, of which there are three. They have amplitudes (or areas) equal to $\pi/2 \parallel \gamma \parallel$, $\pi \parallel \gamma_R \parallel$, $\pi/2 \parallel \gamma \parallel$, and are located at $\mu = -2\omega$, 0, 2ω, respectively. This spectrum is depicted in Fig. 2.4-1a. It is evident that

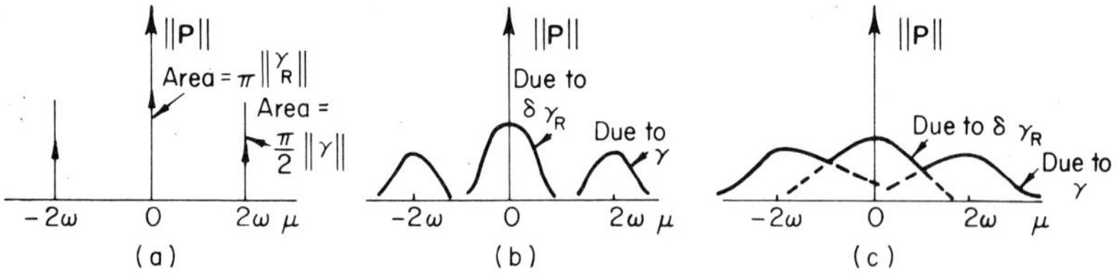

FIG. 2.4-1. Frequency spectra for identification.

to measure γ_R, it is merely necessary to filter ρ with a lowpass filter having high attenuation at $\mu = 2\omega$. Also we note that advantage can be taken of the fact that the output from the filter should not change. The output can be averaged to further surpress the components at $\mu = +2\omega$, and any noise (or zero mean value) present.

Next we consider the case in which system is slowly varying. To avoid confusion as to what is meant by the words slowly varying, the following definition is offered: A slowly-varying system is a varying system in which the augmented transform vector $\gamma(j\omega, t)$ remains essentially constant over every interval of time I which exceeds the period of the slowest natural mode of the system "frozen" at some time $t \in I$; the parameter ω is chosen such that $\parallel \gamma(j\omega) \parallel = \frac{1}{10} \parallel \gamma(j0) \parallel$, where $\gamma(j\omega)$ is the augmented transform vector for the frozen system. Analysis of the spectrum for this case reveals that continuously distributed components have appeared. These components, which will be referred to as subspectra, are tightly grouped about the frequencies at which the impulses had occurred in the stationary case. The subspectrum grouped about $\mu = 0$ corresponds to γ_R, the other two to γ. (Note that the use of the norm to represent the spectra does not allow distinction between Fourier transforms differing only in phase.) A typical spectrum for the slowly-varying case is depicted in Fig. 2.4-1b. With the spectra well separated as shown in the figure, γ_R can be recovered from ρ by filtering with a lowpass filter, just as was done for the stationary system. However, in this case the demands on the filter are more stringent. Neglecting noise, the ideal filter would have uniform response and constant phase shift over the range of frequencies occupied by the subspectrum due to γ_R, and zero response elsewhere. These characteristics define the so-called "cardinal-data hold filter," which, as might be expected, is nonrealizable (for real time operation). The realizable filter used should be selected to approach this characteristic over the range of frequencies for which $\| \mathbf{P} \|$ is not negligibly small.

When noise is present, a realizable approximation of the cardinal-data hold-filter characteristic may no longer yield satisfactory results. (The noise will, in general, consist of a random component plus a component due to the presence of a control signal applied with the test signal at the input.) The optimum filter for this case is the one which most effectively rejects the noise to give the best estimate of γ_R. Ideally the performance criterion defining what is best should take into account Eq. (2.4-31). That is, it is the error in the measurement of $\boldsymbol{\lambda}$ that is of paramount concern.

Even when noise can be neglected, there will generally be some error in the measurement of γ_R. This is partly due to the fact that γ is rarely bandwidth limited in the strictest sense. Consequently, some overlapping of the "tails" of the subspectra occurs. This produces an anomalous component in the output of the filter. Another source of distortion results from the nonideal filter characteristics which must be accepted if on-line identification is to be performed. Both of these effects may be made negligible through proper design, if the system is varying slowly enough.

Finally, the case in which the system is rapidly varying will be considered. The spectrum for this case differs from that of the slowly-varying case primarily in that the distributions of the subspectra are no longer narrow. Instead, they have the spread-out appearance indicated in Fig. 2.4–1c. Clearly if γ_R is to be recovered from $\boldsymbol{\rho}$, the subspectra must overlap as they do in Fig. 2.4–1c. No filter can be used to separate them in this situation. The only recourse is to increase the test signal frequency until the subspectra have been drawn far enough apart to allow the γ_R subspectrum to be separated. This improvement in the status quo is just another manifestation of the well-known principle that more information can be carried over a high carrier frequency than a low one if the available bandwidth is a fixed percentage of the carrier frequency.

It is unfortunate from the viewpoint of the identification problem that nature resists this approach. Two deleterious effects set in as frequency is increased. The first is that, for fixed test-signal amplitude, the amplitude of the spectrum **P** decreases. Of course, this results in decreased signal-to-noise ratio, thereby adding to the noise problem. The second effect is that our solution for $\boldsymbol{\lambda}$ becomes increasingly sensitive to small errors made in the measurement of $\boldsymbol{\gamma}$. It is evident that the second effect aggravates the problems caused by the first, with the net result that $\boldsymbol{\lambda}$ may be difficult to measure accurately for large ω. However, provided the noise level is very low and the measuring equipment very accurate, the choice of a large ω permits the identification of even very rapidly varying systems.

The procedure for determining $\boldsymbol{\gamma}_{\mathrm{I}}$ from $\boldsymbol{\psi}^1$ is a close parallel to that just discussed for determining $\boldsymbol{\gamma}_{\mathrm{R}}$. The only major difference is that $\boldsymbol{\psi}^1$ is multiplied by $\sin \omega t$ instead of $\cos \omega t$. This generates a correlation product which, under favorable circumstances, may be filtered to yield $\boldsymbol{\gamma}_{\mathrm{I}}$. Since no new concepts are involved in this demodulation procedure, attention will now be turned to the task of measuring the vector transforms when the test signal contains sinusoids of more than one frequency.

For the multifrequency case, the test signal is chosen as the sum of sinusoids

$$u_1(t) = \sum_{i=1}^{q} a_i \cos \omega_i t \tag{2.4-37}$$

Calling the response of the augmented state vector to this input $\boldsymbol{\psi}^1(j\omega_i, t)$, it follows from Eq. (2.4-35) and superposition that

$$\boldsymbol{\psi}^1(j\omega_i, t) = \sum_{i=1}^{q} a_i[{}^i\boldsymbol{\gamma}_{\mathrm{R}} \cos \omega_i t - {}^i\boldsymbol{\gamma}_{\mathrm{I}} \sin \omega_i t] \tag{2.4-38}$$

Providing that the spectra of the ${}^i\boldsymbol{\gamma}$ are effectively bandwidth limited, and the ω_i are spaced sufficiently far apart, the spectrum of $\boldsymbol{\psi}^1$ consists of nonoverlapping subspectra such as shown in Fig. 2.4-2. In the figure, $\boldsymbol{\psi}^1(j\omega, j\mu)$ represents the Fourier transform of $\boldsymbol{\psi}^1(j\omega, t)$. With the subspectra well separated as shown, they can be recovered individually through the use of bandpass filters. The outputs of these filters can then be demodulated with the methods discussed previously to give the ${}^i\boldsymbol{\gamma}_{\mathrm{R}}$ and the ${}^i\boldsymbol{\gamma}_{\mathrm{I}}$.

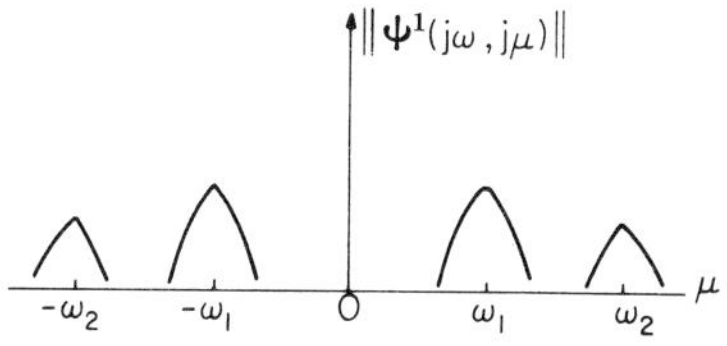

FIG. 2.4-2. Multiple frequency spectra.

It is possible to derive a criterion which specifies the circumstances under which successful measurements of the vector transforms can be made (assuming ideal filters and negligible noise). We write Eq. (2.4-38) in complex notation

$$\psi^1(j\omega_i, t) = \sum_{i=1}^{q} a_i/2[{}^i\gamma e^{j\omega_i t} + ({}^i\gamma e^{j\omega_i t})^*] \tag{2.4-39}$$

where the asterisk denotes the complex conjugate, and taking the Fourier transform of both sides gives the equation

$$\psi^1(j\omega_i, j\mu) = \sum_{i=1}^{q} a_i/2\{\boldsymbol{\Gamma}[j\omega_i, j(\mu - \omega_i)] + \boldsymbol{\Gamma}[-j\omega_i, j(\mu + \omega_i)]\} \tag{2.4-40}$$

Here, $\boldsymbol{\Gamma}$ represents the Fourier transform of γ. If the subspectra do not overlap, the terms inside the summation are disjoint, which implies that the inner product

$$\boldsymbol{\Gamma}^{\dagger}[j\omega_i, j(\mu - \omega_i)]\, \boldsymbol{\Gamma}[+j\omega_k, j(\mu \mp \omega_k)] = 0, \qquad \text{if} \quad k \neq 1 \tag{2.4-41}$$

for $i = 1, 2, \ldots, q$. We may verify that when the condition is satisfied, then not only can each frequency term be separated for the others with a bandpass filter, but also the output of each bandpass filter may be demodulated to give ${}^i\gamma_R$ and ${}^i\gamma_I$.

Example 2.4-1. The problem of identifying the system of Fig. 2.4-3

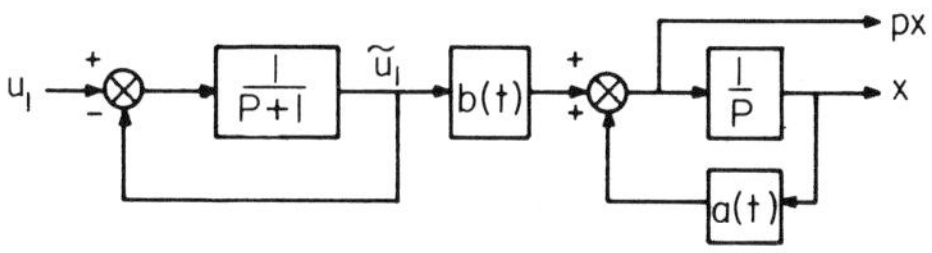

FIG. 2.4-3. Second-order time-varying differential system.

is considered. It is assumed that the test signal is the only input, and that the stationary part of the system is at steady state. The structure of this system suggests the choice of u_1 such that $\tilde{u}_1 = \cos \omega t$. Then the time-varying part of the system may be identified as if it alone were being tested — with no modifications of the preceding theory needed. The proper choice of u_1 is easily verified to be $u_1 = 2 \cos \omega t - \omega \sin \omega t$. Forming the matrix $\mathbf{G}$, and inverting, yields for this case

$$\begin{bmatrix} a(t) \\ b(t) \end{bmatrix} = \begin{bmatrix} h_{\mathrm{I}}^{-1} & f_{n\mathrm{I}} \\ f_{n\mathrm{R}} - h_{\mathrm{R}} & (h_{\mathrm{I}}^{-1} f_{n\mathrm{I}}) \end{bmatrix}$$

The quantities h_{R}, h_{I}, f_{R}, and f_{I} are to be measured by the methods discussed previously. Hence the implementation of the identification scheme takes the form shown in Fig. 2.4–4.

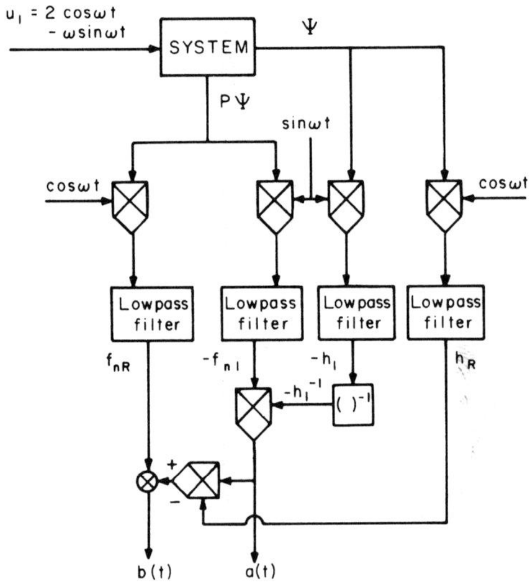

FIG. 2.4-4. Computer simulation of identification technique.

The noise problem in this identification scheme is of some interest and will be briefly mentioned here. There are two principle effects which are desirable: minimization of the effects of noise originating at the input, and of the effects of estimation of γ in the presence of noise when not every element of $\mathbf{x}$ can be measured directly.

Noise at the input of the system will often be the known control input which is "noise" as far as this system identification method is concerned. In some instances, it may be possible to increase the amplitude of the test signal and/or to choose the largest test signal frequency well above the cutoff frequency of the system. Great care must be exercised in attempting to apply either of these remedies because of the obvious harmful side effects.

A more elaborate method of minimizing the effects of noise introduced at the input is based on a model which tracks the primary system as shown in Fig. 2.4-5. In this noise cancellation scheme,

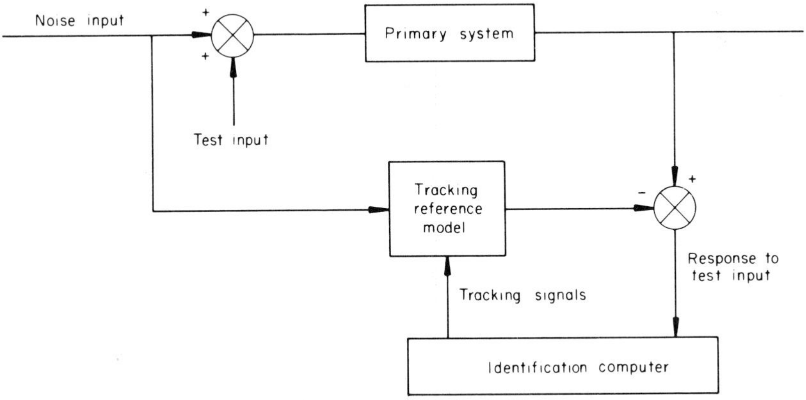

FIG. 2.4-5. Model reference noise cancellation scheme.

the tracking model is adjusted such as to identify the system. In this condition, the system response due to the test input only is input to the identification computer. Thus, we again have a form of model reference identification as discussed in Sect. 2.5.

A scheme for the identification of linear time-varying differential systems has been presented. This method appears promising for on-line identification, and it appears to be eminently satisfactory for off-line use, where physically nonrealizable filter characteristics may be simulated to permit identification rates which approach the theoretical maximum.

A criterion for the proper selection of test signal frequencies was presented. This criterion could be used to determine the credibility of experimental results from a particular identification run.

2.5. Learning Models

One of the conceptually simplest, while at the same time most flexible, methods of system identification is the approach referred to as learning models or model reference. The basic idea of this approach is represented in Fig. 2.5-1. A known input, or class of inputs, is

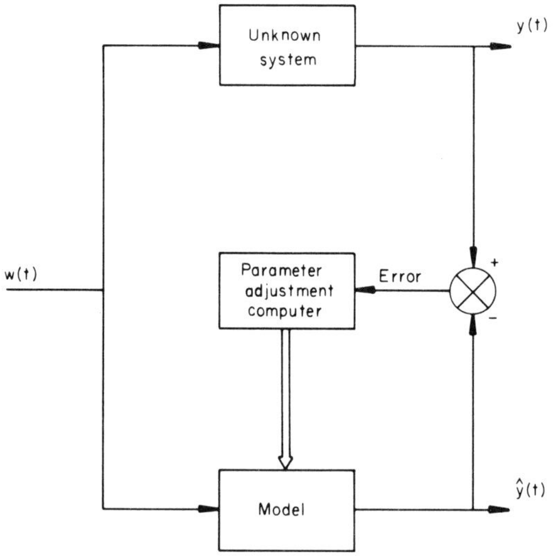

Fig. 2.5-1. Learning model approach to system identification.

applied to the unknown system and a model which is to simulate the system. The difference between the two outputs is then used to adjust the model, and the procedure is repeated. Normally the model is fixed in form, and only a finite number of parameters may be adjusted. This procedure can be used on line by dividing the input signal into finite time records and carrying out an identification after each of these records. The procedure can also be used with noisy observations of the unknown system, although problems associated with estimation, stability of the identification, and nonuniqueness of the identification, complicate the design procedure. If the actual unknown system is expensive to use for experimentation, one may wish to replace it by a stored set of input–output records.

In the application of this procedure, there are several practical problems which must be considered. These include:

1. how to select the model structure,
2. time-scaling possibilities,
3. error criterion,
4. initial conditions, and
5. adjustment strategy.

We shall discuss each of these items in somewhat more detail.

The problem of model selection is an integral part of almost every identification method. In most problems, we have a good deal of knowledge about the physical processes which are represented by the unknown system and lack only knowledge of the specific parameter values. If we have no knowledge of the system structure, we may only guess at various model structures and see if they can adequately match the observed behavior of the unknown system. The first situation is far more desirable, since we may have some "feel" for the parameter ranges and can determine if the answers obtained make physical sense. There are also situations where one purposefully introduces a model which is known to be different than the actual system description. This would be the case if we were seeking the best "linear" model for some system which was known to exhibit nonlinear characteristics. We might want a linear model in order to carry out some other phase of system analysis or design.

If one uses stored input–output records, then there is the possibility of significantly increasing the time scale of the model so that the parameter selection procedure may be completed more rapidly. The model reference identification procedure is ideally suited to hybrid computation, and time scaling in this case may be easily accomplished. Of course, in a digital computer simulation, time scaling is meaningless since there is normally no real-time clock associated with this approach.

Since the learning model optimization is almost always done by a search procedure as opposed to an analytic technique. there is a wide variety of error criteria which can be used without a significant change in complexity. The possibilities include integral square error, integral absolute error, and various time-weighted versions of these, as well as higher-order error criteria. One could also use min–max error criteria, where it would be desired to select parameters which would minimize the maximum error. The criterion which is probably most used is the integral squared error, because it normally leads to smoother error surfaces and hence rapid convergence.

When the identification process is based on finite time records, it is necessary to select initial conditions for the model. If possible, the measurements of the system should be made with known initial conditions, perhaps zero. Then the initial conditions of the model can be selected directly. If this is not possible, then the initial conditions of the model must be treated as additional parameters to be optimally selected.

The technique of parameter adjustment is probably the most difficult part of the model reference technique. Possible sources of difficulty are: (1) existence of multiple local minima or saddle points; (2) extreme sensitivity of some parameter and extreme insensitivities of others; (3) poor convergence for some models; and (4) lack of orthogonality, i.e., the optimal value of one parameter depends on the values of the other parameters. Almost all of the existing search techniques (Wilde, 1964) have been applied at one time or another in learning model identification with varying degrees of success. The two most popular and most successful have been random search and "gradient" techniques. The random search approach is very effective in discarding local minima and obtaining solutions on "bas" error surfaces. The gradient techniques[1] are relatively easy to program and can be quite effective although they seek only local minima. Random starts for a gradient techniques as well as the addition of ridge climbing and edge handling capability generally gives a good algorithm. The application of gradient techniques to learning model identification is discussed in Chap. 4. Pertinent also are the papers by Blandhol and Balchen (1963), Eveleigh (1967), Eykhoff (1968), Margolis and Leondes (1959), Mendel (1968), and Mishkin and Braun (1961).

2.6. Wiener Theory of Nonlinear Systems

Wiener's theory of nonlinear systems is an experimental technique in which unknown system parameters are determined as coefficients of an operator expanded in Hilbert space. The input to a system is expanded in a Laguerre function series. These functions are derived

[1] See Chap. 4.

from the Laguerre polynomials by including the square root of the Laguerre function such that, for the Laguerre polynomial,

$$L_n(t) = \frac{1}{(n-1)!} e^t \frac{d^{n-1}}{dt^{n-1}} [t^{n-1}e^{-t}], \qquad n = 1, 2, \ldots \tag{2.6-1}$$

the nth Laguerre function is given as

$$g_n(t) = \begin{cases} e^{-(t/2)} L_n(t), & t \geqslant 0 \\ 0, & t < 0 \end{cases} \tag{2.6-2}$$

The functions $g_n(t)$ are orthonormal for all $t \in [0, \infty]$. We may expand the past of the input as

$$u(-t) = \sum_{n=1}^{\infty} v_n g_n(t), \qquad t \geqslant 0 \tag{2.6-3}$$

$$v_n = \int_0^{\infty} u(-\tau)\, h_n(\tau)\, d\tau \tag{2.6-4}$$

It is thus possible to obtain a desired Laguerre coefficient by using $x(t)$ as the input to a chain of linear filters. Consider the generation of the Laguerre coefficients for the linear system which results from the Laplace transform of Eq. (2.6-1):

$$G_n(s) = \frac{1}{s + \frac{1}{2}} \left[\frac{s - \frac{1}{2}}{s + \frac{1}{2}}\right]^{n-1} \tag{2.6-5}$$

This result is shown in Fig. 2.6–1. In order to exploit fully the benefits of a Laguerre function expansion of an input signal, the Wiener

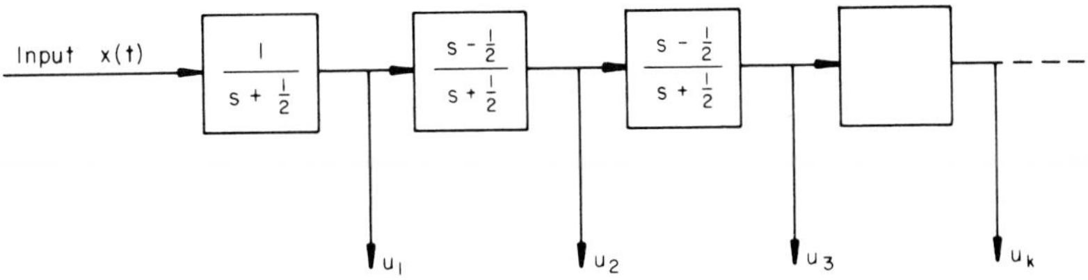

FIG. 2.6-1. Generation of Laguerre coefficients by a linear network.

theory insists that the input used to excite the system response be Gaussian white noise. With this choice, it can be shown that the

Laguerre functions are uncorrelated Gaussian random processes with equal variances. Since Hermite polynomials are orthonormal for all $t \in [-\infty, \infty]$, it is reasonable to expand the system operator in Hermite functions. If $\eta_n(u)$ is the nth Hermite polynomial, Wiener defines the $(n+1)$th Hermite function as

$$H_n(u) = e^{-(n^2/2)}\eta_n(u) \tag{2.6-6}$$

Wiener has shown that the transformation from the Laguerre coefficient input space to the output can be written in terms of Hermite functions as the expansion

$$x(t) = \lim_{p\to\infty} \sum_{i=1}^{\infty} \sum_{j=1}^{\infty} \cdots \sum_{h=1}^{\infty} a_{ij\cdots h} H_i(u_1)\, H_j(u_2) \cdots H_h(u_p) \tag{2.6-7}$$

The coefficients in this expansion $a_{ij\cdots h}$ can be determined by multiplying both sides of the equation by the appropriate products $H_n(u)$ functions and averaging over $t \in [-\infty, \infty]$. However, because of the choice of input signal and expansions, the required averaging can be accomplished by simply crosscorrelating the system output with the Hermite polynomials. For instance, we may show that

$$\begin{aligned} a_{ij\cdots h} &= (2\pi)^{p/2} \lim_{T\to\infty} \frac{1}{2T} \int_{-T}^{T} u(t)\, \eta_i(u_1)\, \eta_j(u_2) \cdots \eta_h(u_p)\, dt \\ &= (2\pi)^{p/2}\, \overrightarrow{u(t)\, v(t)}^{t} \end{aligned} \tag{2.6-8}$$

This equation summarizes the experimental Wiener theory of nonlinear systems. Fig. 2.6-2 is a representation of the computation

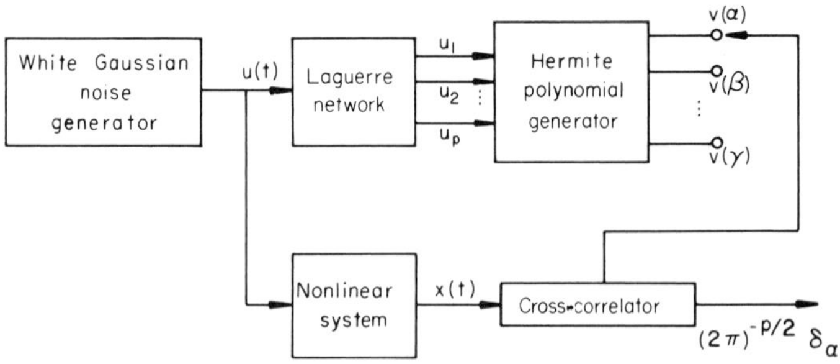

FIG. 2.6-2. Computational arrangement for the evaluation of Wiener coefficients.

required to carry out the various operations. Despite the apparent simplicity of Eq. (2.6-8), there is an infinite number of operations involved in the expression. In order to actually make use of the Wiener theory, it is necessary for us to truncate all limiting operations both with regard to measurement time and also the number of terms taken in the series for $u(t)$. An error analysis of the effect of this truncation would be exceptionally difficult. Further difficulties concern the off-line nature of the procedure, since a special test signal must be applied over a long interval of time. Also, time-varying and unstable systems are not considered. Finally, there is the requirement to convert from the Wiener coefficients to the parameters of the system differential equations, which is often not a simple task. The references which we cite in the bibliography present much more detail on this method of system identification.

2.7. Conclusions

In this chapter, we have examined a number of methods of system identification which have been very effectively applied in the past. The procedures range from the simple numerical deconvolution approach of Sect. 2.2 to the more complex sinusoidal response technique for linear time-varying systems. The remaining chapters of this book examine a number of the so-called modern approaches to system identification. Before we can do this, however, we will need to formulate cost functions for identification, which is the subject of the next chapter.

3

COST FUNCTIONS FOR SYSTEM IDENTIFICATION

3.1. Introduction

In this chapter, we will explore some of the cost functions which may be used for system identification. By cost function for identification, we mean the cost or penalty for not achieving correct identification. For example, if the true value of a parameter to be identified were θ, and we assumed a value $\hat{\theta}$, a suitable cost function might be $(\theta - \hat{\theta})^2$. Generally, the true value of the parameter θ is not known with precision; this is, of course, the primary motivation for identification. Thus some statistical measure of the deviation of $\hat{\theta}$ from θ is more meaningful. We may formulate this error in general terms for the case of a vector parameter $\boldsymbol{\theta}$ as

$$\mathcal{R} = \mathcal{E}\left\{C[\tilde{\boldsymbol{\theta}}(\boldsymbol{Z})] \mid \boldsymbol{Z}\right\} = \int_{-\infty}^{\infty} C[\tilde{\boldsymbol{\theta}}(\boldsymbol{Z})]\, p(\boldsymbol{\theta} \mid \boldsymbol{Z})\, d\boldsymbol{\theta}$$
$$= \int_{-\infty}^{\infty}\int_{-\infty}^{\infty}\cdots\int_{-\infty}^{\infty} C[\tilde{\boldsymbol{\theta}}(\boldsymbol{Z})]\, p(\boldsymbol{\theta} \mid \boldsymbol{Z})\, d\theta_1\, d\theta_2 \cdots d\theta_N \tag{3.1-1}$$

Here, we define $C[\tilde{\boldsymbol{\theta}}(\boldsymbol{Z})]$ as the *cost of an error*. The error is defined as

$$\tilde{\boldsymbol{\theta}}(\boldsymbol{Z}) = \boldsymbol{\theta} - \hat{\boldsymbol{\theta}}(\boldsymbol{Z}) \tag{3.1-2}$$

where $\boldsymbol{\theta}$ is the true value of the parameter and $\hat{\boldsymbol{\theta}}(\boldsymbol{Z})$ is the estimate of the parameter based upon some observation $\boldsymbol{Z}$. Equation (3.1-1) represents the conditioned expected value of the cost of an error in estimating the parameter and is obtained by direct application of the fundamental theorem of expectation. Commonly used error costs are the squared error cost function

$$C[\tilde{\theta}(Z)] = \| \theta - \hat{\theta}(Z) \|_{\mathbf{S}}^2 = [\theta - \hat{\theta}(Z)]^T \mathbf{S} [\theta - \hat{\theta}(Z)] \qquad (3.1\text{-}3)$$

where $\mathbf{S}$ is a nonnegative definite symmetric matrix, and the uniform cost function

$$C[\tilde{\theta}(Z)] = \begin{cases} 1/\epsilon, & \text{if } \| \tilde{\theta}(Z) \| \geqslant \epsilon \\ 0, & \text{if } \| \tilde{\theta}(Z) \| < \epsilon \end{cases} \qquad (3.1\text{-}4)$$

Minimization of Eq. (3.1-1) with the squared error cost function of Eq. (3.1-3) by optimum choice of $\hat{\theta}(Z)$ easily leads to the conclusion that the best estimate of $\hat{\theta}(Z)$ is the conditional mean estimate

$$\hat{\theta}(Z) = \int_{-\infty}^{\infty} \theta p(\theta \mid Z) \, d\theta \qquad (3.1\text{-}5)$$

The error criterion of Eq. (3.1-4) is usually considered for vanishingly small ϵ, such that equivalent to Eq. (3.1-4) is the cost function

$$C[\tilde{\theta}(Z)] = - \prod_{i=1}^{N} \delta_D[\theta_i - \hat{\theta}_i(Z)] \qquad (3.1\text{-}6)$$

By substitution of Eq. (3.1-4) into Eq. (3.1-1) and evaluation of the result as ϵ approaches zero, or by direct use of Eq. (3.1-6) in Eq. (3.1-1), there results the maximum a posteriori cost function

$$\mathscr{R} = -p[\hat{\theta}(Z) \mid Z] \qquad (3.1\text{-}7)$$

in which we desire to minimize $\mathscr{R}$ by choice of $\hat{\theta}(Z)$. This $\hat{\theta}(Z)$ is called the maximum a posteriori (MAP) estimate, since it is the estimate obtained by maximizing the conditional density function $p[\theta \mid Z]$ and is normally obtained from solution of

$$\left. \frac{\partial p(\theta \mid Z)}{\partial \theta} \right|_{\theta = \hat{\theta}_{MAP}(Z)} = 0 \qquad (3.1\text{-}8)$$

We will make frequent use of the MAP estimate in the efforts which follow.

More classical than the maximum a posteriori cost function is the maximum likelihood (ML) cost function in which we desire to maximize the probability density of the observation conditioned upon the parameter $\boldsymbol{\theta}$. This is obtained from

$$\left.\frac{\partial p(Z \mid \boldsymbol{\theta})}{\partial \boldsymbol{\theta}}\right|_{\boldsymbol{\theta}=\hat{\boldsymbol{\theta}}_{\mathrm{ML}}(Z)} = \mathbf{0} \tag{3.1-9}$$

and is the estimate of $\boldsymbol{\theta}$ which most likely caused Z to occur. The MAP and ML estimators are clearly related, since by Bayes' rule

$$p(\boldsymbol{\theta} \mid Z) = p(Z \mid \boldsymbol{\theta})\, p(\boldsymbol{\theta})/p(Z) \tag{3.1-10}$$

Thus the MAP estimator is an ML estimator in which prior knowledge of the parameter, represented by the probability density $p(\boldsymbol{\theta})$, is considered in such a way as to improve the estimate. The ML estimator does not consider any prior knowledge concerning the value of $\boldsymbol{\theta}$, the parameter to be estimated or identified.

We may illustrate the difference between the two approaches by considering a simple problem for which we will have considerable need later. Consider the identification of the optimum N vector state $\mathbf{x}(k)$ which evolves from the linear unforced model

$$\mathbf{x}(k+1) = \boldsymbol{\Phi}(k+1, k)\, \mathbf{x}(k) \tag{3.1-11}$$

Noisy amplitude modulated M vector observations of $\mathbf{x}(k)$ are available of the form

$$\mathbf{z}(k) = \mathbf{H}(k)\, \mathbf{x}(k) + \mathbf{v}(k) \tag{3.1-12}$$

where $\mathbf{v}(k)$ is a zero-mean white Gaussian sequence with

$$\operatorname{cov}\{\mathbf{v}(k), \mathbf{v}(j)\} = \mathbf{V}_{\mathrm{v}}(k)\, \delta_{\mathrm{K}}(k-j), \qquad \operatorname{cov}\{\mathbf{v}(k), \mathbf{x}(j)\} = \mathbf{0} \tag{3.1-13}$$

where δ_{K} represents the Kronecker delta function. First we consider the problem of obtaining the maximum likelihood estimate of $\mathbf{x}(k_0)$ by maximizing the likelihood function

$$p[Z(k_{\mathrm{f}}) \mid \mathbf{x}(k_0)] \tag{3.1-14}$$

with respect to a choice of $\mathbf{x}(k_0)$. Identifying $\mathbf{x}(k_0)$ in this fashion is equivalent to identifying $\mathbf{x}(k)$, for $k_0 \leqslant k \leqslant k_{\mathrm{f}}$, since $\mathbf{x}(k)$ evolves from $\mathbf{x}(k_0)$ according to Eq. (3.1-11). The symbol $Z(k_{\mathrm{f}})$ is used to indicate all $\mathbf{z}(k)$ for $k_0 < k \leqslant k_{\mathrm{f}}$.

We have, for the first two conditional moments,

$$\mathscr{E}\{\mathbf{z}(k) \mid \mathbf{x}(k_0)\} = \mathbf{H}(k)\,\mathbf{x}(k) = \mathbf{H}(k)\,\mathbf{\Phi}(k, k_0)\,\mathbf{x}(k_0) \tag{3.1-15}$$

$$\operatorname{var}\{\mathbf{z}(k) \mid \mathbf{x}(k_0)\} = \mathbf{V}_{\mathrm{v}}(k) \tag{3.1-16}$$

where

$$\mathbf{\Phi}(k, k_0) = \prod_{j=k_0}^{k-1} \mathbf{\Phi}(j+1, j) \tag{3.1-17}$$

The likelihood function or density of $Z(k_{\mathrm{f}})$ conditioned upon $\mathbf{x}(k_0)$ is Gaussian and

$$p[Z(k_{\mathrm{f}}) \mid \mathbf{x}(k_0)] = \prod_{k=k_0+1}^{k_{\mathrm{f}}} \Bigg[\frac{1}{(2\pi)^{M/2}[\det \mathbf{V}_{\mathrm{v}}(k)]^{1/2}} \times \exp\{-0.5[\mathbf{z}(k) - \mathbf{H}(k)\,\mathbf{x}(k)]^{\mathrm{T}}\mathbf{V}_{\mathrm{v}}^{-1}(k)[\mathbf{z}(k) - \mathbf{H}(k)\,\mathbf{x}(k)]\}\Bigg] \tag{3.1-18}$$

We see that maximization of Eq. (3.1-18) is equivalent to minimization of the least squares curve fit type cost function

$$J = \frac{1}{2} \sum_{k=k_0+1}^{k_{\mathrm{f}}} \| \mathbf{z}(k) - \mathbf{H}(k)\,\mathbf{x}(k) \|^2_{\mathbf{V}_{\mathrm{v}}^{-1}(k)} \tag{3.1-19}$$

which must be accomplished with respect to $\mathbf{x}(k_0)$, where

$$\mathbf{x}(k) = \mathbf{\Phi}(k, k_0)\,\mathbf{x}(k_0) \tag{3.1-20}$$

Combining the two foregoing equations, differentiating with respect to $\mathbf{x}(k_0)$, and setting the result equal to zero, easily results in

$$\hat{\mathbf{x}}_{\mathrm{ML}}(k_0) = \mathbf{M}^{-1}(k_{\mathrm{f}}, k_0) \sum_{k+k_0+1}^{k_{\mathrm{f}}} \mathbf{\Phi}^{\mathrm{T}}(k, k_0)\,\mathbf{H}^{\mathrm{T}}(k)\,\mathbf{V}_{\mathrm{v}}^{-1}(k)\,\mathbf{z}(k) \tag{3.1-21}$$

where

$$\mathbf{M}(k_{\mathrm{f}}, k_0) = \sum_{k=k_0+1}^{k_{\mathrm{f}}} \mathbf{\Phi}^{\mathrm{T}}(k, k_0)\,\mathbf{H}^{\mathrm{T}}(k)\,\mathbf{V}_{\mathrm{v}}^{-1}(k)\,\mathbf{H}(k)\,\mathbf{\Phi}(k, k_0) \tag{3.1-22}$$

For a solution of Eq. (3.1-21) to exist, $\mathbf{M}(k_f, k_0)$ must have an inverse. The requirement that $\mathbf{M}(k_f, k_0)$ have an inverse is known as the observability requirement (Sage, 1968).

We may obtain the continuous limit for this problem by allowing the samples to become dense, such that as $k \to \infty$, $k_f T \to t_f$, $k_0 T \to t_0$, and $kT \to t$. The definitions

$$\mathbf{F}(t) \triangleq \lim_{\substack{k\to\infty \\ kT\to t}} [\mathbf{\Phi}(\overline{k+1\, T}, kT) - \mathbf{I}]/T \tag{3.1-23}$$

$$\mathbf{H}(t) \triangleq \lim_{\substack{k\to\infty \\ kT\to t}} \mathbf{H}(kT) \tag{3.1-24}$$

$$\mathbf{\Psi}_v(t) \triangleq \lim_{\substack{k\to\infty \\ kT\to t}} T\mathbf{V}_v(kT) \tag{3.1-25}$$

are used. The probability density function of Eq. (3.1-18) does not exist, since it is infinite in dimension, but the cost function of Eq. (3.1-19) is valid, as the samples become dense, and becomes

$$J' = \frac{1}{2} \int_{t_0}^{t_f} \| \mathbf{z}(t) - \mathbf{H}(t)\, \mathbf{x}(t) \|^2_{\mathbf{\Psi}_v^{-1}(t)}\, dt \tag{3.1-26}$$

The difference equation of Eq. (3.1-11) becomes the differential equation

$$\dot{\mathbf{x}} = \mathbf{F}(t)\, \mathbf{x}(t) \tag{3.1-27}$$

which must be used as a constraint in minimizing Eq. (3.1-26). By writing the solution to Eq. (3.1-27) as

$$\mathbf{x}(t) = \mathbf{\Phi}(t, t_0)\, \mathbf{x}(t_0) \tag{3.1-28}$$

and substituting this relation in Eq. (3.1-26), we differentiate Eq. (3.1-26) with respect to $\mathbf{x}(t_0)$, set the result equal to zero, and have

$$\hat{\mathbf{x}}(t_0) = \mathbf{M}^{-1}(t_f, t_0) \int_{t_0}^{t_f} \mathbf{\Phi}^T(t, t_0)\, \mathbf{H}^T(t)\, \mathbf{z}(t)\, dt \tag{3.1-29}$$

where

$$\mathbf{M}(t_f, t_0) = \int_{t_0}^{t_f} \mathbf{\Phi}^T(t, t_0)\, \mathbf{H}^T(t)\, \mathbf{\Psi}_v^{-1}(t)\, \mathbf{H}(t)\, \mathbf{\Phi}(t, t_0)\, dt \tag{3.1-30}$$

which will have an inverse if the system is observable (Sage, 1968).

We may obtain the density function from which to determine the MAP estimate in terms of that used to obtain the ML estimate by use of Eq. (3.1-10). $p[Z(k_f)]$ is just a constant with respect to the intended maximization, and thus maximization of $p[\mathbf{x}(k_0) \mid Z(k_f)]$ is entirely equivalent to maximization of the unconditional joint density function

$$p[\mathbf{x}(k_0), Z(k_f)] = p[Z(k_f) \mid \mathbf{x}(k_0)]\, p[\mathbf{x}(k_0)] \tag{3.1-31}$$

We see that we need more statistical information to accomplish the intended maximization. Specifically, we need the density (prior density) of $\mathbf{x}(k_0)$. We will assume it to be Gaussian with mean $\boldsymbol{\mu}_{\mathbf{x}_0}$ and variance $\mathbf{V}_{\mathbf{x}_0}$. The joint density function becomes, from Eq. (3.1-18),

$$\begin{aligned} p[\mathbf{x}(k_0), Z(k_f)] = {} & \frac{1}{(2\pi)^{N/2}[\det \mathbf{V}_{\mathbf{x}_0}]^{1/2}} \exp\{-0.5 \,\|\, \mathbf{x}(k_0) - \boldsymbol{\mu}_{\mathbf{x}_0} \|^2_{\mathbf{V}_{\mathbf{x}_0}^{-1}}\} \\ & \times \prod_{k=k_0+1}^{k_f} \frac{1}{(2\pi)^{M/2}[\det \mathbf{V}_{\mathbf{v}}(k)]^{1/2}} \exp\{-0.5 \,\|\, \mathbf{z}(k) \\ & - \mathbf{H}(k)\, \boldsymbol{\Phi}(k, k_0)\, \mathbf{x}(k_0)\|^2_{\mathbf{V}_{\mathbf{v}}^{-1}(k)}\} \end{aligned} \tag{3.1-32}$$

Maximization of this density function is equivalent to minimization of the cost function

$$J = \frac{1}{2} \| \mathbf{x}(k_0) - \boldsymbol{\mu}_{\mathbf{x}_0} \|^2_{\mathbf{V}_{\mathbf{x}_0}^{-1}} + \frac{1}{2} \sum_{k=k_0+1}^{k_f} \| \mathbf{z}(k) - \mathbf{H}(k)\, \boldsymbol{\Phi}(k, k_0)\, \mathbf{x}(k_0) \|^2_{\mathbf{V}_{\mathbf{v}}^{-1}(k)} \tag{3.1-33}$$

Setting the gradient with respect to $\mathbf{x}(k_0)$ of this least squares cost function equal to zero results in the estimator

$$\hat{\mathbf{x}}_{\mathrm{MAP}}(k_0) = [\mathbf{V}_{\mathbf{x}_0}^{-1} + \mathbf{M}(k_f, k_0)]^{-1} \left[\mathbf{V}_{\mathbf{x}_0}^{-1} \boldsymbol{\mu}_{\mathbf{x}_0} + \sum_{k=k_0+1}^{k_f} \boldsymbol{\Phi}^T(k, k_0) \mathbf{H}(k) \mathbf{V}_{\mathbf{v}}^{-1}(k) \mathbf{z}(k)\right] \tag{3.1-34}$$

where $\mathbf{M}(k_f, k_0)$ is defined by Eq. (3.1-22).

It is of interest to determine the error variance for the ML and MAP estimators which are given by

$$\mathrm{var}\{\tilde{\mathbf{x}}_{\mathrm{ML}}(k_0)\} = \mathrm{var}\{\mathbf{x}(k_0) - \hat{\mathbf{x}}_{\mathrm{ML}}(k_0)\} = \mathbf{M}^{-1}(k_f, k_0) \tag{3.1-35}$$

$$\mathrm{var}\{\tilde{\mathbf{x}}_{\mathrm{MAP}}(k_0)\} = \mathrm{var}\{\mathbf{x}(k_0) - \hat{\mathbf{x}}_{\mathrm{MAP}}(k_0)\} = [\mathbf{V}_{\mathbf{x}_0}^{-1} + \mathbf{M}(k_f, k_0)]^{-1} \tag{3.1-36}$$

We see that the error variance in estimation or identification of $\mathbf{x}(k_0)$ is less for the MAP estimator than it is for the ML estimator. Both estimates can easily be shown to be unbiased. These statements are based upon correct prior statistics being used to implement the identification algorithms. If the prior statistics are in error, the ML estimator may well be superior to the MAP estimator. For a complete discussion of error analysis, prior statistics, and associated sensitivity effects, the reader is referred to Chaps. 6 and 8 of Sage and Melsa (1971).

The continuous time MAP estimator is determined by letting the samples become dense and using definitions (3.1-23)–(3.1-25). Equation (3.1-33) becomes

$$J = \frac{1}{2} \| \mathbf{x}(t_0) - \boldsymbol{\mu}_{\mathbf{x}_0} \|^2_{\mathbf{V}^{-1}_{\mathbf{x}_0}} + \int_{t_0}^{t_f} \| \mathbf{z}(t) - \mathbf{H}(t)\, \boldsymbol{\Phi}(t, t_0)\, \mathbf{x}(t_0) \|^2_{\boldsymbol{\Psi}_{\mathbf{v}}^{-1}(t)}\, dt \tag{3.1-37}$$

The MAP estimator becomes

$$\hat{\mathbf{x}}_{\mathrm{MAP}}(t_0) = [\mathbf{V}^{-1}_{\mathbf{x}_0} + \mathbf{M}(t_f, t_0)]^{-1} \left[\mathbf{V}^{-1}_{\mathbf{x}_0} \boldsymbol{\mu}_{\mathbf{x}_0} + \int_{t_0}^{t_f} \boldsymbol{\Phi}^{\mathrm{T}}(t, t_0)\, \mathbf{H}(t)\, \boldsymbol{\Psi}_{\mathbf{v}}^{-1}(t)\, \mathbf{z}(t)\, dt\right] \tag{3.1-38}$$

where $\mathbf{M}(t_f, t_0)$ is defined by Eq. (3.1-30).

The error variances for the two continuous estimators become

$$\operatorname{var}\{\tilde{\mathbf{x}}_{\mathrm{ML}}(t_0)\} = \operatorname{var}\{\mathbf{x}(t_0) - \hat{\mathbf{x}}_{\mathrm{ML}}(t_0)\} = \mathbf{M}^{-1}(t_f, t_0) \tag{3.1-39}$$

$$\operatorname{var}\{\tilde{\mathbf{x}}_{\mathrm{MAP}}(t_0)\} = \operatorname{var}\{\mathbf{x}(t_0) - \hat{\mathbf{x}}_{\mathrm{MAP}}(t_0)\} = [\mathbf{V}^{-1}_{\mathbf{x}_0} + \mathbf{M}(t_f, t_0)]^{-1} \tag{3.1-40}$$

Again, the MAP error variance is less than the ML error variance.

It may appear that this linear estimation model does not apply to the system identification problem, which is often nonlinear. An exception is the identification of the impulse response of a linear system. Also, nonlinear identification problems may be linearized, as in Chap. 6, Quasilinearization. In that event, the methods of this section are directly applicable. One case does arise in which the methods of this section are inapplicable. This occurs whenever there are unknown inputs driving the system. We will now turn our attention to this problem and will consider nonlinear system dynamics.

3.2. Maximum a Posteriori Identification

In this section, we will examine the Bayes maximum a posteriori (MAP) approach to generalized estimation or system identification. Then we will show that many system identification problems may be cast in the framework of maximum a posteriori estimation. We will show that, with Gaussian a priori statistics, the maximum a posteriori estimate is equivalent to an appropriate least squares curve fit estimate. Included also will be a development of the appropriate cost functions and associated two point boundary value problems which may be resolved by the computational methods in Chaps. 4–7.

In the development of this section, emphasis will be placed on a discrete estimation model. The continuous results will be stated in terms of a redefined continuous estimation model.

The discrete message and observation models are given by[1]

$$\mathbf{x}(k+1) = \boldsymbol{\phi}[\mathbf{x}(k), k] + \boldsymbol{\Gamma}[\mathbf{x}(k), k]\, \mathbf{w}(k) \tag{3.2-1}$$

$$\mathbf{z}(k) = \mathbf{h}[\mathbf{x}(k), k] + \mathbf{v}(k) \tag{3.2-2}$$

where

$\mathbf{x}(k)$ = N-dimensional state vector

$\boldsymbol{\phi}[\mathbf{x}(k), k]$ = N-dimensional vector-valued function which includes any known inputs

$\boldsymbol{\Gamma}[\mathbf{x}(k), k]$ = $N \times M$ matrix

$\mathbf{w}(k)$ = M-dimensional plant noise vector

$\mathbf{z}(k)$ = R-dimensional observation vector

$\mathbf{h}[\mathbf{x}(k), k]$ = R-dimensional vector-valued function

$\mathbf{v}(k)$ = R-dimensional observation noise vector

$\mathbf{x}(k)$ is used to represent the generalized state vector at the kth sample time $\mathbf{x}(t_k)$ or $\mathbf{x}(kT_k)$. For the discrete estimation model, $\mathbf{w}(k)$ and $\mathbf{v}(k)$ are assumed to be independent, zero mean, Gauss–Markov white sequences, such that

$$\mathscr{E}\{\mathbf{w}(k)\, \mathbf{w}^{\mathrm{T}}(j)\} = \mathbf{V}_{\mathrm{w}}(k)\, \delta_{\mathrm{K}}(k-j) \tag{3.2-3}$$

$$\mathscr{E}\{\mathbf{v}(k)\, \mathbf{v}^{\mathrm{T}}(j)\} = \mathbf{V}_{\mathrm{v}}(k)\, \delta_{\mathrm{K}}(k-j) \tag{3.2-4}$$

[1] Specific formulation of identification problems in this format will be considered later in this section.

where $\delta_K(k-j)$ is the Kronecker delta function, and $\mathbf{V}_w(k)$ and $\mathbf{V}_v(k)$ are symmetric nonnegative definite $M \times M$ and $R \times R$ covariance matrices, respectively.

A continuous estimation model may often be derived using a nonrigorous limiting procedure such that the continuous model follows directly from the discrete model as the samples become dense; that is, as $t_{k+1} - t_k = T_k$ (the sample period) becomes zero $t_k \to t$. This continuous estimation model is defined by

$$\dot{\mathbf{x}}(t) = \mathbf{f}[\mathbf{x}(t), t] + \mathbf{G}[\mathbf{x}(t), t]\, \mathbf{w}(t) \tag{3.2-5}$$

$$\mathbf{z}(t) = \mathbf{h}[\mathbf{x}(t), t] + \mathbf{v}(t) \tag{3.2-6}$$

where $\mathbf{w}(t)$ and $\mathbf{v}(t)$ are assumed to be zero mean, white and independent, with Gaussian amplitude distributions such that

$$\mathscr{E}\{\mathbf{w}(t)\mathbf{w}^T(\tau)\} = \mathbf{\Psi}_w(t)\delta_D(t-\tau) \tag{3.2-7}$$

$$\mathscr{E}\{\mathbf{v}(t)\mathbf{v}^T(\tau)\} = \mathbf{\Psi}_v(t)\delta_D(t-\tau) \tag{3.2-8}$$

The discrete and continuous models are related by the nonrigorous limits

$$\mathbf{f}[\mathbf{x}(t), t] \triangleq \lim_{\substack{T_k \to 0 \\ t_k \to t}} \frac{1}{T_k}\{\boldsymbol{\phi}[\mathbf{x}(k), k] - \mathbf{x}(k)\} \tag{3.2-9}$$

$$\mathbf{G}[\mathbf{x}(t), t] \triangleq \lim_{\substack{T_k \to 0 \\ t_k \to t}} \frac{1}{T_k}\{\mathbf{\Gamma}[\mathbf{x}(k), k]\} \tag{3.2-10}$$

$$\mathbf{h}[\mathbf{x}(t), t] \triangleq \lim_{\substack{T_k \to 0 \\ t_k \to t}} \mathbf{h}[\mathbf{x}(k), k] \tag{3.2-11}$$

$$\mathbf{\Psi}_w(t) \triangleq \lim_{\substack{T_k \to 0 \\ t_k \to t}} T_k \mathbf{V}_w(k) \tag{3.2-12}$$

$$\mathbf{\Psi}_v(t) \triangleq \lim_{\substack{T_k \to 0 \\ t_k \to t}} T_k \mathbf{V}_v(k) \tag{3.2-13}$$

We observe that the continuous nonlinear differential equation (3.2-5) is quite nonrigorous and should be written as the stochastic differential equation.

$$\mathbf{dx}(t) = \mathbf{f}[\mathbf{x}(t), t]\, dt + \mathbf{G}[\mathbf{x}(t), t]\, \mathbf{du}(t) \tag{3.2-14}$$

where $\mathbf{du}(t)$ is a Wiener process. In a similar manner, results of Eqs. (3.1-9)–(3.1-13) should be obtained rigorously from the stochastic calculus (Sage and Melsa, 1971). The foregoing comments also apply to the limiting procedure used in the remainder of the chapter.

The sequences $\mathbf{x}(k_0)$, $\mathbf{x}(k_1)$,..., $\mathbf{x}(k_f)$, and $\mathbf{z}(k_1)$, $\mathbf{z}(k_2)$,..., $\mathbf{z}(k_f)$ are denoted by $X(k_f)$ and $Z(k_f)$, respectively. In like manner, the continuous values of $\mathbf{x}(t)$ and $\mathbf{z}(t)$ in the interval $[t_0, t_f]$ are denoted by $X(t_f)$ and $Z(t_f)$. $p[X(k_f) \mid Z(k_f)]$ and $p[X(t_f) \mid Z(t_f)]$ denote the conditional probability density function of X given measurements Z. We further assume that $p[\mathbf{x}(k_0)]$ and $p[\mathbf{x}(t_0)]$ are known and are normal with mean $\boldsymbol{\mu}_{\mathbf{x}_0}$ and variance $\mathbf{V}_{\mathbf{x}_0}$.

The best estimate of the generalized state vector $\mathbf{x}$ throughout an interval will, in general, depend on the criteria used to determine the best estimate. Here, the term "best estimate" denotes that estimate derived from maximizing the conditional probability function $p[X \mid Z]$ with respect to X throughout the interval. The resulting estimator is known as the Bayesian maximum likelihood or maximum a posteriori estimator (Sage, 1968; Sage and Melsa, 1971). In the following development, we will derive results for the discrete case, and the continuous results will simply be stated.

Applying Bayes' rule to $p[X(k_f) \mid Z(k_f)]$ results in

$$p[X(k_f) \mid Z(k_f)] = \frac{p[Z(k_f) \mid X(k_f)]\, p[X(k_f)]}{p[Z(k_f)]} \tag{3.2-15}$$

From Eq. (3.2-2) it is clear that if $\mathbf{x}(k)$ is known, $p[\mathbf{z}(k) \mid \mathbf{x}(k)]$ is Gaussian, since $\mathbf{v}(k)$ is Gaussian. If $X(k_f)$ is given, we have

$$p[Z(k_f) \mid X(k_f)] = \prod_{k=k_0+1}^{k_f} \left[\frac{\exp\{-\frac{1}{2}(\mathbf{z}(k) - \mathbf{h}[\mathbf{x}(k), k])^{\mathrm{T}}\mathbf{V}_{\mathbf{v}}^{-1}(k)(\mathbf{z}(k) - \mathbf{h}[\mathbf{x}(k), k])\}}{(2\pi)^{\mathrm{R}/2} \det[\mathbf{V}_{\mathbf{v}}(k)]^{1/2}}\right] \tag{3.2-16}$$

Using the chain rule of probability,

$$p[\boldsymbol{\alpha}, \boldsymbol{\beta}] = p[\boldsymbol{\alpha} \mid \boldsymbol{\beta}]\, p[\boldsymbol{\beta}] \tag{3.2-17}$$

results in

$$p[X(k_f)] = p[\mathbf{x}(k_f) \mid X(k_f - 1)]\, p[\mathbf{x}(k_f - 1) \mid X(k_f - 2)] \cdots p[\mathbf{x}(k_1) \mid \mathbf{x}(k_0)]\, p[\mathbf{x}(k_0)] \tag{3.2-18}$$

Since $\mathbf{w}(k)$ is a white Gauss–Markov sequence, $\mathbf{x}(k)$ is Markov and

$$p[\mathbf{x}(k_f) \mid X(k_f - 1)] = p[\mathbf{x}(k_f) \mid \mathbf{x}(k_f - 1)] \tag{3.2-19}$$

Thus, $p[X(k_f)]$ is composed of Gaussian terms and

$$p[X(k_f)] = p[\mathbf{x}(k_0)] \prod_{k=k_0+1}^{k_f} p[\mathbf{x}(k) \mid \mathbf{x}(k-1)] \tag{3.2-20}$$

where $p[\mathbf{x}(k) \mid \mathbf{x}(k-1)]$ is Gaussian and, from Eq. (3.2-1), has mean $\boldsymbol{\phi}[\mathbf{x}(k-1), k-1]$ and variance

$$\boldsymbol{\Gamma}[\mathbf{x}(k-1), k-1]\, \mathbf{V}_{\mathbf{w}}(k-1)\, \boldsymbol{\Gamma}^{\mathrm{T}}[\mathbf{x}(k-1), k-1].$$

$p[Z(k_f)]$ contains no terms in $\mathbf{x}(k)$, and $Z(k_f)$ is the known conditioning variable for the intended maximization. Thus $p[Z(k_f)]$ can be considered a normalizing constant with respect to the intended maximization. After a modest amount of manipulation, Eq. (3.2-15) may be written in terms of Eqs. (3.2-16) and (3.2-20) as

$$\begin{aligned} p[X(k_f) \mid Z(k_f)] = A \exp\Big\{ &-\frac{1}{2} \sum_{k=k_0+1}^{k_f} \| \mathbf{z}(k) - \mathbf{h}[\mathbf{x}(k), k] \|^2_{\mathbf{V}_{\mathbf{v}}^{-1}(k)} \\ &-\frac{1}{2} \sum_{k=k_0+1}^{k_f} \| \mathbf{x}(k) - \boldsymbol{\phi}[\mathbf{x}(k-1), k-1] \|^2_{\boldsymbol{\Omega}^{-1}(k-1)} \\ &-\frac{1}{2} \| \mathbf{x}(k_0) - \boldsymbol{\mu}_{\mathbf{x}}(k_0) \|^2_{\mathbf{V}_{\tilde{\mathbf{x}}_0}^{-1}} \Big\} \end{aligned} \tag{3.2-21}$$

where we assume[2] that A is not a function of $\mathbf{x}(k)$ and

$$\boldsymbol{\Omega}(k) = \boldsymbol{\Gamma}[\mathbf{x}(k), k]\, \mathbf{V}_{\mathbf{w}}(k)\, \boldsymbol{\Gamma}^{\mathrm{T}}[\mathbf{x}(k), k] \tag{3.2-22}$$

It is now clear that maximizing Eq. (3.2-21) with respect to a choice of $X(k_f)$ is equivalent to minimizing

$$\begin{aligned} J = \frac{1}{2} \| \mathbf{x}(k_0) - \boldsymbol{\mu}_{\mathbf{x}}(k_0) \|^2_{\mathbf{V}_{\tilde{\mathbf{x}}_0}^{-1}} &+ \frac{1}{2} \sum_{k=k_0}^{k_f-1} \| \mathbf{z}(k+1) - \mathbf{h}[\mathbf{x}(k+1), k+1] \|^2_{\mathbf{V}_{\mathbf{v}}^{-1}(k+1)} \\ &+ \frac{1}{2} \sum_{k=k_0}^{k_f-1} \| \mathbf{w}(k) \|^2_{\mathbf{V}_{\mathbf{w}}^{-1}(k)} \end{aligned} \tag{3.2-23}$$

[2] This is true if $\boldsymbol{\Gamma}$ is not a function of $\mathbf{x}$ and is a useful result in general.

Similarly, maximization of $p[X(t_f) \mid Z(t_f)]$ is equivalent to minimizing

$$J' = \frac{1}{2} \| \mathbf{x}(t_0) - \boldsymbol{\mu}_{\mathbf{x}}(t_0) \|^2_{\mathbf{V}^{-1}_{\tilde{\mathbf{x}}_0}} + \frac{1}{2} \int_{t_0}^{t_f} \{ \| \mathbf{z}(t) - \mathbf{h}[\mathbf{x}(t), t] \|^2_{\boldsymbol{\Psi}^{-1}_{\mathbf{v}}(t)} + \| \mathbf{w}(t) \|^2_{\boldsymbol{\Psi}^{-1}_{\mathbf{w}}(t)} \} \, dt \qquad (3.2\text{-}24)$$

subject to the differential equality constraint of Eq. (3.2-5). This relation, Eq. (3.2-24), is a least squares curve fit cost function and, if the prior variances are chosen correctly and appropriate Gaussian assumptions on $\mathbf{x}(k_0)$, $\mathbf{v}(k)$ and $\mathbf{w}(k)$ hold, is equivalent to the maximum a posteriori cost function.

Equation (3.2-23) is of the form that suggests application of the discrete maximum principle or the discrete Euler–Lagrange (Sage, 1968) equations. The Hamiltonian is defined as

$$H[\mathbf{x}(k), \mathbf{w}(k), \boldsymbol{\lambda}(k+1), k] = \frac{1}{2} \| \mathbf{z}(k+1) - \ell[\mathbf{x}(k), \mathbf{w}(k), k+1] \|^2_{\mathbf{V}^{-1}_{\mathbf{v}}(k+1)} + \frac{1}{2} \| \mathbf{w}(k) \|^2_{\mathbf{V}^{-1}_{\mathbf{w}}(k)} + \boldsymbol{\lambda}^T(k+1) \, \boldsymbol{\phi}[\mathbf{x}(k), k] + \boldsymbol{\lambda}^T(k+1] \, \boldsymbol{\Gamma}[\mathbf{x}(k), k] \, \mathbf{w}(k) \qquad (3.2\text{-}25)$$

where

$$\ell[\mathbf{x}(k), \mathbf{w}(k), k+1] \triangleq \mathbf{h}\{\boldsymbol{\phi}[\mathbf{x}(k), k] + \boldsymbol{\Gamma}[\mathbf{x}(k), k] \, \mathbf{w}(k), k+1\} = \mathbf{h}[\mathbf{x}(k+1), k+1] \qquad (3.2\text{-}26)$$

The canonic equations and boundary conditions are given by

$$\hat{\mathbf{x}}(k+1 \mid k_f) = \left. \frac{\partial H}{\partial \boldsymbol{\lambda}(k+1)} \right|_{\mathbf{x}(k)=\hat{\mathbf{x}}(k \mid k_f)}, \qquad \boldsymbol{\lambda}(k_0 \mid k_0) = \mathbf{V}^{-1}_{\tilde{\mathbf{x}}_0} [\mathbf{x}(k_0) - \boldsymbol{\mu}_{\mathbf{x}}(k_0)] \qquad (3.2\text{-}27)$$

$$\boldsymbol{\lambda}(k \mid k_f) = \left. \frac{\partial H}{\partial \mathbf{x}(k)} \right|_{\mathbf{x}(k)=\hat{\mathbf{x}}(k \mid k_f)}, \qquad \boldsymbol{\lambda}(k_f \mid k_f) = \mathbf{0} \qquad (3.2\text{-}28)$$

$$\left. \frac{\partial H}{\partial \mathbf{w}(k)} \right|_{\mathbf{w}(k)=\hat{\mathbf{w}}(k)} = \mathbf{0} \qquad (3.2\text{-}29)$$

These canonic equations and the associated boundary conditions specify a nonlinear two-point boundary value problem (TPBVP), the solution to which yields a fixed interval smoothing estimate.

After a considerable amount of algebraic manipulation, we obtain the canonic equations from the foregoing:

$$\hat{\mathbf{x}}(k+1 \mid k_f) = \boldsymbol{\phi}[\hat{\mathbf{x}}(k \mid k_f), k] - \boldsymbol{\Gamma}[\hat{\mathbf{x}}(k \mid k_f), k]\mathbf{V}_w(k)\boldsymbol{\Gamma}^T[\hat{\mathbf{x}}(k \mid k_f), k]\boldsymbol{\Psi}^{-1}\boldsymbol{\lambda}(k \mid k_f) \tag{3.2-30}$$

$$\boldsymbol{\lambda}(k+1 \mid k_f) = \boldsymbol{\Psi}^{-1}\boldsymbol{\lambda}(k \mid k_f) + \frac{\partial \mathbf{h}^T[\hat{\mathbf{x}}(k+1 \mid k_f)]}{\partial \hat{\mathbf{x}}(k+1 \mid k_f)} \mathbf{V}_v^{-1}(k+1) \times [\mathbf{z}(k+1) - \mathbf{h}[\hat{\mathbf{x}}(k+1 \mid k_f), k+1]] \tag{3.2-31}$$

where

$$\boldsymbol{\Psi} = \frac{\partial \boldsymbol{\phi}^T[\hat{\mathbf{x}}(k \mid k_f), k]}{\partial \hat{\mathbf{x}}(k \mid k_f)} + \frac{\partial[\boldsymbol{\Gamma}[\hat{\mathbf{x}}(k \mid k_f), k]\,\hat{\mathbf{w}}(k)]^T}{\partial \hat{\mathbf{x}}(k \mid k_f)}$$
$$\hat{\mathbf{w}}(k) = -\mathbf{V}_w(k)\,\boldsymbol{\Gamma}^T[\hat{\mathbf{x}}(k \mid k_f), k]\,\boldsymbol{\Psi}^{-1}\boldsymbol{\lambda}(k \mid k_f) \tag{3.2-32}$$

The terms in $\hat{\mathbf{x}}(k+1 \mid k_f)$ and $\boldsymbol{\lambda}(k+1 \mid k_f)$ involve expressions quadratic in $\boldsymbol{\lambda}$. When using the invariant imbedding procedure, terms of powers greater than the first in $\boldsymbol{\lambda}$ can be eliminated. Thus for solution by the invariant imbedding procedure, we may use the equivalent expression

$$\boldsymbol{\Psi} = \frac{\partial \boldsymbol{\phi}^T[\hat{\mathbf{x}}(k \mid k_f), k]}{\partial \hat{\mathbf{x}}(k \mid k_f)} \tag{3.2-33}$$

These relations are solved subject to the two-point boundary conditions

$$\boldsymbol{\lambda}(k_0 \mid k_0) = -\mathbf{V}_{\tilde{\mathbf{x}}_0}^{-1}[\hat{\mathbf{x}}(k_0) - \boldsymbol{\mu}_x(k_0)], \qquad \boldsymbol{\lambda}(k_f \mid k_f) = \mathbf{0} \tag{3.2-34}$$

The two-point boundary value problem for the continuous case may be obtained by letting the samples become dense in Eqs. (3.2-30) and (3.2-31) or by using the continuous maximum principle (Sage, 1968) to minimize the cost function of Eq. (3.2-24) subject to the differential equality constraint of Eq. (3.2-5). If we use the latter approach, we define the Hamiltonian

$$H[\mathbf{x}(t), \mathbf{w}(t), \boldsymbol{\lambda}(t), t] = \tfrac{1}{2} \| \mathbf{z}(t) - \mathbf{h}[\mathbf{x}(t), t]\|^2_{\boldsymbol{\Psi}_v^{-1}(t)} + \tfrac{1}{2} \| \mathbf{w}(t)\|^2_{\boldsymbol{\Psi}_w^{-1}(t)} + \boldsymbol{\lambda}^T(t)\{\mathbf{f}[\mathbf{x}(t), t] + \mathbf{G}[\mathbf{x}(t), t]\,\mathbf{w}(t)\} \tag{3.2-35}$$

and obtain the canonic equations from

$$\dot{\hat{\mathbf{x}}} = \frac{\partial H}{\partial \boldsymbol{\lambda}}, \qquad \boldsymbol{\lambda}(t_0) = \mathbf{V}_{\tilde{\mathbf{x}}_0}^{-1}(t)[\hat{\mathbf{x}}(t_0) - \boldsymbol{\mu}_{\mathbf{x}}(t_0)]$$

$$\dot{\boldsymbol{\lambda}} = -\frac{\partial H}{\partial \hat{\mathbf{x}}}, \qquad \boldsymbol{\lambda}(t_f) = \mathbf{0} \tag{3.2-36}$$

$$\frac{\partial H}{\partial \hat{\mathbf{w}}} = \mathbf{0}$$

Either procedure leads to the two-point boundary value problem

$$\dot{\hat{\mathbf{x}}} = \mathbf{f}[\hat{\mathbf{x}}(t), t] - \mathbf{G}[\hat{\mathbf{x}}(t), t]\, \boldsymbol{\Psi}_{\mathrm{w}}(t)\, \mathbf{G}^{\mathrm{T}}[\hat{\mathbf{x}}(t), t]\, \boldsymbol{\lambda}(t) \tag{3.2-37}$$

$$\dot{\boldsymbol{\lambda}} = \frac{\partial \mathbf{h}^{\mathrm{T}}[\hat{\mathbf{x}}(t), t]}{\partial \hat{\mathbf{x}}(t)}\, \boldsymbol{\Psi}_{\mathrm{v}}^{-1}(t)\{\mathbf{z}(t) - \mathbf{h}[\hat{\mathbf{x}}(t), t]\} - \frac{\partial \mathbf{f}^{\mathrm{T}}[\hat{\mathbf{x}}(t), t]}{\partial \hat{\mathbf{x}}(t)}\, \boldsymbol{\lambda}(t)$$
$$+ \frac{\partial\{\boldsymbol{\lambda}^{\mathrm{T}}(t)\, \mathbf{G}[\hat{\mathbf{x}}(t), t]\, \boldsymbol{\Psi}_{\mathrm{w}}(t)\, \mathbf{G}^{\mathrm{T}}[\hat{\mathbf{x}}(t), t]\}}{\partial \hat{\mathbf{x}}(t)}\, \boldsymbol{\lambda}(t) \tag{3.2-38}$$

with the initial and terminal conditions

$$\boldsymbol{\lambda}(t_0) = -\mathbf{V}_{\mathbf{x}_0}^{-1}(t)[\hat{\mathbf{x}}(t_0) - \boldsymbol{\mu}_{\mathbf{x}}(t_0)], \qquad \boldsymbol{\lambda}(t_f) = \mathbf{0} \tag{3.2-39}$$

The state variable can and should be written here as $\hat{\mathbf{x}}(t \mid t_f)$ to indicate that, if the two point boundary value problem is actually solved, the smoothing solution or estimate of $\mathbf{x}$ with observation through time t_f is obtained. We will, in the next four chapters, obtain both smoothing solutions $\hat{\mathbf{x}}(t \mid t_f)$ and sequential or filtering solutions $\hat{\mathbf{x}}(t \mid t)$ to the two-point boundary value problem of Eqs. (3.2-37)–(3.2-39). It is now of interest to relate the assumed message and observation models to the system identification problem. We will accomplish this only for the continuous problem. However, a similar procedure holds for the discrete problem.

Consider the generalized estimation and identification problem in which the message model is

$$\dot{x} = \mathbf{f}[x(t), \mathbf{a}, t] + \mathbf{G}[x(t), \mathbf{b}, t]\, \mathbf{w}(t) + \mathbf{c} \tag{3.2-40}$$

The observation model is

$$\mathbf{z}(t) = \mathbf{h}[x(t), \mathbf{d}, t] + \mathbf{e} + \mathbf{v}(t) \tag{3.2-41}$$

Here **a**, **b**, **c**, **d**, and **e** are constant parameters which are to be identified. Since they are constants, the differential relations

$$\dot{\mathbf{a}} = 0, \qquad \dot{\mathbf{b}} = 0, \qquad \dot{\mathbf{c}} = 0, \qquad \dot{\mathbf{d}} = 0, \qquad \dot{\mathbf{e}} = 0 \tag{3.2-42}$$

are valid. This model is sufficiently general to represent a vast variety of identification situations: **c** may represent an unknown mean value of the plant noise input; **e** may represent the unknown mean value of the measurement noise; **b** may be used to indicate certain unknown variance terms in the plant noise input; **a** and **d** represent unknowns associated with the message and observation models.

By defining the generalized state vector

$$\mathbf{x}^{\mathrm{T}} = [x^{\mathrm{T}} \ \mathbf{a}^{\mathrm{T}} \ \mathbf{b}^{\mathrm{T}} \ \mathbf{c}^{\mathrm{T}} \ \mathbf{d}^{\mathrm{T}} \ \mathbf{c}^{\mathrm{T}}] \tag{3.2-43}$$

it is clear that our identification problem has been cast into the framework of Eqs. (3.2-5) and (3.2-6). To obtain maximum a posteriori identification by minimization of Eq. (3.2-24), obtained by solution of the TPBVP of Eqs. (3.2-27)–(3.2-39), it is necessary that the random constant parameters **a**, **b**, **c**, **d**, and **e** have a Gaussian probability density function with known mean and variance. If these requirements are not satisfied, solution of the TPBVP still guarantees a least squares curve fit with the cost function of Eq. (3.2-24).

Four problems of interest in system identification are not included in the models of Eqs. (3.2-40)–(3.2-42):

1. unknown measurement noise variance problems,
2. colored plant and/or measurement noise problems,
3. time-varying unknown parameter problems, and
4. correlated plant and measurement noise problems.

The cost functions of the next two sections will allow us to solve system identification problems in which measurement noise variances are unknown. Colored plant noise problems may be easily resolved by adjoining the state vector such that the plant noise for the adjoined state vector is white. Colored measurement noise problems may be resolved by differentiating the observation vector **z** a sufficient number of times such that white noise appears as the measurement noise in the differentiated observations. Sage and Melsa (1971) present detailed discussions of the colored plant and measurement noise problems.

Time-varying unknown parameter problems may be treated by

regarding the unknown parameter as a stochastic process satisfying the Markov model

$$\dot{\mathbf{a}} = \mathbf{A}\mathbf{a}(t) + \mathbf{B}\boldsymbol{\eta}(t) \tag{3.2-44}$$

where $\boldsymbol{\eta}(t)$ is Gaussian white noise with a known mean and variance coefficient and $\mathbf{a}(t)$ is the unknown time varying parameter. In order to obtain maximum a posteriori identification, it is necessary that the prior distribution of the parameter $\mathbf{a}(t)$ be Gaussian with known mean $\boldsymbol{\mu}_{\mathbf{a}}(t_0)$ and variance $\mathbf{V}_{\mathbf{a}}(t_0)$.

Finally, many identification problems are similar to that illustrated in Fig. 3.2-1 in which a noise corrupted version of the input plant

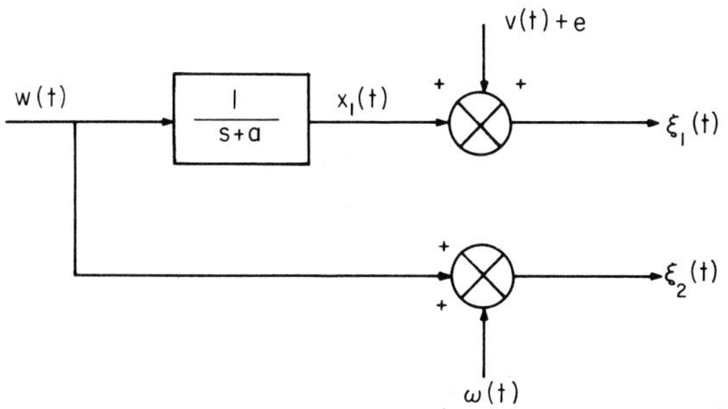

FIG. 3.2-1. A simple identification problem.

noise may be observed. In this case, it is clear that the discrete model of Eqs. (3.2-1)–(3.2-4) and the continuous model of Eqs. (3.2-5)–(3.2-7) are valid, except that the plant and augmented measurement noise vector must be regarded as correlated with

$$\mathscr{E}\{\mathbf{w}(k)\,\mathbf{v}^{\mathrm{T}}(j)\} = \mathbf{V}_{\mathbf{wv}}(k)\,\delta_{\mathrm{K}}(k-j) \tag{3.2-45}$$

$$\mathscr{E}\{\mathbf{w}(t)\,\mathbf{v}^{\mathrm{T}}(\tau)\} = \boldsymbol{\Psi}_{\mathbf{wv}}(t)\,\delta_{\mathrm{D}}(t-\tau) \tag{3.2-46}$$

$$\boldsymbol{\Psi}_{\mathbf{wv}}(t) = \lim_{\substack{t_k \to t \\ T_k \to 0}} T_k \mathbf{V}_{\mathbf{wv}}(t_k) \tag{3.2-47}$$

This alters the equivalent cost functions for the maximum a posteriori identification. The cost function equivalent to Eq. (3.2-23) becomes

$$J = \frac{1}{2}\|\mathbf{x}(k_0) - \boldsymbol{\mu}_{\mathbf{x}}(k_0)\|^2_{\mathbf{V}^{-1}_{\tilde{\mathbf{x}}_0}} + \frac{1}{2}\sum_{k=k_0}^{k_f-1}\|\mathbf{y}(k)\|^2_{\mathbf{Y}^{-1}(k)} + \frac{1}{2}\|\mathbf{w}(k_0)\|^2_{\mathbf{V}^{-1}_{\overline{\mathbf{w}}}(k_0)} \tag{3.2-48}$$

where

$$\mathbf{y}(k) = \begin{bmatrix} \mathbf{w}(k+1) \\ \mathbf{z}(k+1) - \mathbf{h}[\mathbf{x}(k+1), k+1] \end{bmatrix} \tag{3.2-49}$$

$$\mathbf{Y}(k) = \begin{bmatrix} \mathbf{V}_{\mathrm{w}}(k+1) & \mathbf{V}_{\mathrm{wv}}(k+1) \\ \mathbf{V}_{\mathrm{vw}}(k+1) & \mathbf{V}_{\mathrm{v}}(k+1) \end{bmatrix} \tag{3.2-50}$$

It is easily verified that

$$\mathbf{Y}^{-1}(k) = \begin{bmatrix} \Xi_{11}(k) & \Xi_{12}(k) \\ \Xi_{12}^{T}(k) & \Xi_{22}(k) \end{bmatrix} \tag{3.2-51}$$

where

$$\Xi_{11}(k) = [\mathbf{V}_{\mathrm{w}}(k+1) - \mathbf{V}_{\mathrm{wv}}(k+1)\, \mathbf{V}_{\mathrm{v}}^{-1}(k+1)\, \mathbf{V}_{\mathrm{vw}}(k+1)]^{-1} \tag{3.2-52}$$

$$\Xi_{12}(k) = -\Xi_{11}(k)\, \mathbf{V}_{\mathrm{wv}}(k+1)\, \mathbf{V}_{\mathrm{v}}^{-1}(k+1) \tag{3.2-53}$$

$$\Xi_{22}(k) = [\mathbf{V}_{\mathrm{v}}(k+1) - \mathbf{V}_{\mathrm{vw}}(k+1)\, \mathbf{V}_{\mathrm{w}}^{-1}(k+1)\, \mathbf{V}_{\mathrm{wv}}(k+1)]^{-1} \tag{3.2-54}$$

For the continuous case, the cost function of Eq. (3.2-24) becomes

$$J' = \frac{1}{2} \| \mathbf{x}(t_0) - \boldsymbol{\mu}_{\mathrm{x}}(t_0) \|^2_{\mathbf{V}_{\bar{\mathbf{x}}_0}^{-1}} + \frac{1}{2} \int_{t_0}^{t_{\mathrm{f}}} \| \mathbf{y}(t) \|^2_{\mathbf{Y}^{-1}(t)}\, dt \tag{3.2-55}$$

where

$$\mathbf{y}(t) = \begin{bmatrix} \mathbf{w}(t) \\ \mathbf{z}(t) - \mathbf{h}[\mathbf{x}(t), t] \end{bmatrix} \tag{3.2-56}$$

$$\mathbf{Y}(t) = \begin{bmatrix} \boldsymbol{\Psi}_{\mathrm{w}}(t) & \boldsymbol{\Psi}_{\mathrm{wv}}(t) \\ \boldsymbol{\Psi}_{\mathrm{vw}}(t) & \boldsymbol{\Psi}_{\mathrm{v}}(t) \end{bmatrix} \tag{3.2-57}$$

Since

$$\mathbf{Y}^{-1}(t) = \begin{bmatrix} \Xi_{11}(t) & \Xi_{12}(t) \\ \Xi_{12}^{\mathrm{T}}(t) & \Xi_{22}(t) \end{bmatrix} \tag{3.2-58}$$

where

$$\Xi_{11}(t) = [\boldsymbol{\Psi}_{\mathrm{w}}(t) - \boldsymbol{\Psi}_{\mathrm{wv}}(t)\, \boldsymbol{\Psi}_{\mathrm{v}}^{-1}(t)\, \boldsymbol{\Psi}_{\mathrm{vw}}(t)]^{-1} \tag{3.2-59}$$

$$\Xi_{12}(t) = -\Xi_{11}(t)\, \boldsymbol{\Psi}_{\mathrm{wv}}(t)\, \boldsymbol{\Psi}_{\mathrm{v}}^{-1}(t) \tag{3.2-60}$$

$$\Xi_{22}(t) = [\boldsymbol{\Psi}_{\mathrm{v}}(t) - \boldsymbol{\Psi}_{\mathrm{vw}}(t)\, \boldsymbol{\Psi}_{\mathrm{w}}^{-1}(t)\, \boldsymbol{\Psi}_{\mathrm{wv}}(t)]^{-1} \tag{3.2-61}$$

it follows that the cost function of Eq. (3.2-55) becomes

$$J' = \frac{1}{2}\| \mathbf{x}(t_0) - \boldsymbol{\mu}_x(t_0)\|^2_{\mathbf{V}^{-1}_{\tilde{\mathbf{x}}_0}} + \frac{1}{2}\int_{t_0}^{t_f} \{\| \mathbf{z}(t) - \mathbf{h}[\mathbf{x}(t), t]\|^2_{\Xi_{22}(t)}$$

$$+ 2\mathbf{w}^T(t)\, \Xi_{12}(t)[\mathbf{z}(t) - \mathbf{h}[\mathbf{x}(t), t]] + \| \mathbf{w}(t)\|^2_{\Xi_{11}(t)}\}\, dt \qquad (3.2\text{-}62)$$

When this cost function is minimized subject to the equality constraint of Eq. (3.2-5) by direct use of the maximum principle, the canonic equations (TPBVP)

$$\dot{\hat{\mathbf{x}}} = \mathbf{f}[\hat{\mathbf{x}}(t), t] - \mathbf{G}[\hat{\mathbf{x}}(t), t]\{\boldsymbol{\Psi}_w(t) - \boldsymbol{\Psi}_{wv}(t)\boldsymbol{\Psi}_v^{-1}(t)\boldsymbol{\Psi}_{vw}(t)\}\mathbf{G}^T[\hat{\mathbf{x}}(t), t]\boldsymbol{\lambda}(t)$$

$$- \mathbf{G}[\hat{\mathbf{x}}(t), t]\boldsymbol{\Psi}_{wv}(t)\boldsymbol{\Psi}_v^{-1}(t)\{\mathbf{z}(t) - \mathbf{h}[\hat{\mathbf{x}}(t), t]\} \qquad (3.2\text{-}63)$$

$$\dot{\boldsymbol{\lambda}} = \frac{\partial \mathbf{h}^T[\hat{\mathbf{x}}(t), t]}{\partial \hat{\mathbf{x}}(t)} \boldsymbol{\Psi}_v^{-1}(t)\{\mathbf{z}(t) - \mathbf{h}[\hat{\mathbf{x}}(t), t]\} - \frac{\partial \mathbf{f}^T[\hat{\mathbf{x}}(t), t]}{\partial \hat{\mathbf{x}}(t)} \boldsymbol{\lambda}(t)$$

$$+ \frac{\partial \mathbf{h}^T[\hat{\mathbf{x}}(t), t]}{\partial \hat{\mathbf{x}}(t)} \boldsymbol{\Psi}_v^{-1}(t)\boldsymbol{\Psi}_{vw}(t)\mathbf{G}^T[\hat{\mathbf{x}}(t), t]\boldsymbol{\lambda}(t)$$

$$- \frac{\partial\{\boldsymbol{\lambda}^T(t)\mathbf{G}[\hat{\mathbf{x}}(t), t]\}}{\partial \hat{\mathbf{x}}(t)} \boldsymbol{\Psi}_{wv}(t)\boldsymbol{\Psi}_v^{-1}(t)\{\mathbf{z}(t) - \mathbf{h}[\hat{\mathbf{x}}(t), t]\}$$

$$+ \frac{\partial\{\boldsymbol{\lambda}^T(t)\mathbf{G}[\hat{\mathbf{x}}(t), t][\boldsymbol{\Psi}_w(t) - \boldsymbol{\Psi}_{wv}(t)\boldsymbol{\Psi}_v^{-1}(t)\boldsymbol{\Psi}_{vw}(t)]\mathbf{G}^T[\hat{\mathbf{x}}(t), t]\}}{\partial \hat{\mathbf{x}}(t)} \boldsymbol{\lambda}(t) \qquad (3.2\text{-}64)$$

result after a modest amount of algebraic effort. These canonic equations are solved subject to the two-point boundary conditions

$$\boldsymbol{\lambda}(t_0) = \mathbf{V}_{\tilde{\mathbf{x}}_0}[\hat{\mathbf{x}}(0) - \boldsymbol{\mu}_x(0)], \qquad \boldsymbol{\lambda}(t_f) = \mathbf{0} \qquad (3.2\text{-}65)$$

Depending upon which computational technique we use to solve the TPBVP, we obtain filtering or smoothing estimates of the system states and parameters.

Example 3.2-1. We consider the identification of the parameter a and the unknown mean value of the measurement noise e for the first-order system shown in Fig. 3.2-1. The message and observation models are

$$\dot{x}_1 = -ax_1(t) + w(t)$$
$$\xi_1(t) = x_1(t) + v(t) + e$$
$$\xi_2(t) = w(t) + \omega(t)$$

where we use

$$\dot{a} = \dot{e} = 0$$

to constrain the unknown random parameters to be constant. It is assumed that $w(t)$, $v(t)$, and $\omega(t)$ are zero mean white Gaussian, uncorrelated, with known variance coefficients.

The appropriate terms for the MAP cost function of Eq. (3.2-62) and the canonic equations of Eqs. (3.2-63) and (3.2-64) become

$$\mathbf{x}(t) = \begin{bmatrix} x_1(t) \\ a \\ e \end{bmatrix} = \begin{bmatrix} x_1(t) \\ x_2(t) \\ x_3(t) \end{bmatrix}, \qquad \mathbf{f}[\mathbf{x}(t), t] = \begin{bmatrix} -x_1(t)\, x_2(t) \\ 0 \\ 0 \end{bmatrix}$$

$$\mathbf{G}[\mathbf{x}(t), t] = \begin{bmatrix} 1 \\ 0 \\ 0 \end{bmatrix}, \qquad \mathbf{w}(t) = w(t), \qquad \mathbf{h}[\mathbf{x}(t), t] = \begin{bmatrix} x_1(t) + x_3(t) \\ 0 \end{bmatrix}$$

$$\mathbf{v}(t) = \begin{bmatrix} v(t) \\ \omega(t) + w(t) \end{bmatrix} = \begin{bmatrix} v_1(t) \\ v_2(t) \end{bmatrix}, \qquad \boldsymbol{\Psi}_{\mathbf{w}}(t) = \Psi_w(t), \qquad \mathbf{z}(t) = \begin{bmatrix} \xi_1(t) \\ \xi_2(t) \end{bmatrix}$$

$$\boldsymbol{\Psi}_{\mathbf{v}}(t) = \begin{bmatrix} \Psi_v(t) & 0 \\ 0 & \Psi_\omega(t) + \Psi_w(t) \end{bmatrix}, \qquad \boldsymbol{\Psi}_{\mathbf{wv}}(t) = [0 \quad \Psi_w(t)]$$

$$\boldsymbol{\mu}_{\mathbf{x}}(k_0) = \begin{bmatrix} \mu_{x_1}(t_0) \\ \mu_a \\ \mu_e \end{bmatrix}, \qquad \mathbf{V}_{\tilde{\mathbf{x}}}(t_0) = \mathbf{V}_{\mathbf{x}}(t_0) = \begin{bmatrix} V_{x_1}(t_0) & 0 & 0 \\ 0 & V_a & 0 \\ 0 & 0 & V_e \end{bmatrix}$$

It is instructive to detail the canonic equations for this specific problem. They are

$$\dot{\hat{x}}_1 = -\hat{x}_1(t)\, \hat{x}_2(t) - \frac{\Psi_w(t)\, \Psi_\omega(t)}{\Psi_w(t) + \Psi_\omega(t)} \lambda_1(t) + \frac{\Psi_w(t)}{\Psi_w(t) + \Psi_w(t)} \xi_2(t)$$

$$\dot{\hat{x}}_2 = 0$$

$$\dot{\hat{x}}_3 = 0$$

$$\dot{\lambda}_1 = \frac{1}{\Psi_v(t)} [\xi_1(t) - \hat{x}_1(t) - \hat{x}_3(t)] + \hat{x}_2(t)\, \lambda_1(t)$$

$$\dot{\lambda}_2 = \hat{x}_1(t)\, \lambda_2(t)$$

$$\dot{\lambda}_3 = \frac{1}{\Psi_v(t)} [\xi_1(t) - \hat{x}_1(t) - \hat{x}_3(t)]$$

The boundary conditions are

$$\lambda_1(t_0) = V_{x1}(t_0)[\hat{x}_1(t_0) - \mu_{x1}(t_0)], \qquad \lambda_1(t_f) = 0$$

$$\lambda_2(t_0) = V_a[\hat{x}_2(t_0) - \mu_a], \qquad \lambda_2(t_f) = 0$$

$$\lambda_3(t_0) = V_e[\hat{x}_3(t_0) - \mu_e], \qquad \lambda_3(t_f) = 0$$

The TPBVP is decidely nonlinear, and therefore there is no hope of an analytical solution. By detailing the canonic equations, we are able to see the specific way in which the various terms, in particular the prior statistics, enter the TPBVP. Of great interest later will be the way in which these prior statistics $[V_{x_1}(t_0), V_a, V_e, \mu_{x_1}(t_0), \mu_a, \mu_e]$ affect the accuracy and speed with which we are able to identify the system parameters.

3.3. Maximum a Posteriori Identification with Unknown Prior Statistics

If the prior statistics of the plant and measurement noise, or some components of them, are unknown, we will not be able to solve the previous TPBVP for system identification. In many cases, unknown prior means may be treated as unknown constants to be identified. These unknown constants are added to the plant and observation equations and the TPBVP obtained as in the last section. Alternatively, they may be considered by the method of this section. Unknown prior variances present a considerably more difficult problem however, in that these variances enter into the nonexponential part of the MAP cost function of Eq. (3.2-21). To insure compatibility with MAP identification, it is convenient to assume a distribution for the unknown prior statistics such that we may use Bayes' rule to obtain, for stage invariant prior statistics,

$$p[\mathbf{X}(k_f), \boldsymbol{\mu}_w, \mathbf{V}_w, \boldsymbol{\mu}_v, \mathbf{V}_v \mid \mathbf{Z}(k_f)] = \frac{p[\mathbf{Z}(k_f) \mid \mathbf{X}(k_f), \boldsymbol{\mu}_w, \mathbf{V}_w, \boldsymbol{\mu}_v, \mathbf{V}_v] \times p[\mathbf{X}(k_f) \mid \boldsymbol{\mu}_w, \mathbf{V}_w, \boldsymbol{\mu}_v, \mathbf{V}_v]\, p[\boldsymbol{\mu}_w]\, p[\mathbf{V}_w]\, p[\boldsymbol{\mu}_v]\, p[\mathbf{V}_v]}{p[\mathbf{Z}(k_f)]} \tag{3.3-1}$$

where we assume that the prior statistics are statistically independent. Again, the joint density $p[\mathbf{Z}(k_f)]$ does not influence the optimization, since it does not explicitly depend upon the state variable $\mathbf{X}(k)$ and the prior statistics $\boldsymbol{\mu}_w, \mathbf{V}_w, \boldsymbol{\mu}_v, \mathbf{V}_v$ which are the negotiable variables for the optimization of the posterior density. Thus the MAP estimator may be found by maximization of the unconditional density

$$p[X(k_f), Z(k_f), \boldsymbol{\mu}_w, \mathbf{V}_w, \boldsymbol{\mu}_v, \mathbf{V}_v] = p[Z(k_f) \mid X(k_f), \boldsymbol{\mu}_w, \mathbf{V}_w, \boldsymbol{\mu}_v, \mathbf{V}_v] p[X(k_f) \mid \boldsymbol{\mu}_w, \mathbf{V}_w, \boldsymbol{\mu}_v, \mathbf{V}_v] \times p[\boldsymbol{\mu}_w] p[\mathbf{V}_w] p[\boldsymbol{\mu}_v] p[\mathbf{V}_v] \tag{3.3-2}$$

Often it is convenient to assume that the distribution of the prior statistic is uniform (Sage and Husa, 1969) such that for each component of the prior statistic $\boldsymbol{\mu}_w$

$$p[\mu_{wi}] = \begin{cases} \dfrac{1}{\mu_{wi\,\max} - \mu_{wi\,\min}}, & \mu_{w\,\max} \leqslant \mu_w \leqslant \mu_{w\,\min} \\ 0, & \text{otherwise} \end{cases} \tag{3.3-3}$$

A similar expression would be assumed for each of the other prior statistics. Assuming that the estimates of the prior statistics are such that the inequalities in the uniform densities are not violated, the estimate of the states, parameters, and prior statistics are determined by maximizing the mixed expression

$$p[Z(k_f) \mid X(k_f), \boldsymbol{\mu}_w, \mathbf{V}_w, \boldsymbol{\mu}_v, \mathbf{V}_v] \, p[X(k_f) \mid \boldsymbol{\mu}_w, \mathbf{V}_w, \boldsymbol{\mu}_v, \mathbf{V}_v] \tag{3.3-4}$$

For the case where the plant and measurement noise are uncorrelated, this is precisely the same expression obtained in Eq. (3.2-21). Thus is it desired to maximize

$$J = \frac{1}{[\det \mathbf{V}_x(k_0)]^{1/2}\{[\det \boldsymbol{\Gamma}\mathbf{V}_w\boldsymbol{\Gamma}^T][\det \mathbf{V}_v]\}^{(k_f-k_0)/2}} \times \exp\{-0.5 \| \mathbf{x}(k_0) - \boldsymbol{\mu}_x(k_0)\|^2_{\mathbf{V}_x^{-1}(k_0)} - 0.5 \sum_{k=k_0+1}^{k_f} \| \mathbf{z}(k) - \boldsymbol{\mu}_v - \mathbf{h}[\mathbf{x}(k), k]\|^2_{\mathbf{V}_v^{-1}} - 0.5 \sum_{k=k_0}^{k_f-1} \| \mathbf{w}(k) - \boldsymbol{\mu}_w \|^2_{\mathbf{V}_w^{-1}} \tag{3.3-5}$$

with respect to a choice of $X(k_f)$, $\boldsymbol{\mu}_w$, $\mathbf{V}_w$, $\boldsymbol{\mu}_v$, and $\mathbf{V}_v$. This MAP cost function is developed for the message and observation model with Gaussian white noises $\mathbf{w}(k)$ and $\mathbf{v}(k)$:

$$\mathbf{x}(k+1) = \boldsymbol{\phi}[\mathbf{x}(k), k] + \boldsymbol{\Gamma}\mathbf{w}(k) \tag{3.3-6}$$

$$\mathbf{z}(k) = \mathbf{h}[\mathbf{x}(k), k] + \mathbf{v}(k) \tag{3.3-7}$$

and (unknown) prior statistics

$$\boldsymbol{\mu}_{\mathrm{w}} = \mathscr{E}\{\mathbf{w}(k)\}, \qquad \mathbf{V}_{\mathrm{w}} = \operatorname{var}\{\mathbf{w}(k)\}$$
$$\boldsymbol{\mu}_{\mathrm{v}} = \mathscr{E}\{\mathbf{v}(k)\}, \qquad \mathbf{V}_{\mathrm{v}} = \operatorname{var}\{\mathbf{v}(k)\} \tag{3.3-8}$$

and (known) prior statistics

$$\boldsymbol{\mu}_{\mathrm{x}}(k_0) = \mathscr{E}\{\mathbf{x}(k_0)\}, \qquad \mathbf{V}_{\mathrm{v}}(k_0) = \operatorname{var}\{\mathbf{x}(k_0)\}$$
$$\operatorname{cov}\{\mathbf{w}(k), \mathbf{v}(k)\} = \mathbf{0}, \qquad \operatorname{cov}\{\mathbf{x}(k_0), \mathbf{w}(k)\} = \mathbf{0}, \qquad \operatorname{cov}\{\mathbf{x}(k), \mathbf{v}(k)\} = \mathbf{0} \tag{3.3-9}$$

We recall that in this formulation unknown parameters, which are to be identified, form part of the generalized state vector $X(k)$. Any of the components of $\boldsymbol{\mu}_{\mathrm{w}}$, $\mathbf{V}_{\mathrm{w}}$, $\boldsymbol{\mu}_{\mathrm{v}}$, and $\mathbf{V}_{\mathrm{v}}$ which are known are treated as knowns are; thus the cost function of Eq. (3.5-5) is not maximized with respect to them. Maximization of Eq. (3.3-5) with respect to choice of the prior statistics is a standard problem in the matrix calculus. There results

$$\boldsymbol{\Gamma}\hat{\boldsymbol{\mu}}_{\mathrm{w}}(k_{\mathrm{f}} \mid k_{\mathrm{f}}) = \frac{1}{k_{\mathrm{f}} - k_0} \sum_{k=k_0+1}^{k_{\mathrm{f}}} \hat{\mathbf{x}}(k \mid k_{\mathrm{f}}) - \boldsymbol{\phi}[\hat{\mathbf{x}}(k-1 \mid k_{\mathrm{f}}), k-1] \tag{3.3-10}$$

$$\begin{aligned}\boldsymbol{\Gamma}\hat{\mathbf{V}}_{\mathrm{w}}(k_{\mathrm{f}} \mid k_{\mathrm{f}})\boldsymbol{\Gamma}^{\mathrm{T}} = \frac{1}{k_{\mathrm{f}} - k_0} \sum_{k=k_0+1}^{k_{\mathrm{f}}} &\{\hat{\mathbf{x}}(k \mid k_{\mathrm{f}}) - \boldsymbol{\phi}[\hat{\mathbf{x}}(k-1 \mid k_{\mathrm{f}}), k-1] \\ &- \boldsymbol{\Gamma}\hat{\boldsymbol{\mu}}_{\mathrm{w}}(k_{\mathrm{f}} \mid k_{\mathrm{f}})\}\{\hat{\mathbf{x}}(k \mid k_{\mathrm{f}}) - \boldsymbol{\phi}[\hat{\mathbf{x}}(k-1 \mid k_{\mathrm{f}}), k-1] \\ &- \boldsymbol{\Gamma}\hat{\boldsymbol{\mu}}_{\mathrm{w}}(k_{\mathrm{f}} \mid k_{\mathrm{f}})\}^{\mathrm{T}}\end{aligned} \tag{3.3-11}$$

$$\hat{\boldsymbol{\mu}}_{\mathrm{v}}(k_{\mathrm{f}} \mid k_{\mathrm{f}}) = \frac{1}{k_{\mathrm{f}} - k_0} \sum_{k=k_0+1}^{k_{\mathrm{f}}} \mathbf{z}(k) - \mathbf{h}[\hat{\mathbf{x}}(k \mid k_{\mathrm{f}}), k] \tag{3.3-12}$$

$$\begin{aligned}\hat{\mathbf{V}}_{\mathrm{v}}(k_{\mathrm{f}} \mid k_{\mathrm{f}}) = \frac{1}{k_{\mathrm{f}} - k_0} \sum_{k=k_0+1}^{k_{\mathrm{f}}} &\{\mathbf{z}(k) - \mathbf{h}[\hat{\mathbf{x}}(k \mid k_{\mathrm{f}}), k] \\ &- \hat{\boldsymbol{\mu}}_{\mathrm{v}}(k_{\mathrm{f}} \mid k_{\mathrm{f}})\}\{\mathbf{z}(k) - \mathbf{h}[\hat{\mathbf{x}}(k \mid k_{\mathrm{f}}), k] - \hat{\boldsymbol{\mu}}_{\mathrm{v}}(k_{\mathrm{f}} \mid k_{\mathrm{f}})\}^{\mathrm{T}}\end{aligned} \tag{3.3-13}$$

where $\hat{\mathbf{x}}(k \mid k_{\mathrm{f}})$ is the smoothing solution to the identification problem which results from maximization of Eq. (3.3-5) with respect to $X(k_{\mathrm{f}})$ using for prior statistics the estimates $\boldsymbol{\Gamma}\hat{\boldsymbol{\mu}}_{\mathrm{w}}(k_{\mathrm{f}} \mid k_{\mathrm{f}})$, $\boldsymbol{\Gamma}\hat{\mathbf{V}}_{\mathrm{w}}(k_{\mathrm{f}} \mid k_{\mathrm{f}})\,\boldsymbol{\Gamma}^{\mathrm{T}}$, $\hat{\boldsymbol{\mu}}_{\mathrm{v}}(k_{\mathrm{f}} \mid k_{\mathrm{f}})$, and $\hat{\mathbf{V}}_{\mathrm{v}}(k_{\mathrm{f}} \mid k_{\mathrm{f}})$. Approximate smoothing solution for the class of problems posed here are presented in Chap. 9 of Sage and Melsa (1971). Unfortunately, the algorithms for $\hat{\mathbf{x}}(k \mid k_{\mathrm{f}})$, when coupled with those for estimating the prior statistics [Eqs. (3.3-10)–(3.3-13)] may not be processed easily. We will use the smoothed prior statistics estimates in Eqs. (3.3-10)–(3.3-13) to develop estimation algorithms by the gradient and stochastic approximation methods of the next two chapters. We now turn our attention to formulating the identification problem such that the quasilinearization and invariant imbedding methods of Chaps. 6 and 7 may be used to overcome the TPBVP.

It is convenient to minimize the negative of the natural logarithm of the MAP cost function of Eq. (3.5-5) which is equivalent to minimizing

$$
\begin{aligned}
J = {} & 0.5 \| \mathbf{x}(k_0) - \boldsymbol{\mu}_x(k_0)\|^2_{\mathbf{V}_{\mathbf{x}}^{-1}(k_0)} \\
& + 0.5 \sum_{k=k_0}^{k_f-1} \{\| \mathbf{z}(k+1) - \boldsymbol{\mu}_v(k+1) - \mathbf{h}[\mathbf{x}(k+1), k+1]\|^2_{\mathbf{V}_{\mathbf{v}}^{-1}(k+1)} \\
& + 0.5 \| \mathbf{w}(k) - \boldsymbol{\mu}_w(k)\|^2_{\mathbf{V}_{\mathbf{w}}^{-1}(k)} \\
& + 0.5 \ln[\det \boldsymbol{\Gamma} \mathbf{V}_w(k) \boldsymbol{\Gamma}^T] + 0.5 \ln[\det \mathbf{V}_v(k+1)]\}
\end{aligned} \tag{3.3-14}
$$

We wish to accomplish this subject to the message model constraint of Eq. (3.3-6) and the constant prior statistics constraints implied by Eq. (3.3-7):

$$\mathbf{x}(k+1) = \boldsymbol{\phi}[\mathbf{x}(k), k] + \boldsymbol{\Gamma}\mathbf{w}(k) \tag{3.3-15}$$

$$\boldsymbol{\mu}_w(k+1) = \boldsymbol{\mu}_w(k) \tag{3.3-16}$$

$$\mathbf{V}_w(k+1) = \mathbf{V}_w(k) \tag{3.3-17}$$

$$\boldsymbol{\mu}_v(k+1) = \boldsymbol{\mu}_v(k) \tag{3.3-18}$$

$$\mathbf{V}_v(k+1) = \mathbf{V}_v(k) \tag{3.3-19}$$

The discrete maximum principle or the discrete Euler–Lagrange equations may be applied directly to this problem. We define the Hamiltonian:

$$
\begin{aligned}
& H[\mathbf{x}(k), \mathbf{w}(k), \boldsymbol{\lambda}(k+1), \boldsymbol{\gamma}_w(k+1), \boldsymbol{\Xi}_w(k+1), \boldsymbol{\gamma}_v(k+1), \boldsymbol{\Xi}_v(k+1), \\
& \qquad \boldsymbol{\mu}_w(k), \mathbf{V}_w(k), \boldsymbol{\mu}_v(k), \mathbf{V}_v(k), k] \\
& = \tfrac{1}{2} \| \mathbf{z}(k+1) - \boldsymbol{\mu}_v(k+1) - \ell[\mathbf{x}(k), \mathbf{w}(k), k+1]\|^2_{\mathbf{V}_{\mathbf{v}}^{-1}(k+1)} \\
& \quad + \tfrac{1}{2} \| \mathbf{w}(k) - \boldsymbol{\mu}_w(k)\|^2_{\mathbf{V}_{\mathbf{w}}^{-1}(k)} + \tfrac{1}{2} \ln\{\det \boldsymbol{\Gamma} \mathbf{V}_w(k) \boldsymbol{\Gamma}^T\} \\
& \quad + \tfrac{1}{2} \ln\{\det \mathbf{V}_v(k+1)\} + \boldsymbol{\lambda}^T(k+1)\{\boldsymbol{\phi}[\mathbf{x}(k), k] + \boldsymbol{\Gamma}\mathbf{w}(k)\} \\
& \quad + \boldsymbol{\gamma}_w^T(k+1)\boldsymbol{\mu}_w(k) + \operatorname{tr}\{\boldsymbol{\Xi}_w(k+1) \boldsymbol{\Gamma} \mathbf{V}_w(k) \boldsymbol{\Gamma}^T\} + \boldsymbol{\gamma}_v^T(k+1)\boldsymbol{\mu}_v(k) \\
& \quad + \operatorname{tr}\{\boldsymbol{\Xi}_v(k+1)\mathbf{V}_v(k+1)\}
\end{aligned} \tag{3.3-20}
$$

where $\boldsymbol{\lambda}$, $\boldsymbol{\gamma}_w$, and $\boldsymbol{\gamma}_v$ are vector Lagrange multipliers, and $\boldsymbol{\Xi}_w$ and $\boldsymbol{\Xi}_v$ are matrix Lagrange multipliers.

The canonic equations become

$$\hat{\mathbf{x}}(k+1 \mid k_f) = \boldsymbol{\phi}[\hat{\mathbf{x}}(k \mid k_f), k] - \boldsymbol{\Gamma}\hat{\mathbf{V}}_w(k \mid k_f)\boldsymbol{\Gamma}^T \left[\frac{\partial \boldsymbol{\phi}[\hat{\mathbf{x}}(k \mid k_f), k]}{\partial \hat{\mathbf{x}}(k \mid k_f)}\right]^{-T} \times \boldsymbol{\lambda}(k \mid k_f) + \boldsymbol{\Gamma}\hat{\boldsymbol{\mu}}_w(k \mid k_f) \tag{3.3-21}$$

$$\boldsymbol{\lambda}(k+1 \mid k_f) = \left[\frac{\partial \boldsymbol{\phi}[\hat{\mathbf{x}}(k \mid k_f), k]}{\partial \hat{\mathbf{x}}(k \mid k_f)}\right]^{-T} \boldsymbol{\lambda}(k \mid k_f) + \frac{\partial \mathbf{h}^T[\hat{\mathbf{x}}(k+1 \mid k_f), k+1]}{\partial \hat{\mathbf{x}}(k+1 \mid k_f)} \cdot \hat{\mathbf{V}}_v^{-1}(k+1 \mid k_f) \times \{\mathbf{z}(k+1) - \hat{\boldsymbol{\mu}}_v(k \mid k_f) - \mathbf{h}[\hat{\mathbf{x}}(k+1 \mid k_f), k+1]\} \tag{3.3-22}$$

$$\hat{\boldsymbol{\mu}}_w(k+1 \mid k_f) = \hat{\boldsymbol{\mu}}_w(k \mid k_f) \tag{3.3-23}$$

$$\boldsymbol{\gamma}_w(k+1 \mid k_f) = \boldsymbol{\gamma}_w(k \mid k_f) - \boldsymbol{\Gamma}^T \left[\frac{\partial \boldsymbol{\phi}[\hat{\mathbf{x}}(k \mid k_f), k]}{\partial \hat{\mathbf{x}}(k \mid k_f)}\right]^{-T} \boldsymbol{\lambda}(k \mid k_f) \tag{3.3-24}$$

$$\boldsymbol{\Gamma}\hat{\mathbf{V}}_w(k+1 \mid k_f)\boldsymbol{\Gamma}^T = \boldsymbol{\Gamma}\hat{\mathbf{V}}_w(k \mid k_f)\boldsymbol{\Gamma}^T \tag{3.3-25}$$

$$\boldsymbol{\Xi}_w(k+1 \mid k_f) = \boldsymbol{\Xi}_w(k \mid k_f) - 0.5[\boldsymbol{\Gamma}\hat{\mathbf{V}}_w(k \mid k_f)\boldsymbol{\Gamma}^T]^{-1} - 0.5\boldsymbol{\Gamma}\hat{\mathbf{V}}_w(k \mid k_f)\boldsymbol{\Gamma}^T \times \left[\frac{\partial \boldsymbol{\phi}[\hat{\mathbf{x}}(k \mid k_f), k]}{\partial \hat{\mathbf{x}}(k \mid k_f)}\right]^{-T} \boldsymbol{\lambda}(k \mid k_f)\, \boldsymbol{\lambda}^T(k \mid k_f) \times \left[\frac{\partial \boldsymbol{\phi}[\hat{\mathbf{x}}(k \mid k_f), k]}{\partial \hat{\mathbf{x}}(k \mid k_f)}\right] \boldsymbol{\Gamma}\hat{\mathbf{V}}_w(k \mid k_f)\boldsymbol{\Gamma}^T \tag{3.3-26}$$

$$\hat{\boldsymbol{\mu}}_v(k+1 \mid k_f) = \hat{\boldsymbol{\mu}}_v(k \mid k_f) \tag{3.3-27}$$

$$\boldsymbol{\gamma}_v(k+1 \mid k_f) = \boldsymbol{\gamma}_v(k \mid k_f) + \hat{\mathbf{V}}_v^{-1}(k \mid k_f)\{\mathbf{z}(k+1) - \hat{\boldsymbol{\mu}}_v(k \mid k_f) - \mathbf{h}[\hat{\mathbf{x}}(k+1 \mid k_f), k+1]\} \tag{3.3-28}$$

$$\hat{\mathbf{V}}_v(k+1 \mid k_f) = \hat{\mathbf{V}}_v(k \mid k_f) \tag{3.3-29}$$

$$\boldsymbol{\Xi}_v(k+1 \mid k_f) = \boldsymbol{\Xi}_v(k \mid k_f) + 0.5\hat{\mathbf{V}}_v^{-1}(k \mid k_f)\{\mathbf{z}(k+1) - \hat{\boldsymbol{\mu}}_v(k \mid k_f) - \mathbf{h}[\hat{\mathbf{x}}(k+1 \mid k_f), k+1]\}\{\mathbf{z}(k+1) - \hat{\boldsymbol{\mu}}_v(k \mid k_f) - \mathbf{h}[\hat{\mathbf{x}}(k+1 \mid k_f), k+1]\}^T\hat{\mathbf{V}}_v^{-1}(k \mid k_f) - 0.5\hat{\mathbf{V}}_v^{-1}(k \mid k_f) \tag{3.3-30}$$

For the important case where the plant and measurement noise are correlated, the cost function of Eq. (3.3-14) becomes

$$J = 0.5 \| \mathbf{x}(k_0) - \boldsymbol{\mu}_x(k_0) \|^2_{\mathbf{V}_x^{-1}(k_0)} + 0.5 \| \mathbf{w}(k_0) - \boldsymbol{\mu}_w(k_0) \|^2_{\mathbf{V}_w^{-1}(k_0)}$$

$$+ 0.5 \sum_{k=k_0}^{k_f-1} \{ \| \mathbf{y}(k) \|^2_{\mathbf{Y}^{-1}(k)} + 0.5 \ln[\det \mathbf{Y}(k)] \} \tag{3.3-31}$$

where

$$\mathbf{y}(k) = \begin{bmatrix} \boldsymbol{\Gamma}[\mathbf{w}(k+1) - \boldsymbol{\mu}_w(k+1)] \\ \mathbf{z}(k+1) - \mathbf{h}[\mathbf{x}(k+1), k+1] \end{bmatrix} \tag{3.3-32}$$

$$\mathbf{Y}(k) = \begin{bmatrix} \boldsymbol{\Gamma}\mathbf{V}_w(k+1)\boldsymbol{\Gamma}^T & \boldsymbol{\Gamma}\mathbf{V}_{wv}(k+1) \\ \mathbf{V}_{vw}(k+1)\boldsymbol{\Gamma}^T & \mathbf{V}_v(k+1) \end{bmatrix} \tag{3.3-33}$$

This cost function is maximized subject to the constraint relations

$$\begin{aligned} \mathbf{x}(k+1) &= \boldsymbol{\phi}[\mathbf{x}(k), k] + \boldsymbol{\Gamma}\mathbf{w}(k) \\ \boldsymbol{\mu}_w(k+1) &= \boldsymbol{\mu}_w(k) \\ \boldsymbol{\mu}_v(k+1) &= \boldsymbol{\mu}_v(k) \\ \mathbf{Y}(k+1) &= \mathbf{Y}(k) \end{aligned} \tag{3.3-34}$$

Again, optimization theory may be used to obtain the TPBVP. The results are extremely cumbersome in the general case and so will not be given here. For any specific problem, obtaining the TPBVP is a straightforward but possibly very tedious exercise. We shall present some specific examples in the next four chapters dealing with computational methods using this TPBVP.

Example 3.3-1. We consider the simple system with an initial known Gaussian distribution

$$\mathbf{x}(k+1) = \mathbf{x}(k)$$

where the scalar observation is corrupted by zero mean Gaussian white noise with an unknown variance

$$z(k) = \mathbf{H}(k)\,\mathbf{x}(k) + v(k), \qquad k = 1, 2, \ldots, k_f$$

Thus, this example is representative of estimating the value of a constant signal in the presence of additive white Gaussian noise of unknown mean and variance. We desire to obtain the identification of $\mathbf{x}$, and V_v by maximization of the density function

$$p[X(k_f), Z(k_f) \mid V_v] = p[Z(k_f) \mid X(k_f), V_v]\, p[X(k_f) \mid V_v]$$

We have simply

$$E\{z(k) \mid \mathbf{x}(k), V_v\} = \mathbf{H}(k)\,\mathbf{x}(k)$$

$$\operatorname{var}\{z(k) \mid \mathbf{x}(k), V_v\} = V_v$$

$$\operatorname{var}\{\mathbf{x}(k+1) \mid \mathbf{x}(k)\} = \mathbf{0}$$

such that

$$p[X(k_f), Z(k_f) \mid V_v] = \frac{1}{(2\pi)^{1/2}(\det \mathbf{V}_x)^{1/2}} \exp\{-0.5 \| \mathbf{x}(0) - \boldsymbol{\mu}_x \|^2_{\mathbf{V}_x^{-1}}\} \times \prod_{k=1}^{k_f} \frac{1}{(2\pi)^{1/2} V_v^{1/2}} \exp\{-0.5 V_v^{-1}[z(k) - \mathbf{H}(k)\,\mathbf{x}(k)]^2\}$$

The equivalent cost function to be minimized is

$$J = 0.5 \| \mathbf{x}(0) - \boldsymbol{\mu}_x \|^2_{\mathbf{V}_x^{-1}} + 0.5 \sum_{k=1}^{k_f} V_v^{-1}[z(k) - \mathbf{H}(k)\,\mathbf{x}(k)]^2 + 0.5 k_f \ln V_v$$

For convenience, we will now assume $\mathbf{V}_x = \infty$. This is equivalent to asserting that we have total prior uncertainty concerning the parameter $\mathbf{x}$. We differentiate with respect to each of the unknowns, $\mathbf{x}(k)$ and V_v, and set the result equal to zero to obtain

$$\hat{\mathbf{x}}(k_f \mid k_f) = \left[\sum_{k=1}^{k_f} \mathbf{H}^T(k)\,\mathbf{H}(k)\right]^{-1} \sum_{k=1}^{k_f} \mathbf{H}^T(k)\, z(k)$$

$$\hat{V}_v(k_f \mid k_f) = \frac{1}{k_f} \sum_{k=1}^{k_f} [z(k) - \mathbf{H}(k)\,\hat{\mathbf{x}}(k_f \mid k_f)]^2$$

For this particularly simple example, the estimate of $\mathbf{x}$ is determined without knowledge of V_v. The estimate $\hat{\mathbf{x}}(k_f \mid k_f)$ is then used to determine the estimate $\hat{V}_v(k_f \mid k_f)$.

A sequential or recursive solution is desirable. For this simple example, we may obtain the sequential estimation algorithms by induction. First we define, in accordance with Eq. (3.1-22),

$$\mathbf{M}(k_\mathrm{f}) = \mathbf{M}(k_\mathrm{f}\,, 0) = \sum_{k=1}^{k_\mathrm{f}} \mathbf{H}^\mathrm{T}(k)\,\mathbf{H}(k)$$

and then note that

$$\mathbf{M}(k_\mathrm{f}) = \sum_{k=1}^{k_\mathrm{f}-1} \mathbf{H}^\mathrm{T}(k)\,\mathbf{H}(k) + \mathbf{H}^\mathrm{T}(k_\mathrm{f})\,\mathbf{H}(k_\mathrm{f}) = \mathbf{M}(k_\mathrm{f} - 1) + \mathbf{H}^\mathrm{T}(k_\mathrm{f})\,\mathbf{H}(k_\mathrm{f})$$

Use of the matrix inversion lemma yields

$$\begin{aligned}\mathbf{M}^{-1}(k_\mathrm{f}) = {} & \mathbf{M}^{-1}(k_\mathrm{f} - 1) - \mathbf{M}^{-1}(k_\mathrm{f} - 1)\mathbf{H}^\mathrm{T}(k_\mathrm{f})[1 + \mathbf{H}(k_\mathrm{f})\mathbf{M}^{-1}(k_\mathrm{f} - 1)\mathbf{H}^\mathrm{T}(k_\mathrm{f})]^{-1} \\ & \times \mathbf{H}(k_\mathrm{f})\,\mathbf{M}^{-1}(k_\mathrm{f} - 1)\end{aligned}$$

By combining the relations for two successive stages of the estimator for $\mathbf{x}$:

$$\hat{\mathbf{x}}(k_\mathrm{f} \mid k_\mathrm{f}) = \mathbf{M}^{-1}(k_\mathrm{f}) \sum_{k=1}^{k_\mathrm{f}-1} \mathbf{H}^\mathrm{T}(k)\,z(k)$$

$$\hat{\mathbf{x}}(k_\mathrm{f} - 1 \mid k_\mathrm{f} - 1) = \mathbf{M}^{-1}(k_\mathrm{f} - 1) \sum_{k=1}^{k_\mathrm{f}} \mathbf{H}^\mathrm{T}(k)\,z(k)$$

we obtain

$$\hat{\mathbf{x}}(k_\mathrm{f} \mid k_\mathrm{f}) = \mathbf{M}^{-1}(k_\mathrm{f})\,\mathbf{M}(k_\mathrm{f} - 1)\,\hat{\mathbf{x}}(k_\mathrm{f} - 1 \mid k_\mathrm{f} - 1) + \mathbf{M}^{-1}(k_\mathrm{f})\,\mathbf{H}^\mathrm{T}(k_\mathrm{f})\,z(k_\mathrm{f})$$

Again using the matrix inversion for $\mathbf{M}^{-1}(k_\mathrm{f})$, we have finally, where we drop the f subscript and use for convenience $\hat{\mathbf{x}}(k) = \hat{\mathbf{x}}(k \mid k)$,

$$\hat{\mathbf{x}}(k) = \hat{\mathbf{x}}(k - 1) + \frac{\mathbf{M}^{-1}(k - 1)\,\mathbf{H}^\mathrm{T}(k)}{1 + \mathbf{H}(k)\,\mathbf{M}^{-1}(k - 1)\,\mathbf{H}^\mathrm{T}(k)}\,[z(k) - \mathbf{H}(k)\,\hat{\mathbf{x}}(k - 1)]$$

In precisely the same way, the sequential estimator $V_v(k)$ is obtained. We have

$$\hat{V}_v(k) = \frac{1}{k}\left[(k - 1)\,\hat{V}_v(k - 1) + \frac{[z(k) - \mathbf{H}(k)\,\hat{\mathbf{x}}(k - 1)]^2}{1 + \mathbf{H}(k)\,\mathbf{M}^{-1}(k - 1)\,\mathbf{H}^\mathrm{T}(k)}\right]$$

These two estimation or identification algorithms are processed together with the computationally efficient algorithm for $\mathbf{M}^{-1}(k)$:

$$\begin{aligned}\mathbf{M}^{-1}(k) = {} & \mathbf{M}^{-1}(k - 1) - \mathbf{M}^{-1}(k - 1)\,\mathbf{H}^\mathrm{T}(k) \\ & \times [1 + \mathbf{H}(k)\,\mathbf{M}^{-1}(k)\,\mathbf{H}^\mathrm{T}(k)]^{-1}\mathbf{H}(k)\,\mathbf{M}^{-1}(k - 1)\end{aligned}$$

Unfortunately, for problems more difficult than this simple one we are not able to obtain the sequential or nonsequential solution in as simple a fashion as we have here. For instance, if the prior variance of $\mathbf{x}$ is not infinite, the smoothed (nonsequential) estimator for $\mathbf{x}$ is

$$\hat{\mathbf{x}}(k_f \mid k_f) = \left[\sum_{k=1}^{k_f} \mathbf{H}^T(k)\hat{V}_v(k_f \mid k_f)\mathbf{H}(k)\right]^{-1}\left[\mathbf{V}_\mathbf{x}^{-1}\boldsymbol{\mu}_\mathbf{x} + \sum_{k=1}^{k_f} \mathbf{H}^T(k)\hat{V}_v^{-1}(k_f \mid k_f)z(k)\right]$$

and this may not be solved for $\hat{\mathbf{x}}(k_f \mid k_f)$ without first solving for $\hat{V}_v(k_f \mid k_f)$, which depends upon $\hat{\mathbf{x}}(k_f \mid k_f)$. We will resolve this solution difficulty by the computational techniques of the next chapters.

In concluding this example, we wish to obtain the TPBVP which minimizes the cost function J. Thus, we wish to minimize

$$J = 0.5 \| \mathbf{x}(0) - \boldsymbol{\mu}_\mathbf{x} \|^2_{\mathbf{V}_\mathbf{x}^{-1}} + 0.5k_f \ln V_v(k) + 0.5 \sum_{k=1}^{k_f} V_v^{-1}(k)[z(k) - \mathbf{H}(k)\mathbf{x}(k)]^2$$

subject to the constraint relations

$$\mathbf{x}(k+1) = \mathbf{x}(k), \qquad V_v(k+1) = V_v(k)$$

The canonic equations become, from Eqs. (3.3-21)–(3.3-30),

$$\begin{aligned}
\hat{\mathbf{x}}(k+1 \mid k_f) &= \hat{\mathbf{x}}(k \mid k_f) \\
\boldsymbol{\lambda}(k+1 \mid k_f) &= \boldsymbol{\lambda}(k \mid k_f) + \mathbf{H}^T(k+1)\,\hat{V}_v^{-1}(k+1 \mid k_f) \\
&\quad \times [\mathbf{z}(k+1) - \mathbf{H}(k+1)\,\hat{\mathbf{x}}(k+1 \mid k_f)] \\
\hat{V}_v(k+1 \mid k_f) &= \hat{V}_v(k \mid k_f) \\
\Xi_v(k+1 \mid k_f) &= \Xi_v(k \mid k_f) - \frac{0.5}{\hat{V}_v(k+1 \mid k_f)} + \frac{0.5}{\hat{V}_v{}^2(k+1 \mid k_f)} \\
&\quad \times [z(k+1) - \mathbf{H}(k+1)\,\hat{\mathbf{x}}(k+1 \mid k_f)]
\end{aligned}$$

These must be solved with the two-point boundary conditions

$$\begin{aligned}
\boldsymbol{\lambda}(0 \mid k_f) &= \mathbf{V}_\mathbf{x}^{-1}[\hat{\mathbf{x}}(0 \mid k_f) - \boldsymbol{\mu}_\mathbf{x}], &\qquad \boldsymbol{\lambda}(k_f \mid k_f) &= \mathbf{0} \\
\Xi_v(0 \mid k_f) &= 0, & \Xi_v(k_f \mid k_f) &= 0
\end{aligned}$$

We will now turn our attention to the formulation of system identification problems using maximum likelihood techniques.

3.4. Maximum Likelihood Identification

In our previous development of cost functions for system identification, we assumed that we wished to estimate or identify system parameters *and states*. We now wish to relax the requirement that the system states formally be identified. Thus we wish to maximize $p[\boldsymbol{\theta} \mid Z(k_f)]$ or $p[\boldsymbol{\theta} \mid Z(t_f)]$ for maximum a posteriori identification, where $\boldsymbol{\theta}$ represents the constant unknown parameters (not including the system states) with which the maximization is to be conducted. For maximum likelihood identification, we maximize $p[Z(k_f) \mid \boldsymbol{\theta}]$ or $p[Z(t_f) \mid \boldsymbol{\theta}]$ with respect to $\boldsymbol{\theta}$. If the distribution of the unknown parameter $\boldsymbol{\theta}$ is uniform, or if there is large uncertainty (large $\mathbf{V}_\theta$) in the prior statistics, we have seen in the last section that maximum likelihood and maximum a posteriori identification of the system parameters are equivalent. In this section, we will only consider maximum likelihood estimation of system parameters (including prior statistics). If prior statistics are available concerning unknown system parameters, these may be incorporated into the cost functions of this section: However, we will not develop this case here. As in our previous efforts, discrete systems will be considered first. The continuous results will then follow. For simplicity, we will consider first the case where the plant noise is not observed, and then we will remove this restriction.

We consider the nonlinear message and observation models of Eqs. (3.2-1) and (3.2-2):

$$\mathbf{x}(k+1) = \boldsymbol{\phi}[\mathbf{x}(k), k, \mathbf{a}] + \boldsymbol{\Gamma}[\mathbf{x}(k), k, \mathbf{b}]\, \mathbf{w}(k) \tag{3.4-1}$$

$$\mathbf{z}(k) = \mathbf{h}[\mathbf{x}(k), k, \mathbf{c}] + \mathbf{v}(k) \tag{3.4-2}$$

where $\mathbf{a}$, $\mathbf{b}$, and $\mathbf{c}$ are unknown parameter vectors which are to be identified. In addition, the zero-mean white uncorrelated Gaussian plant and measurement noises may have variance matrix components which must be identified. We will let the symbol $\boldsymbol{\theta}$ represent all of the unknown constants which are to be identified. We will not explicitely use the symbols, $\mathbf{a}$, $\mathbf{b}$, and $\mathbf{c}$ which represent the unknown parameters in $\boldsymbol{\phi}$, $\boldsymbol{\Gamma}$, and $\mathbf{h}$ in the development to follow but will constantly remember that in $\boldsymbol{\phi}$, $\boldsymbol{\Gamma}$, and $\mathbf{h}$ contain these parameters.

We desire to maximize the expression $p[Z(k_f) \mid \boldsymbol{\theta}]$, which may be rewritten, using the chain rule for probabilities, as

$$p[Z(k_f) \mid \boldsymbol{\theta}] = \prod_{k=k_1}^{k_f} p[\mathbf{z}(k) \mid Z(k-1), \boldsymbol{\theta}] \tag{3.4-3}$$

Since there is assumed to be no observation at stage k_0

$$p[\mathbf{z}(k_1) \mid \mathbf{z}(k_0), \boldsymbol{\theta}] = p[\mathbf{z}(k_1) \mid \boldsymbol{\theta}] \tag{3.4-4}$$

We define the conditional moments

$$\hat{\mathbf{h}}[\mathbf{x}(k), k \mid \boldsymbol{\theta}] \triangleq \mathscr{E}\{\mathbf{z}(k) \mid \mathbf{Z}(k-1), \boldsymbol{\theta}\} = \mathscr{E}\{\mathbf{h}[\mathbf{x}(k), k] \mid \mathbf{Z}(k-1), \boldsymbol{\theta}\} \tag{3.4-5}$$

$$\mathbf{V}_{\mathbf{z}}(k \mid k-1, \boldsymbol{\theta}) \triangleq \operatorname{var}\{\mathbf{z}(k) \mid \mathbf{Z}(k-1), \boldsymbol{\theta}\} = \mathbf{V}_{\hat{\mathbf{z}}}(k \mid k-1, \boldsymbol{\theta}) + \mathbf{V}_{\mathbf{v}}(k) \tag{3.4-6}$$

$$\mathbf{V}_{\hat{\mathbf{z}}}(k \mid k-1, \boldsymbol{\theta}) \triangleq \operatorname{var}\{\mathbf{h}[\mathbf{x}(k), k] \mid \mathbf{Z}(k-1), \boldsymbol{\theta}\} \tag{3.4-7}$$

The density functions in Eq. (3.4-3) are, in general, non-Gaussian. However, it is possible to obtain a "pseudo-Bayes" density function by assuming that the conditional densities in Eq. (3.4-3) are Gaussian, such that

$$\begin{aligned} p[\mathbf{z}(k) \mid \mathbf{Z}(k-1), \boldsymbol{\theta}] = {} & \frac{1}{(2\pi)^{R/2}[\det \mathbf{V}_{\mathbf{z}}(k \mid k-1, \boldsymbol{\theta})]^{1/2}} \\ & \times \exp\{-0.5[\mathbf{z}(k) - \hat{\mathbf{h}}[\mathbf{x}(k), k \mid \boldsymbol{\theta}]]^{\mathrm{T}} \\ & \times \mathbf{V}_{\mathbf{z}}^{-1}(k \mid k-1, \boldsymbol{\theta})[\mathbf{z}(k) - \hat{\mathbf{h}}[\mathbf{x}(k), k \mid \boldsymbol{\theta}]]\} \end{aligned} \tag{3.4-8}$$

With this pseudo-Bayes assumption, the likelihood function becomes

$$\begin{aligned} p[\mathbf{Z}(k_f) \mid \boldsymbol{\theta}] = {} & \prod_{k=k_1}^{k_f} \frac{1}{(2\pi)^{R/2}[\det \mathbf{V}_{\mathbf{z}}(k \mid k-1, \boldsymbol{\theta})]^{1/2}} \\ & \times \exp\{-\frac{1}{2} \| \tilde{\mathbf{z}}(k \mid k-1, \boldsymbol{\theta}) \|^2_{\mathbf{V}_{\mathbf{z}}^{-1}(k \mid k-1, \boldsymbol{\theta})}\} \end{aligned} \tag{3.4-9}$$

where $\tilde{\mathbf{z}}(k \mid k-1, \boldsymbol{\theta})$ is the "innovations" process (Sage and Melsa, 1971)

$$\begin{aligned} \tilde{\mathbf{z}}(k \mid k-1, \boldsymbol{\theta}) & \triangleq \mathbf{z}(k) - \hat{\mathbf{h}}[\mathbf{x}(k), k \mid \boldsymbol{\theta}] \\ & = \mathbf{z}(k) - \mathscr{E}\{\mathbf{z}(k) \mid \mathbf{Z}(k-1), \boldsymbol{\theta}\} \end{aligned} \tag{3.4-10}$$

which represents the "new information" added by the observation $\mathbf{z}(k)$. It will often be more convenient to minimize the negative of the logarithm of Eq. (3.4-9) rather than to maximize Eq. (3.4-9). Thus maximum likelihood identification, for the discrete model considered, is obtained by minimizing the cost function

$$J = \frac{1}{2} \sum_{k=k_1}^{k_f} \ln \det\{\mathbf{V}_z(k \mid k-1, \boldsymbol{\theta})\} + \| \tilde{\mathbf{z}}(k \mid k-1, \boldsymbol{\theta}) \|^2_{\mathbf{V}_{\tilde{z}}^{-1}(k \mid k-1, \boldsymbol{\theta})} \tag{3.4-11}$$

with respect to $\boldsymbol{\theta}$. By using results from the Ito calculus, it is possible to show that, as the samples become dense, minimization of Eq. (3.4-11) or maximization of Eq. (3.4-9) is not a meaningful operation. This is due to two factors. The variance of the plant and measurement noise is infinite in the continuous case. Also, the density function of Eq. (3.4-9) is infinite dimensional in the variable $Z(t_f)$ in the continuous case. There does not exist the possibility of identifying $\mathbf{V}_w$ or $\mathbf{V}_v$ in the continuous case, since it is infinite. If we disregard the possibility of identifying $\mathbf{V}_v$, we may reformulate the identification problem as one of maximizing the likelihood ratio

$$L[Z(k_f) \mid \boldsymbol{\theta}] = \frac{p[Z(k_f) \mid \mathscr{H}_1, \boldsymbol{\theta}]}{p[Z(k_f) \mid \mathscr{H}_0]} \tag{3.4-12}$$

where $\mathscr{H}_1$ represents the hypothesis that

$$\mathbf{z}(k) = \mathbf{h}[\mathbf{x}(k), k] + \mathbf{v}(k) \tag{3.4-13}$$

and $\mathscr{H}_0$ represents the hypothesis that

$$\mathbf{z}(k) = \mathbf{v}(k) \tag{3.4-14}$$

It is an easy matter to show that

$$p[\mathbf{z}(k) \mid Z(k-1), \mathscr{H}_0] = \frac{1}{(2\pi)^{R/2}[\det \mathbf{V}_v(k)]^{1/2}} \exp\{-\frac{1}{2} \| \mathbf{z}(k) \|^2_{\mathbf{V}_v^{-1}(k)}\} \tag{3.4-15}$$

such that the likelihood ratio becomes

$$L[Z(k_f) \mid \boldsymbol{\theta}] = \prod_{k=k_1}^{k_f} \frac{\det[\mathbf{V}_v(k)]^{1/2}}{[\det \mathbf{V}_{\hat{z}}(k \mid k-1, \boldsymbol{\theta}) + \mathbf{V}_v(k)]^{1/2}} \exp\{\mathscr{S}[Z(k_f) \mid \boldsymbol{\theta}]\} \tag{3.4-16}$$

where $\mathscr{S}$ is the sufficient statistic

$$\mathscr{S}[Z(k_f) \mid \boldsymbol{\theta}] = -\frac{1}{2} \sum_{k=k_1}^{k_f} \{\| \tilde{\mathbf{z}}(k \mid k-1, \boldsymbol{\theta}) \|^2_{\mathbf{V}_{\tilde{z}}^{-1}(k \mid k-1, \boldsymbol{\theta})} - \| \mathbf{z}(k) \|^2_{\mathbf{V}_v^{-1}(k)}\} \tag{3.4-17}$$

If identification of $\mathbf{V}_v$ is not involved, maximization of Eq. (3.4-16) with respect to $\boldsymbol{\theta}$ is equivalent to maximization of Eq. (3.4-11) with respect to $\boldsymbol{\theta}$. We are now in a position to let the samples become dense.

It is possible to show (McLendon and Sage, 1970; Sage and Melsa, 1971) that, as the samples become dense, maximization of Eq. (3.4-16) is equivalent to minimization of

$$J = -\frac{1}{2} ⨍_{t_0}^{t_f} \{\mathbf{dU}^T(t)\, \boldsymbol{\Psi}_v^{-1}(t)\, \hat{\mathbf{h}}[\mathbf{x}(t), t \mid \boldsymbol{\theta}] + \hat{\mathbf{h}}^T[\mathbf{x}(t), t \mid \boldsymbol{\theta}]\, \boldsymbol{\Psi}_v^{-1}(t)\, \mathbf{dU}(t)\}$$

$$+ \frac{1}{2} \int_{t_0}^{t_f} \hat{\mathbf{h}}^T[\mathbf{x}(t), t \mid \boldsymbol{\theta}]\, \boldsymbol{\Psi}_v^{-1}(t)\, \hat{\mathbf{h}}[\mathbf{x}(t), t \mid \boldsymbol{\theta}]\, dt \tag{3.4-18}$$

where

$$\mathbf{dU}(t) = \mathbf{z}(t)\, dt \tag{3.4-19}$$

represents the observations, and ⨍ represents Ito integration. As in the discrete case, we have

$$\hat{\mathbf{h}}[\mathbf{x}(t), t \mid \boldsymbol{\theta}] = \mathscr{E}\{\mathbf{h}[\mathbf{x}(t), t] \mid \mathbf{Z}(t), \boldsymbol{\theta}\} \tag{3.4-20}$$

Unfortunately, we will not be able to determine the cost functions of Eqs. (3.4-9) and (3.4-11) due to the terms $\hat{\mathbf{h}}[\mathbf{x}(k), k \mid \boldsymbol{\theta}]$ and $\mathbf{V}_z(k \mid k-1, \boldsymbol{\theta})$ which are, in general, impossible to obtain exactly. Perhaps the most reasonable means for obtaining these terms is to approximate $\mathbf{h}[\mathbf{x}(k), k]$ in a Taylor series about $\mathbf{x}(k) = \hat{\mathbf{x}}(k \mid k-1, \boldsymbol{\theta})$, where the conditional mean estimate is defined as

$$\hat{\mathbf{x}}(k \mid k-1, \boldsymbol{\theta}) = \mathscr{E}\{\mathbf{x}(k) \mid \mathbf{Z}(k-1), \boldsymbol{\theta}\} \tag{3.4-21}$$

This gives (to a first-order approximation)

$$\mathbf{h}[\mathbf{x}(k), k] = \mathbf{h}[\hat{\mathbf{x}}(k \mid k-1, \boldsymbol{\theta}), k] + \left[\frac{\partial \mathbf{h}^T[\hat{\mathbf{x}}(k \mid k-1, \boldsymbol{\theta}), k]}{\partial \hat{\mathbf{x}}(k \mid k-1, \boldsymbol{\theta})}\right]^T [\mathbf{x}(k) - \hat{\mathbf{x}}(k \mid k-1, \boldsymbol{\theta})] \tag{3.4-22}$$

Substituting this expression into Eqs. (3.4-5)–(3.4-7) gives the following first-order approximations to the required terms:

$$\hat{\mathbf{h}}[\mathbf{x}(k), k \mid \boldsymbol{\theta}] \simeq \mathbf{h}[\hat{\mathbf{x}}(k \mid k-1, \boldsymbol{\theta}), k] \tag{3.4-23}$$

$$\mathbf{V}_{\hat{\mathbf{z}}}(k \mid k-1, \boldsymbol{\theta}) \cong \left[\frac{\partial \mathbf{h}^{\mathrm{T}}[\hat{\mathbf{x}}(k \mid k-1, \boldsymbol{\theta}), k]}{\partial \hat{\mathbf{x}}(k \mid k-1, \boldsymbol{\theta})}\right]^{\mathrm{T}} \mathbf{V}_{\tilde{\mathbf{x}}}(k \mid k-1, \boldsymbol{\theta}) \times \left[\frac{\partial \mathbf{h}^{\mathrm{T}}[\hat{\mathbf{x}}(k \mid k-1, \boldsymbol{\theta}), k]}{\partial \hat{\mathbf{x}}(k \mid k-1, \boldsymbol{\theta})}\right] \quad (3.4\text{-}24)$$

where

$$\mathbf{V}_{\tilde{\mathbf{x}}}(k \mid k-1, \boldsymbol{\theta}) = \operatorname{var}\{\hat{\mathbf{x}}(k \mid k-1, \boldsymbol{\theta})\} = \operatorname{var}\{\mathbf{x}(k) \mid \boldsymbol{Z}(k-1, \boldsymbol{\theta}\} \quad (3.4\text{-}25)$$

$$\tilde{\mathbf{z}}(k \mid k-1, \boldsymbol{\theta}) \cong \mathbf{z}(k) - \mathbf{h}[\hat{\mathbf{x}}(k \mid k-1, \boldsymbol{\theta}), k] \quad (3.4\text{-}26)$$

$$\mathbf{V}_{\mathbf{z}}(k \mid k-1, \boldsymbol{\theta}) \cong \mathbf{V}_{\mathbf{v}}(k) + \mathbf{V}_{\hat{\mathbf{z}}}(k \mid k-1, \boldsymbol{\theta}) \quad (3.4\text{-}27)$$

The first-order approximation to the cost function is then

$$J = \frac{1}{2} \sum_{k=k_1}^{k_{\mathrm{f}}} \ln \det\{\mathbf{V}_{\mathbf{z}}(k \mid k-1, \boldsymbol{\theta})\} + \| \tilde{\mathbf{z}}(k, k-1, \boldsymbol{\theta}) \|^2_{\mathbf{V}_{\mathbf{z}}^{-1}(k \mid k-1, \boldsymbol{\theta})} \quad (3.4\text{-}11)$$

where the various terms needed in the foregoing are given by Eqs. (3.4-23)–(3.4-27). We have thus obtained the interesting and fundamental result that the one-stage prediction estimate of $\mathbf{x}(k)$, $\hat{\mathbf{x}}(k \mid k-1, \boldsymbol{\theta})$, is needed in order to obtain the maximum likelihood cost function, even though we did not specify initially that estimation of $\mathbf{x}(k)$ was required. In order to actually implement this cost function, it is necessary that we obtain algorithms to obtain $\hat{\mathbf{x}}(k \mid k-1, \boldsymbol{\theta})$ and $\mathbf{V}_{\tilde{\mathbf{x}}}(k \mid k-1, \boldsymbol{\theta})$ from $\boldsymbol{Z}(k-1)$.

In general, the conditional mean estimation algorithms, which can be derived for the nonlinear message model of Eq. (3.4-1), are infinite dimensional, and in order to obtain realizable algorithms, approximations must be made. One approximation that can be employed is to assume Gaussian statistics for the density functions $p[\mathbf{x}(k) \mid \boldsymbol{Z}(k-1), \boldsymbol{\theta}]$ and $p[\mathbf{z}(k) \mid \boldsymbol{Z}(k-1), \boldsymbol{\theta}]$, and thereby obtain a "pseudo-Bayes" approximate algorithm in the same manner that the "pseudo-Bayes" approximate likelihood ratio was obtained in the last section.

To facilitate determination of the desired estimation algorithms, the conditional expected value and variance theorem for a Gaussian random variable will be used. This theorem states that if the vector random variables $\boldsymbol{\alpha}$ and $\boldsymbol{\beta}$ are each individually Gaussian, then the conditional expected value and variance of $\boldsymbol{\alpha}$ given $\boldsymbol{\beta}$ are given by

$$\mathscr{E}\{\boldsymbol{\alpha} \mid \boldsymbol{\beta}\} = \boldsymbol{\mu}_{\alpha} + \mathbf{V}_{\alpha\beta}\mathbf{V}_{\beta}^{-1}(\boldsymbol{\beta} - \boldsymbol{\mu}_{\beta}) \quad (3.4\text{-}28)$$

$$\operatorname{var}\{\boldsymbol{\alpha} \mid \boldsymbol{\beta}\} = \mathbf{V}_{\alpha} - \mathbf{V}_{\alpha\beta}\mathbf{V}_{\beta}^{-1}\mathbf{V}_{\beta\alpha} \quad (3.4\text{-}29)$$

where

$$\begin{aligned} \boldsymbol{\mu}_\alpha &= \mathscr{E}\{\boldsymbol{\alpha}\} \\ \mathbf{V}_\alpha &= \text{var}\{\boldsymbol{\alpha}\} = \mathscr{E}\{(\boldsymbol{\alpha} - \boldsymbol{\mu}_\alpha)(\boldsymbol{\alpha} - \boldsymbol{\mu}_\alpha)^T\} \\ \mathbf{V}_{\alpha\beta} &= \text{cov}\{\boldsymbol{\alpha}, \boldsymbol{\beta}\} = \mathscr{E}\{(\boldsymbol{\alpha} - \boldsymbol{\mu}_\alpha)(\boldsymbol{\beta} - \boldsymbol{\mu}_\beta)^T\} \end{aligned} \tag{3.4-30}$$

So, if the random variables $\mathbf{x}(k)$ and $\mathbf{z}(k)$ conditioned upon $Z(k-1)$ are each individually Gaussian, there results

$$\begin{aligned} \hat{\mathbf{x}}(k \mid \boldsymbol{\theta}) = \mathscr{E}\{\mathbf{x}(k) \mid Z(k), \boldsymbol{\theta}\} &= \mathscr{E}\{(k) \mid Z(k-1), \mathbf{z}(k), \boldsymbol{\theta}\} = \mathscr{E}\{\mathbf{x}(k) \mid Z(k-1), \boldsymbol{\theta}\} \\ &+ \text{cov}\{\mathbf{x}(k), \mathbf{z}(k) \mid Z(k-1), \boldsymbol{\theta}\} \text{ var}^{-1}\{\mathbf{z}(k) \mid Z(k-1), \boldsymbol{\theta}\} \\ &\times [\mathbf{z}(k) - \mathscr{E}\{\mathbf{z}(k) \mid Z(k-1), \boldsymbol{\theta}\}] \end{aligned} \tag{3.4-31}$$

such that the conditional mean estimate becomes

$$\begin{aligned} \hat{\mathbf{x}}(k \mid \theta) = \hat{\mathbf{x}}(k \mid k-1, \boldsymbol{\theta}) &+ \text{cov}\{\mathbf{x}(k), \mathbf{z}(k) \mid Z(k-1), \boldsymbol{\theta}\} \\ &\times \text{var}^{-1}\{\mathbf{z}(k) \mid Z(k-1), \boldsymbol{\theta}\}[\mathbf{z}(k) - \hat{\mathbf{h}}[\mathbf{x}(k), k \mid \boldsymbol{\theta}]] \end{aligned} \tag{3.4-32}$$

The companion relation to the conditional expectation theorem, the conditional variance theorem of Eq. (3.4-29) yields

$$\begin{aligned} \text{var}\{\mathbf{x}(k) \mid Z(k), \boldsymbol{\theta}\} = \text{var}\{\mathbf{x}(k) \mid Z(k-1), \boldsymbol{\theta}\} &- \text{cov}\{\mathbf{x}(k), \mathbf{z}(k) \mid Z(k-1), \boldsymbol{\theta}\} \\ &\times \text{var}^{-1}\{\mathbf{z}(k) \mid Z(k-1), \boldsymbol{\theta}\} \text{ cov}\{\mathbf{z}(k), \mathbf{x}(k) \mid Z(k-1), \boldsymbol{\theta}\} \end{aligned} \tag{3.4-33}$$

Also

$$\mathbf{V}_{\tilde{\mathbf{x}}}(k \mid \boldsymbol{\theta}) = \text{var}\{\mathbf{x}(k) - \hat{\mathbf{x}}(k) \mid Z(k), \boldsymbol{\theta}\} = \text{var}\{\mathbf{x}(k) \mid Z(k), \boldsymbol{\theta}\} \tag{3.4-34}$$

$$\mathbf{V}_{\tilde{\mathbf{x}}}(k \mid k-1, \boldsymbol{\theta}) = \text{var}\{\mathbf{x}(k) - \hat{\mathbf{x}}(k) \mid Z(k-1), \boldsymbol{\theta}\} = \text{var}\{\mathbf{x}(k) \mid Z(k-1), \boldsymbol{\theta}\} \tag{3.4-35}$$

since the conditional mean estimator is unbiased. Thus the conditional variance theorem has led to an approximate relation for the error variance in filtering:

$$\begin{aligned} \mathbf{V}_{\tilde{\mathbf{x}}}(k \mid \boldsymbol{\theta}) = \mathbf{V}_{\tilde{\mathbf{x}}}(k \mid k-1, \boldsymbol{\theta}) &- \text{cov}\{\mathbf{x}(k), \mathbf{z}(k) \mid Z(k-1), \boldsymbol{\theta}\} \\ &\times \text{var}^{-1}\{\mathbf{z}(k) \mid Z(k-1), \boldsymbol{\theta}\} \text{ cov}\{\mathbf{z}(k), \mathbf{x}(k) \mid Z(k-1), \boldsymbol{\theta}\} \end{aligned} \tag{3.4-36}$$

Equations (3.4-32) and (3.4-36) represent, therefore, approximate algorithms for discrete nonlinear conditional mean filtering and the associated error variance. In order to implement these algorithms, it is necessary to determine expressions for $\text{cov}\{\mathbf{x}(k), \mathbf{z}(k) \mid Z(k-1), \boldsymbol{\theta}\}$

and $\operatorname{var}\{\mathbf{z}(k) \mid Z(k-1), \boldsymbol{\theta}\}$. Since the problem is basically nonlinear, these relations may not be determined exactly, so further approximations must be made. The order of the approximations will determine the resulting filter algorithms.

A first order solution for these needed variances may be obtained by expanding $\boldsymbol{\phi}$ and $\mathbf{h}$ in a first-order Taylor series about the filtered and one-stage prediction solution to obtain

$$\boldsymbol{\phi}[\mathbf{x}(k), k] \cong \boldsymbol{\phi}[\hat{\mathbf{x}}(k \mid \boldsymbol{\theta}), \mathbf{h}] + \left[\frac{\partial \boldsymbol{\phi}[\hat{\mathbf{x}}(k \mid \boldsymbol{\theta}), k]}{\partial \hat{\mathbf{x}}(k \mid \boldsymbol{\theta})}\right]^{\mathrm{T}} [\mathbf{x}(k) - \hat{\mathbf{x}}(k \mid \boldsymbol{\theta})] \qquad (3.4\text{-}37)$$

$$\begin{aligned}\mathbf{h}[\mathbf{x}(k), k] &\cong \mathbf{h}[\hat{\mathbf{x}}(k \mid k-1, \boldsymbol{\theta}), k] \\ &\quad + \left[\frac{\partial \mathbf{h}^{\mathrm{T}}[\hat{\mathbf{x}}(k \mid k-1, \boldsymbol{\theta}), k]}{\partial \hat{\mathbf{x}}(k \mid k-1, \boldsymbol{\theta})}\right]^{\mathrm{T}} [\mathbf{x}(k) - \hat{\mathbf{x}}(k \mid k-1, \boldsymbol{\theta})] \qquad (3.4\text{-}38)\end{aligned}$$

The first-order approximate[3] of $\boldsymbol{\Gamma}\mathbf{V}_{\mathrm{w}}\boldsymbol{\Gamma}^{\mathrm{T}}$ about the filtered estimate $\hat{\mathbf{x}}(k \mid \boldsymbol{\theta})$ is then used:

$$\begin{aligned}&\boldsymbol{\Gamma}[\mathbf{x}(k), k]\mathbf{V}_{\mathrm{w}}(k)\boldsymbol{\Gamma}^{\mathrm{T}}[\mathbf{x}(k), k] \\ &\quad \cong \boldsymbol{\Gamma}[\hat{\mathbf{x}}(k \mid \boldsymbol{\theta}), k]\mathbf{V}_{\mathrm{w}}(k)\boldsymbol{\Gamma}^{\mathrm{T}}[\hat{\mathbf{x}}(k \mid \boldsymbol{\theta}), k] \\ &\qquad + [\mathbf{x}(k) - \hat{\mathbf{x}}(k \mid \boldsymbol{\theta})] \left\{\left[\frac{\partial}{\partial \hat{\mathbf{x}}(k \mid \boldsymbol{\theta})}\right]^{\mathrm{T}} \boldsymbol{\Gamma}[\hat{\mathbf{x}}(k \mid \boldsymbol{\theta}), k]\mathbf{V}_{\mathrm{w}}(k)\boldsymbol{\Gamma}^{\mathrm{T}}[\hat{\mathbf{x}}(k \mid \boldsymbol{\theta}), k]\right\} \\ &\hspace{30em} (3.4\text{-}39)\end{aligned}$$

such that we are able to evaluate (approximately) the needed variances. They are:

$$\begin{aligned}\operatorname{cov}\{\mathbf{x}(k), \mathbf{z}(k) \mid Z(k-1), \boldsymbol{\theta}\} &= \operatorname{cov}\{\hat{\mathbf{x}}(k \mid k-1, \boldsymbol{\theta}), \mathbf{h}[\mathbf{x}(k), k] \mid Z(k-1), \boldsymbol{\theta}\} \\ &\cong \mathbf{V}_{\tilde{\mathbf{x}}}(k \mid k-1, \boldsymbol{\theta}) \frac{\partial \mathbf{h}^{\mathrm{T}}[\hat{\mathbf{x}}(k \mid k-1, \boldsymbol{\theta}), k]}{\partial \hat{\mathbf{x}}(k \mid k-1, \boldsymbol{\theta})} \qquad (3.4\text{-}40)\end{aligned}$$

$$\begin{aligned}\operatorname{var}\{\mathbf{z}(k) \mid Z(k-1), \boldsymbol{\theta}\} &= \operatorname{var}\{\mathbf{h}[\mathbf{x}(k), k] \mid Z(k-1), \boldsymbol{\theta}\} + \operatorname{var}\{\mathbf{v}(k)\} \\ &= \left[\frac{\partial \mathbf{h}^{\mathrm{T}}[\hat{\mathbf{x}}(k \mid k-1, \boldsymbol{\theta}), k]}{\partial \hat{\mathbf{x}}(k \mid k-1, \boldsymbol{\theta})}\right]^{\mathrm{T}} \mathbf{V}_{\tilde{\mathbf{x}}}(k \mid k-1, \boldsymbol{\theta}) \\ &\quad \times \left[\frac{\partial \mathbf{h}^{\mathrm{T}}[\hat{\mathbf{x}}(k \mid k-1, \boldsymbol{\theta}), k]}{\partial \hat{\mathbf{x}}(k \mid k-1, \boldsymbol{\theta})}\right] + \mathbf{V}_{\mathrm{v}}(k) \qquad (3.4\text{-}41)\end{aligned}$$

[3] Alternatively, the zeroth order approximation $\boldsymbol{\Gamma}[\mathbf{x}(k), k] = \boldsymbol{\Gamma}[\hat{\mathbf{x}}(k \mid \boldsymbol{\theta}), k]$ could be used to obtain the same result.

To complete the derivation of this first-order conditional mean filter, it is necessary to determine the prior error variance. This is easily obtained from the series expansion as

$$
\begin{aligned}
&\mathbf{V}_{\tilde{\mathbf{x}}}(k \mid k-1, \boldsymbol{\theta}) \\
&\quad = \operatorname{var}\{\tilde{\mathbf{x}}(k) \mid \mathbf{Z}(k-1), \boldsymbol{\theta}\} = \operatorname{var}\{\mathbf{x}(k) \mid \mathbf{Z}(k-1), \boldsymbol{\theta}\} \\
&\quad = \operatorname{var}\{\boldsymbol{\phi}[\mathbf{x}(k-1), k-1] + \boldsymbol{\Gamma}[\mathbf{x}(k-1), k-1]\mathbf{w}(k-1) \mid \mathbf{Z}(k-1), \boldsymbol{\theta}\} \\
&\quad \cong \frac{\partial \boldsymbol{\phi}[\hat{\mathbf{x}}(k-1 \mid \boldsymbol{\theta}), k-1]}{\partial \hat{\mathbf{x}}(k-1 \mid \boldsymbol{\theta})} \mathbf{V}_{\tilde{\mathbf{x}}}(k-1 \mid \boldsymbol{\theta}) \frac{\partial \boldsymbol{\phi}^T[\hat{\mathbf{x}}(k-1 \mid \boldsymbol{\theta}), k-1]}{\partial \hat{\mathbf{x}}(k-1 \mid \boldsymbol{\theta})} \\
&\quad\quad + \boldsymbol{\Gamma}[\hat{\mathbf{x}}(k-1 \mid \boldsymbol{\theta}), k-1]\mathbf{V}_{\mathbf{w}}(k-1)\boldsymbol{\Gamma}^T[\hat{\mathbf{x}}(k-1 \mid \boldsymbol{\theta}), k-1] \qquad (3.4\text{-}42)
\end{aligned}
$$

The one-stage prediction estimate that is required for the likelihood function of Eq. (3.4-9) is easily obtained in terms of the estimate $\hat{\mathbf{x}}(k \mid \boldsymbol{\theta})$ by substituting the expansion of Eq. (3.4-37) into the message model of Eq. (3.4-1) and taking the expected value of the resulting expression conditioned upon $\mathbf{Z}(k)$ and $\boldsymbol{\theta}$. This yields

$$\hat{\mathbf{x}}(k+1 \mid k, \boldsymbol{\theta}) = \boldsymbol{\phi}[\hat{\mathbf{x}}(k \mid \boldsymbol{\theta}), k] \qquad (3.4\text{-}43)$$

Thus a complete set of first order nonlinear discrete conditional mean estimation algorithms has been discerned. These are summarized in Table 3.4-1 which includes the algorithms necessary for the maximum likelihood cost function determination by this first-order method. Higher-order estimation approximations (Sage and Melsa, 1971) may be developed, but this will not be accomplished here.

We note that these conditional mean (approximate) filter algorithms are just the extended Kalman filter algorithms. If the message and observation models are linear, they become the standard algorithms for the linear Kalman filter. Thus, minimization of the cost function of Table 3.4-1 together with the constraints of the associated difference equations in the table leads to the final maximum likelihood identification algorithms. These we will develop in the next four chapters. We now present a simple example to indicate the salient features of our discussions.

Example 3.4-1. A simple example will be used to indicate the procedure for maximum likelihood identification cost function determination. The scalar message and observation model with constant V_v and V_w:

$$x(k+1) = \Phi x(k) + w(k)$$

$$z(k) = x(k) + v(k)$$

TABLE 3.4-1

FIRST-ORDER MAXIMUM LIKELIHOOD IDENTIFICATION ALGORITHMS (COST FUNCTION)

Message model	$\mathbf{x}(k+1) = \boldsymbol{\phi}[\mathbf{x}(k), k] + \boldsymbol{\Gamma}[\mathbf{x}(k), k]\mathbf{w}(k)$
Observation model	$\mathbf{z}(k) = \mathbf{h}[\mathbf{x}(k), k] + \mathbf{v}(k)$
Prior statistics	$\mathscr{E}\{\mathbf{w}(k)\} = \mathbf{0} \qquad \operatorname{cov}\{\mathbf{w}(k), \mathbf{w}(j)\} = \mathbf{V}_{\mathbf{w}}(k)\delta_{\mathrm{K}}(k-j)$ $\mathscr{E}\{\mathbf{v}(k)\} = \mathbf{0} \qquad \operatorname{cov}\{\mathbf{v}(k), \mathbf{v}(j)\} = \mathbf{V}_{\mathbf{v}}(k)\delta_{\mathrm{K}}(k-j)$ $\mathscr{E}\{\mathbf{x}(0)\} = \boldsymbol{\mu}_{\mathbf{x}}(0) = \boldsymbol{\mu}_{\mathbf{x}_0} \qquad \operatorname{var}\{\mathbf{x}(0)\} = \mathbf{V}_{\tilde{\mathbf{x}}}(0) = \mathbf{V}_{\mathbf{x}}\{0\} = \mathbf{V}_{\mathbf{x}_0}$ $\operatorname{cov}\{\mathbf{w}(k), \mathbf{v}(k)\} = \operatorname{cov}\{\mathbf{x}(k), \mathbf{v}(k)\} = \operatorname{cov}\{\mathbf{x}(k), \mathbf{w}(j)\} = \mathbf{0}, \quad j \geqslant k$
Conditional mean filter algorithms	$\hat{\mathbf{x}}(k+1 \mid \boldsymbol{\theta}) = \hat{\mathbf{x}}(k+1 \mid k, \boldsymbol{\theta}) + \mathbf{K}(k+1)[\mathbf{z}(k+1) - \mathbf{h}[\hat{\mathbf{x}}(k+1 \mid k, \boldsymbol{\theta}), k+1]]$
One-stage prediction algorithm	$\hat{\mathbf{x}}(k+1 \mid k, \boldsymbol{\theta}) = \boldsymbol{\phi}[\hat{\mathbf{x}}(k \mid \boldsymbol{\theta}), k]$
Filter gain algorithm	$\mathbf{K}(k+1) = \mathbf{V}_{\tilde{\mathbf{x}}}(k+1 \mid \boldsymbol{\theta}) \dfrac{\partial \mathbf{h}^T[\hat{\mathbf{x}}(k+1 \mid k, \boldsymbol{\theta}), k+1]}{\partial \hat{\mathbf{x}}(k+1 \mid k, \boldsymbol{\theta})} \mathbf{V}_{\mathbf{v}}^{-1}(k+1)$
Filter prior error variance algorithm	$\mathbf{V}_{\tilde{\mathbf{x}}}(k+1 \mid k, \boldsymbol{\theta}) = \dfrac{\partial \boldsymbol{\phi}[\hat{\mathbf{x}}(k \mid \boldsymbol{\theta}), k]}{\partial \hat{\mathbf{x}}(k \mid \boldsymbol{\theta})} \mathbf{V}_{\tilde{\mathbf{x}}}(k) \dfrac{\partial \boldsymbol{\phi}^T[\hat{\mathbf{x}}(k \mid \boldsymbol{\theta}), k]}{\partial \hat{\mathbf{x}}(k \mid \boldsymbol{\theta})} + \boldsymbol{\Gamma}[\hat{\mathbf{x}}(k \mid \boldsymbol{\theta}), k]\mathbf{V}_{\mathbf{w}}(k)\boldsymbol{\Gamma}^{\mathrm{T}}[\hat{\mathbf{x}}(k \mid \boldsymbol{\theta}), k]$
Filter error variance algorithm	$\mathbf{V}_{\tilde{\mathbf{x}}}(k+1 \mid \boldsymbol{\theta}) = \mathbf{V}_{\tilde{\mathbf{x}}}(k+1 \mid k, \boldsymbol{\theta}) - \mathbf{V}_{\tilde{\mathbf{x}}}(k+1 \mid k, \boldsymbol{\theta}) \times \left[\dfrac{\partial \mathbf{h}^{\mathrm{T}}[\hat{\mathbf{x}}(k+1 \mid k, \boldsymbol{\theta}), k+1]}{\partial \hat{\mathbf{x}}(k+1 \mid k, \boldsymbol{\theta})}\right] \times \left[\left[\dfrac{\partial \mathbf{h}^{\mathrm{T}}[\hat{\mathbf{x}}(k+1 \mid k, \boldsymbol{\theta}), k+1]}{\partial \hat{\mathbf{x}}(k+1 \mid k, \boldsymbol{\theta})}\right]^{\mathrm{T}} \mathbf{V}_{\tilde{\mathbf{x}}}(k+1 \mid k, \boldsymbol{\theta}) \times \left[\dfrac{\partial \mathbf{h}^{\mathrm{T}}[\hat{\mathbf{x}}(k+1 \mid k, \boldsymbol{\theta}), k+1]}{\partial \hat{\mathbf{x}}(k+1 \mid k, \boldsymbol{\theta})}\right] + \mathbf{V}_{\mathbf{v}}(k+1)\right]^{-1} \times \left[\dfrac{\partial \mathbf{h}^T[\hat{\mathbf{x}}(k+1 \mid k, \boldsymbol{\theta}), k+1]}{\partial \hat{\mathbf{x}}(k+1 \mid k, \boldsymbol{\theta})}\right]^{\mathrm{T}} \mathbf{V}_{\tilde{\mathbf{x}}}(k+1 \mid k, \boldsymbol{\theta})$

TABLE 3.4-1 (*continued*)

Maximum likelihood identification cost function	$J = \frac{1}{2}\sum_{k=k_1}^{k_f} \ln\det \mathbf{V}_{\mathbf{z}}(k \mid k-1, \theta) + \lVert \mathbf{z}(k) - \mathbf{h}[\hat{\mathbf{x}}(k \mid k-1, \boldsymbol{\theta}), k] \rVert^2_{\mathbf{V}_{\mathbf{z}}^{-1}(k\mid k-1,\boldsymbol{\theta})}$
	$\mathbf{V}_{\mathbf{z}}(k \mid k-1, \boldsymbol{\theta}) = \mathbf{V}_{\mathbf{v}}(k) + \left[\frac{\partial \mathbf{h}^T[\hat{\mathbf{x}}(k \mid k-1, \boldsymbol{\theta}), k]}{\partial \hat{\mathbf{x}}(k \mid k-1, \boldsymbol{\theta})}\right]^T \times \mathbf{V}_{\tilde{\mathbf{x}}}(k \mid k-1, \boldsymbol{\theta}) \left[\frac{\partial \mathbf{h}^T[\hat{\mathbf{x}}(k \mid k-1, \boldsymbol{\theta}), k]}{\partial \hat{\mathbf{x}}(k \mid k-1, \boldsymbol{\theta})}\right]$
Filter initial conditions	$\hat{\mathbf{x}}(0 \mid 0, \theta) = \boldsymbol{\mu}_{\tilde{\mathbf{x}}0}, \quad \mathbf{V}_{\tilde{\mathbf{x}}}(0 \mid 0, \theta) = \mathbf{V}_{\mathbf{x}0}$

is considered. For this or any linear message and observation model, algorithms of Table 3.4-1 are exact and not just good approximations. The cost function to be minimized, Eq. (3.4-21), becomes

$$J = \frac{1}{2}\sum_{k=k_1}^{k_f} \left\{ \ln\det[V_v + V_{\tilde{x}}(k \mid k-1, \Phi]) + \frac{[z(k) - \hat{x}(k \mid k-1, \Phi)]^2}{V_v + V_x(k \mid k-1, \Phi)} \right\}$$

The one-stage prediction algorithms of Table 3.4-1, which are again exact since this is a linear system, become

$$\hat{x}(k+1 \mid k, \Phi) = \Phi\hat{x}(k \mid k-1, \Phi) + \frac{\Phi V_{\tilde{x}}(k \mid k-1, \Phi)}{V_{\tilde{x}}(k \mid k-1, \Phi) + V_v}[z(k) - \hat{x}(k \mid k-1, \Phi)]$$

$$V_{\tilde{x}}(k+1 \mid k, \Phi) = \frac{\Phi^2 V_v V_{\tilde{x}}(k \mid k-1, \Phi)}{V_{\tilde{x}}(k \mid k-1, \Phi) + V_v} + V_w$$

In order to determine the optimum value of Φ, the transition matrix of the model, we minimize the cost function J with respect to Φ while constraining the difference equations in $\hat{x}(k+1 \mid k, \Phi)$ and $V_{\tilde{x}}(k+1 \mid k, \Phi)$ to be satisfied. Since there is but a single parameter to be adjusted, it is a relatively simple task (on a computer) to evaluate the cost function for as many values of Φ as desired. Figures 3.4-1–3.4-3 illustrate the variation in J versus Φ for several different ratios

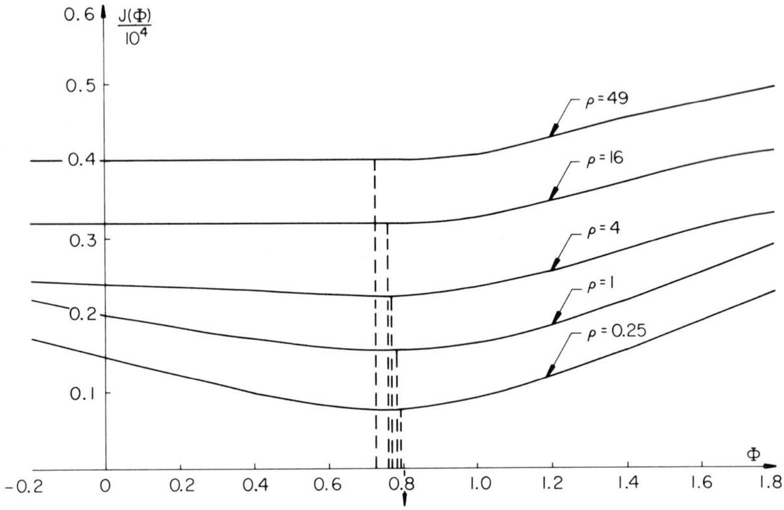

FIG. 3.4-1. Cost versus Φ for different values of V_v/V_w, Example 3.4-1. $x_0 = 0.0$, $V_v/V_w = \rho$, $k_f = 800$.

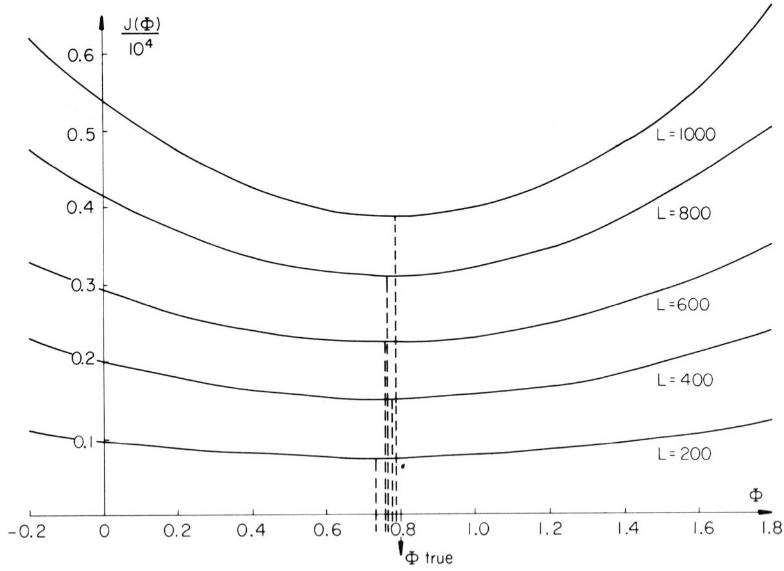

FIG. 3.4-2. Cost versus Φ for different values of L, Example 3.4-1. L = number of observations, $V_v = 1.0$, $V_w = 16.0$, $x_0 = 0.5$.

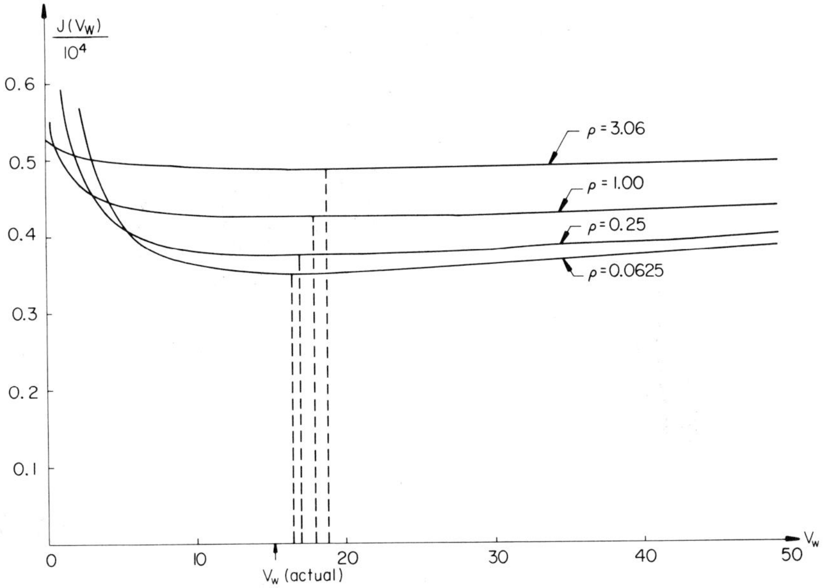

FIG. 3.4-3. Cost function versus V_w for different V_v/V_w, Example 3.4-1. $\rho = V_v/V_w$, k_f = number of observations = 900, $x_0 = 0.0$.

of V_v/V_w. It should be noted that with a finite sample size (finite k_f, $k_0 = 0$) the minimum of the cost function may, for particular noise sequences, not always be at the true value of Φ. For "low" plant noise and a "long" observation sequence, the minimum J will occur very near the true value of Φ. As the record is shortened or the measurement noise increased, the identification error becomes greater. Also, as we might expect, the prior statistics of $x(k_0)$ affect the identification. Clearly, if $V_x(k_0)$ is zero and there is no plant or measurement noise, we have the simple sort of identification problem considered in Chap. 2. Figure 3.4-4 indicates cost function variation versus $\sqrt{V_v}$ for different sample size.

These approximative algorithms need to be modified slightly to consider the case where the plant and measurement noise are correlated. Equations (3.4-9) and (3.4-11) are still valid cost functions. However, the expectations involved are different from the uncorrelated case. We have

$$\begin{aligned}\hat{\mathbf{h}}[\mathbf{x}(k), k \mid \boldsymbol{\theta}] &= \mathscr{E}\{\mathbf{z}(k) \mid Z(k-1), \boldsymbol{\theta}\} \\ &= \mathscr{E}\{\mathbf{h}[\mathbf{x}(k), k] \mid Z(k-1), \boldsymbol{\theta}\}\end{aligned}$$

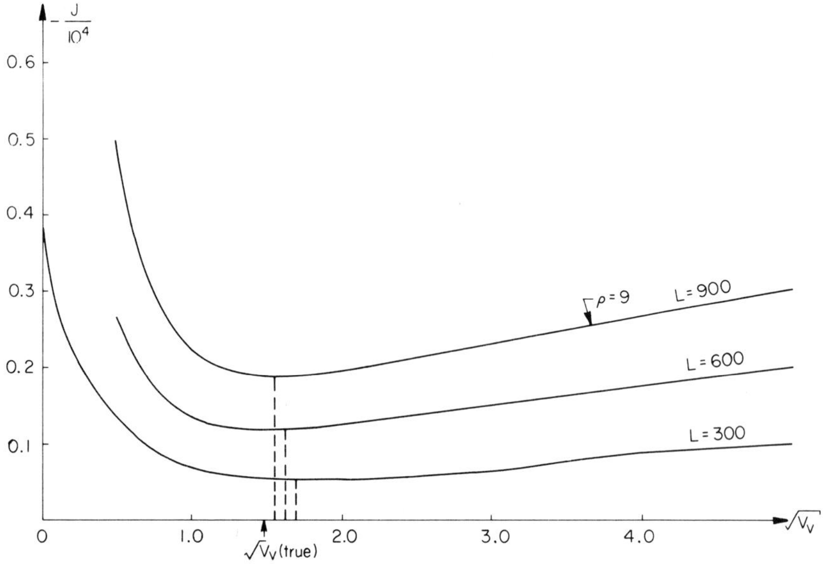

FIG. 3.4-4. Cost function versus $\sqrt{V_v}$ for different L. $V_v/V_w = \rho = 9.0$, L = number of observations, $x_0 = 0.0$.

By employing the Taylor series expansion of Eq. (3.4-22) we see that we have need for $\mathbf{h}[\hat{\mathbf{x}}(k \mid k-1, \boldsymbol{\theta}])$ which is, of course, still defined as in Eq. (3.4-21). The one-stage prediction algorithm of Eq. (3.4-43) is not valid now since

$$\mathbf{x}(k+1) = \boldsymbol{\phi}[\mathbf{x}(k), k] + \boldsymbol{\Gamma}[\mathbf{x}(k), k]\mathbf{w}(k)$$

and the first-order approximation is

$$\hat{\mathbf{x}}(k+1 \mid k, \boldsymbol{\theta}) = \boldsymbol{\phi}[\hat{\mathbf{x}}(k \mid \boldsymbol{\theta}), k] + \boldsymbol{\Gamma}[\hat{\mathbf{x}}(k), k]\mathscr{E}\{\mathbf{w}(k) \mid \boldsymbol{Z}(k), \boldsymbol{\theta}\} \quad (3.4\text{-}44)$$

which becomes

$$\begin{aligned}\hat{\mathbf{x}}(k+1 \mid k, \boldsymbol{\theta}) = {} & \boldsymbol{\phi}[\hat{\mathbf{x}}(k \mid \boldsymbol{\theta}), k] + \boldsymbol{\Gamma}[\hat{\mathbf{x}}(k), k]\mathbf{V}_{\mathrm{wv}}(k)\mathbf{V}_{\mathrm{z}}^{-1}(k \mid k-1, \boldsymbol{\theta}) \\ & \times \{\mathbf{z}(k) - \mathbf{h}[\hat{\mathbf{x}}(k \mid k-1, \boldsymbol{\theta}), k]\} \qquad (3.4\text{-}45)\end{aligned}$$

by application of the conditional expectation theorem of Eq. (3.4-28), and the relations

$$\operatorname{var}\{\mathbf{w}(k) \mid \boldsymbol{Z}(k), \boldsymbol{\theta}\} = \mathbf{V}_{\mathrm{w}}(k) - \mathbf{V}_{\mathrm{wv}}(k)\mathbf{V}_{\mathrm{z}}^{-1}(k \mid k-1)\mathbf{V}_{\mathrm{vw}}(k) \quad (3.4\text{-}46)$$

In a similar way, other algorithms for the first-order conditional mean filter need to be modified. This is easily accomplished and the resulting approximate algorithms are detailed in Table 3.4-2.

TABLE 3.4-2

FIRST-ORDER MAXIMUM LIKELIHOOD IDENTIFICATION ALGORITHMS (CORRELATED PLANT AND MEASUREMENT NOISE)

Conditional mean filter algorithms	$\hat{\mathbf{x}}(k+1 \mid \boldsymbol{\theta}) = \hat{\mathbf{x}}(k+1 \mid k, \boldsymbol{\theta}) + \mathbf{K}(k+1)\{\mathbf{z}(k+1) - \mathbf{h}[\hat{\mathbf{x}}(k+1 \mid k, \boldsymbol{\theta}), k+1]\}$
One-stage prediction algorithm	$\hat{\mathbf{x}}(k+1 \mid k, \boldsymbol{\theta}) = \boldsymbol{\phi}[\hat{\mathbf{x}}(k \mid \boldsymbol{\theta}), k] + \mathbf{K}_p(k)\{\mathbf{z}(k) - \mathbf{h}[\hat{\mathbf{x}}(k \mid k-1, \boldsymbol{\theta}), k]\}$
Identification gain algorithm	$\mathbf{K}(k+1) = \mathbf{V}_{\tilde{\mathbf{x}}}(k+1 \mid k, \boldsymbol{\theta}) \dfrac{\partial \mathbf{h}^T[\hat{\mathbf{x}}(k+1 \mid k, \boldsymbol{\theta}), k+1]}{\partial \hat{\mathbf{x}}(k+1 \mid k, \boldsymbol{\theta})} \times \mathbf{V}_{\mathbf{z}}^{-1}(k+1 \mid k, \boldsymbol{\theta})$
Identification prior error variance	$\mathbf{V}_{\tilde{\mathbf{x}}}(k+1 \mid k, \boldsymbol{\theta}) = \left[\dfrac{\partial \boldsymbol{\phi}[\hat{\mathbf{x}}(k \mid \boldsymbol{\theta}), k]}{\partial \hat{\mathbf{x}}(k \mid \boldsymbol{\theta})}\right] \mathbf{V}_{\tilde{\mathbf{x}}}(k) \left[\dfrac{\partial \boldsymbol{\phi}^T[\hat{\mathbf{x}}(k \mid \boldsymbol{\theta}), k]}{\partial \hat{\mathbf{x}}(k \mid \boldsymbol{\theta})}\right] + \boldsymbol{\Gamma}[\hat{\mathbf{x}}(k \mid \boldsymbol{\theta}), k]\mathbf{V}_{\mathbf{w}}(k)\boldsymbol{\Gamma}^T[\hat{\mathbf{x}}(k \mid \boldsymbol{\theta}), k] - \mathbf{K}_p(k)\mathbf{V}_{\mathbf{z}}^{-1}(k \mid k-1, \boldsymbol{\theta})\mathbf{K}_{\mathbf{p}}{}^T(k) - \dfrac{\partial \boldsymbol{\phi}[\hat{\mathbf{x}}(k \mid \boldsymbol{\theta}), k]}{\partial \hat{\mathbf{x}}(k \mid \boldsymbol{\theta})} \mathbf{K}(k)\mathbf{V}_{\mathbf{vw}}(k)\boldsymbol{\Gamma}^T[\hat{\mathbf{x}}(k \mid \boldsymbol{\theta}), k] - \boldsymbol{\Gamma}[\hat{\mathbf{x}}(k \mid \boldsymbol{\theta}), k]\mathbf{V}_{\mathbf{wv}}(k)\mathbf{K}^T(k) \dfrac{\partial \boldsymbol{\phi}^T[\mathbf{x}(k \mid \boldsymbol{\theta}), k]}{\partial \hat{\mathbf{x}}(k \mid \boldsymbol{\theta})}$
Filter error variance algorithm	$\mathbf{V}_{\tilde{\mathbf{x}}}(k+1 \mid \boldsymbol{\theta}) = \left\{\mathbf{I} - \mathbf{K}(k+1)\left[\dfrac{\partial \mathbf{h}^T[\hat{\mathbf{x}}(k+1 \mid k, \boldsymbol{\theta}), k+1]}{\partial \hat{\mathbf{x}}(k+1 \mid k, \boldsymbol{\theta})}\right]^T\right\} \times \mathbf{V}_{\tilde{\mathbf{x}}}(k+1 \mid k, \boldsymbol{\theta})$
One-stage prediction gain algorithm	$\mathbf{K}_p(k) = \boldsymbol{\Gamma}[\hat{\mathbf{x}}(k \mid \boldsymbol{\theta}), k]\mathbf{V}_{\mathbf{wv}}(k)\mathbf{V}_{\mathbf{z}}^{-1}(k+1 \mid k, \boldsymbol{\theta})$

TABLE 3.4-2 (*continued*)

Maximum likelihood identification cost function	$J = \frac{1}{2}\sum_{k=k_1}^{k_f} \ln\det \mathbf{V}_{\mathbf{z}}(k \mid k-1, \boldsymbol{\theta}) + \lVert \mathbf{z}(k) - \mathbf{h}[\hat{\mathbf{x}}(k \mid k-1, \boldsymbol{\theta}), k] \rVert^2_{\mathbf{V}_{\mathbf{z}}^{-1}(k\mid k-1,\boldsymbol{\theta})}$ $\mathbf{V}_{\mathbf{z}}(k \mid k-1, \boldsymbol{\theta}) = \mathbf{V}_{\mathbf{v}}(k) + \left[\frac{\partial \mathbf{h}^T[\hat{\mathbf{x}}(k \mid k-1, \boldsymbol{\theta}), k]}{\partial \hat{\mathbf{x}}(k \mid k \mid k-1, \boldsymbol{\theta})}\right]^T \times \mathbf{V}_{\tilde{\mathbf{x}}}(k \mid k-1, \boldsymbol{\theta}) \left[\frac{\partial \mathbf{h}^T[\hat{\mathbf{x}}(k \mid k-1, \boldsymbol{\theta}), k]}{\partial \hat{\mathbf{x}}(k \mid k-1, \boldsymbol{\theta})}\right]$
Filter initial conditions	$\hat{\mathbf{x}}(0 \mid 0, \boldsymbol{\theta}) = \boldsymbol{\mu}_{\mathbf{x}_0}, \quad \mathbf{V}_{\tilde{\mathbf{x}}}(0 \mid 0, \boldsymbol{\theta}) = \mathbf{V}_{\mathbf{x}_0}$

Example 3.4-2. We will reconsider Example 3.4-1, except that we now assume a second observation

$$z_2(k) = w(k) + v_2(k)$$

such that the plant and observation noise are correlated with

$$\mathbf{z}(k) = \begin{bmatrix} z(k) \\ z_2(k) \end{bmatrix} = \begin{bmatrix} x(k) + v(k) \\ w(k) + v_2(k) \end{bmatrix}$$

$$\mathbf{h}[x(k), k] = \begin{bmatrix} x(k) \\ 0 \end{bmatrix}, \qquad \mathbf{v}(k) = \begin{bmatrix} v(k) \\ w(k) + v_2(k) \end{bmatrix}$$

$$\mathbf{V}_{w\mathbf{v}} = [0 \quad V_w], \qquad \mathbf{V}_{\mathbf{v}} = \begin{bmatrix} V_v & 0 \\ 0 & V_w + V_{v_2} \end{bmatrix}$$

Use of the algorithms of Table 3.4-2 leads immediately to the cost function and associated equality constraints

$$J = \frac{1}{2}\sum_{k=k_1}^{k_f} \left\{ \ln\det[V_v + V_{\tilde{x}}(k \mid k+1, \Phi) + \frac{[z(k) - \hat{x}(k \mid k-1, \Phi)]^2}{V_v + V_{\tilde{x}}(k \mid k-1, \Phi)} \right\}$$

$$\hat{x}(k+1 \mid k, \Phi) = \Phi\hat{x}(k \mid k-1, \Phi) + \frac{V_w}{V_w + V_{v_2}} z_2(k) + \frac{\Phi V_{\tilde{x}}(k \mid k-1, \Phi)}{V_{\tilde{x}}(k \mid k-1, \Phi) + V_v} [z(k) - \hat{x}(k \mid k-1, \Phi)]$$

$$V_{\tilde{x}}(k+1 \mid k, \Phi) = \frac{\Phi^2 V_v V_{\tilde{x}}(k-1 \mid k, \Phi)}{V_{\tilde{x}}(k \mid k-1, \Phi) + V_v} + V_w - \frac{V_w^2}{V_w + V_{v_2}}$$

Figure 3.4-5 indicates the variation of the cost function J versus Φ for several different values of V_{v_2}. As is apparent, if $V_{v_2} = \infty$, there is no use to be made of $z_2(k)$, and the algorithms of this example degenerate to those of Example 3.4-1.

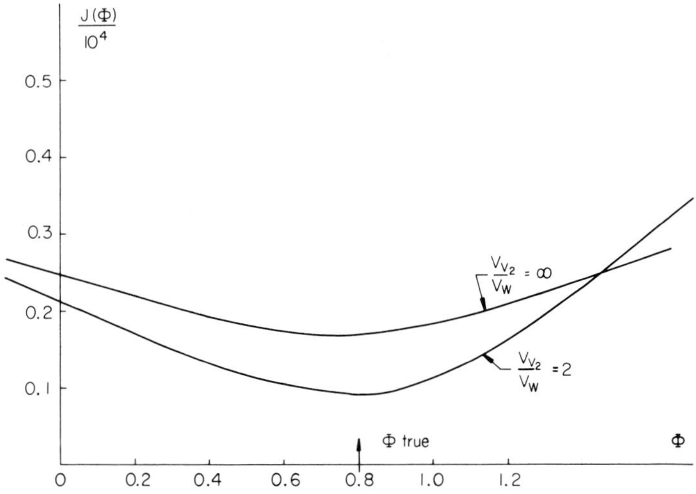

FIG. 3.4-5. Cost function versus Φ for two different values of V_{v2}, Example 3.4-2. $V_{v1}/V_w = 1$, $x_0 = 0$, $k_f = 800$.

3.5. SUMMARY

In this chapter, we have examined and formulated the cost functions for discrete system identification for the maximum a posteriori and maximum likelihood criteria. In several instances, the TPBVP for system identification were derived and generalization to the continuous case made. We will now turn our attention to computational methods for determination of parameters and states in system identification problems.

4

GRADIENT TECHNIQUES FOR SYSTEM IDENTIFICATION

4.1. Introduction

We will now turn our attention to computational methods for system identification. In this chapter, we will consider a class of direct methods known as gradient methods. These methods are called direct since the attempt is to minimize the cost function at each iteration. In an indirect method such as the quasilinearization method of Chap. 6, the attempt is to solve the two-point boundary value problem resulting from the use of optimization theory at each stage in the computation.

We will consider static or single-stage decision problems and will then proceed to the multistage and continuous cases. In addition to considering the first-order or basic gradient method, we will also consider second-order series expansion of various nonlinear terms in order to obtain what are called second-order gradient or second variation methods. These methods have the advantage over the first-order gradient method of much more rapid convergence. They suffer the disadvantage of being considerably more complex and of possessing a much more narrow range of initial conditions for which convergence will occur than the first-order gradient method. Finally, we will consider the conjugate gradient method, which generates conjugate directions of search and therefore minimizes a positive definite quadratic function of m variables in m steps. This method possesses many of the advantages of the first- and second-order gradient methods.

We will illustrate the gradient techniques by several system identification examples. The most accessible tutorial works concerning gradient techniques are the texts by Sage (1968) and Bryson and Ho (1969). In each of these texts, the gradient method is considered primarily for the solution of optimal control problems. Bekey and Karplus (1968) use some first-order gradient techniques in their text concerned with hybrid computation. Many journal papers are available on this subject, but the majority of these again concern optimal control rather than the specific problem of system identification. Many of these are cited in the bibliography.

4.2. The Gradient Technique for Single-Stage Identification Problems

The classical gradient or steepest descent method for extremization problems is a direct method which has been in existence for a number of years. Fundamentally it is concerned with finding the extremum of the scalar function of an M vector

$$J = \theta(\mathbf{u}) \tag{4.2-1}$$

Elementary calculus techniques show that the first necessary requirement for an extremum is

$$\left.\frac{dJ}{d\mathbf{u}}\right|_{\mathbf{u}=\hat{\mathbf{u}}} = \mathbf{0} = \left.\frac{d\theta(\mathbf{u})}{d\mathbf{u}}\right|_{\mathbf{u}=\hat{\mathbf{u}}} \tag{4.2-2}$$

The vector $dJ/d\mathbf{u}$ is often spoken of as the *gradient vector* and setting $d\theta(\mathbf{u})/d\mathbf{u} = \mathbf{0}$ results in a set of M algebraic equations which may be quite difficult to solve for the optimum vector $\hat{\mathbf{u}}$. If this is the case, we may choose instead to develop an iterative method based upon expanding $\theta(\mathbf{u})$ in a Taylor's series about an assumed vector $\mathbf{u}^i$. We have

$$J = \theta(\mathbf{u}) = \theta(\mathbf{u}^i) + \left[\frac{d\theta(\mathbf{u}^i)}{d\mathbf{u}^i}\right]^{\mathrm{T}}(\mathbf{u} - \mathbf{u}^i) + \frac{1}{2}(\mathbf{u} - \mathbf{u}^i)^{\mathrm{T}} \frac{d^2\theta(\mathbf{u}^i)}{(d\mathbf{u}^i)^2}(\mathbf{u} - \mathbf{u}^i) + \cdots \tag{4.2-3}$$

Let us truncate this series after the linear term in $\mathbf{u}$ such that we have

$$J = \theta(\mathbf{u}) \cong \theta(\mathbf{u}^i) + \left[\frac{d\theta(\mathbf{u}^i)}{d\mathbf{u}^i}\right]^{\mathrm{T}}(\mathbf{u} - \mathbf{u}^i) \tag{4.2-4}$$

Now suppose that we evaluate the cost function J with $\mathbf{u} = \mathbf{u}^i$ and then calculate

$$\Delta J^i = \theta(\mathbf{u}) - \theta(\mathbf{u}^i) \cong \left[\frac{d\theta(\mathbf{u}^i)}{d\mathbf{u}^i}\right]^{\mathrm{T}} \Delta\mathbf{u}^i \tag{4.2-5}$$

where

$$\Delta J^i = J - J^i, \qquad \Delta\mathbf{u}^i = \mathbf{u} - \mathbf{u}^i \tag{4.2-6}$$

We wish to pick $\Delta\mathbf{u}^i$ such that ΔJ^i is as negative as possible in order to make the *steepest descent* towards a minimum of J. It is clear from Eq. (4.2-5) that, in order to minimize ΔJ^i subject to the constraint $(\Delta\mathbf{u}^i)^{\mathrm{T}}\Delta\mathbf{u}^i = \alpha$, we should choose

$$\Delta\mathbf{u}^i = -K^i \frac{d\theta(\mathbf{u}^i)}{d\mathbf{u}^i} \tag{4.2-7}$$

where K^i is a positive scalar, possibly a function of the iteration index i, which is chosen with respect to such factors as convergence, rate of descent, etc. The computation of the optimum vector $\hat{\mathbf{u}}$ proceeds in a straightforward fashion. We assume a ${}_1\mathbf{u}^i$ and then calculate $d\theta(\mathbf{u}^i)/d\mathbf{u}^i$. $\Delta\mathbf{u}^i$ is then computed in accordance with Eq. (4.2-7) and $\mathbf{u}^{i+1}$ determined from

$$\mathbf{u}^{i+1} = \mathbf{u}^i + \Delta\mathbf{u}^i = \mathbf{u}^i - K^i \frac{d\theta(\mathbf{u}^i)}{d\mathbf{u}^i} \tag{4.2-8}$$

The computation is repeated until there is no significant change in $\mathbf{u}$ from iteration to iteration. There is no assurance that the computation will converge unless K^i is suitably restricted. A simple example will illustrate the point.

Example 4.2-1. We desire to solve the set of linear algebraic equations $\mathbf{Au} = \mathbf{b}$ for $\mathbf{u}$. Clearly $\mathbf{u} = \mathbf{A}^{-1}\mathbf{b}$, but this may be a computationally difficult problem if the dimension of $\mathbf{u}$ is large. We know that the nonnegative cost function, where $\mathbf{R}$ is a positive semidefinite weighting matrix,

$$J = \theta(\mathbf{u}) = \tfrac{1}{2}\tilde{\mathbf{e}}^{\mathrm{T}}\mathbf{R}\tilde{\mathbf{e}} = \tfrac{1}{2}(\mathbf{Au} - \mathbf{b})^{\mathrm{T}}\mathbf{R}(\mathbf{Au} - \mathbf{b})$$

will reach its minimum value of zero only if $\mathbf{u} = \mathbf{A}^{-1}\mathbf{b}$; thus minimizing J is equivalent to solving $\mathbf{Au} = \mathbf{b}$. We use Eq. (4.2-8) and have

$$\mathbf{u}^{i+1} = \mathbf{u}^i - K^i\mathbf{A}^{\mathrm{T}}\mathbf{R}(\mathbf{Au}^i - \mathbf{b})$$

The error in the computation is related to the decrease of J from iteration to iteration. We have

$$J^i = \tfrac{1}{2} \| \mathbf{Au}^i - \mathbf{b} \|_{\mathbf{R}}^2$$

$$J^{i+1} = \tfrac{1}{2} \| \mathbf{Au}^i - K^i\mathbf{AA}^{\mathrm{T}}\mathbf{R}(\mathbf{Au}^i - \mathbf{b}) - \mathbf{b} \|_{\mathbf{R}}^2$$

$$= \tfrac{1}{2} \| \mathbf{Au}^i - \mathbf{b} \|_{(\mathbf{I}-K^i\mathbf{AA}^{\mathrm{T}}\mathbf{R})^{\mathrm{T}}\mathbf{R}(\mathbf{I}-K^i\mathbf{AA}^{\mathrm{T}}\mathbf{R})}^2$$

For stability, we require $J^{i+1} < J^i$ and thus require that the matrix $2\mathbf{I} - K^i\mathbf{A}^{\mathrm{T}}\mathbf{RA}$ be positive definite.

The error in the computation is actually $\mathbf{e}^i = \mathbf{u} - \mathbf{u}^i$, and we have

$$\mathbf{e}^{i+1} = \mathbf{Q}^i\mathbf{e}^i, \qquad \| \mathbf{e}^{i+1} \|^2 = \| \mathbf{e}^i \|_{\mathbf{Q}^{i\mathrm{T}}\mathbf{Q}^i}^2$$

where

$$\mathbf{Q}^i = \mathbf{I} - K^i\mathbf{A}^{\mathrm{T}}\mathbf{RA}$$

In order to make $\| \mathbf{e}^{i+1} \|^2 < \| \mathbf{e}^i \|^2$, we must require that the matrix $2\mathbf{I} - K^i\mathbf{A}^{\mathrm{T}}\mathbf{RA}$ be positive definite. This sets an upper bound on the K^i. Unfortunately, convergence requirements are not easily established for problems of much greater complexity than the simple one examined here.

Satisfaction of Eq. (4.2-2) insures only that a relative extremum of the cost function exists. To determine whether the obtained value of $\hat{\mathbf{u}}$ maximizes or minimizes the cost function, the second derivative or second variation of the cost function must be examined; however, we will not pursue this topic here (Sage, 1968).

Minimization of cost function subject to equality constraints is a topic of much more interest here than the simple extremization problem considered thus far. Thus we wish to consider the minimization of

$$J = \theta(\mathbf{x}, \mathbf{u}) \tag{4.2-9}$$

by choice of $\mathbf{u}$, subject to the equality constraint relating $\mathbf{x}$ and $\mathbf{u}$

$$\mathbf{f}(\mathbf{x}, \mathbf{u}) = \mathbf{0} \tag{4.2-10}$$

where $\mathbf{u}$ is an independent M vector and $\mathbf{f}$ and $\mathbf{x}$ are N vectors. This may be accomplished by defining a quantity

$$H(\mathbf{x}, \mathbf{u}, \boldsymbol{\lambda}) = \theta(\mathbf{x}, \mathbf{u}) + \boldsymbol{\lambda}^{\mathrm{T}}\mathbf{f}(\mathbf{x}, \mathbf{u}) \tag{4.2-11}$$

where $\boldsymbol{\lambda}$ is a Lagrange multiplier. We then solve the algebraic equations

$$\frac{\partial H}{\partial \mathbf{u}} = \frac{\partial H}{\partial \mathbf{x}} = \frac{\partial H}{\partial \boldsymbol{\lambda}} = \mathbf{0} \tag{4.2-12}$$

or

$$\frac{\partial \theta}{\partial \mathbf{u}} + \frac{\partial \mathbf{f}^{\mathrm{T}}}{\partial \mathbf{u}} \boldsymbol{\lambda} = \mathbf{0} \tag{4.2-13}$$

$$\frac{\partial \theta}{\partial \mathbf{x}} + \frac{\partial \mathbf{f}^{\mathrm{T}}}{\partial \mathbf{x}} \boldsymbol{\lambda} = \mathbf{0} \tag{4.2-14}$$

$$\mathbf{f}(\mathbf{x}, \mathbf{u}) = \mathbf{0} \tag{4.2-15}$$

Often this will be tedious to do, so we attempt an iterative procedure. If we choose a $\mathbf{u}^i$, we may solve Eq. (4.2-10) to obtain $\mathbf{f}(\mathbf{x}^i, \mathbf{u}^i) = \mathbf{0}$. If we expand H about this solution, we obtain

$$H(\mathbf{x}, \mathbf{u}, \boldsymbol{\lambda}) = \theta(\mathbf{x}^i, \mathbf{u}^i) + \left[\frac{\partial \theta(\mathbf{x}^i, \mathbf{u}^i)}{\partial \mathbf{x}^i}\right]^{\mathrm{T}} (\mathbf{x} - \mathbf{x}^i) + \left[\frac{\partial \theta(\mathbf{x}^i, \mathbf{u}^i)}{\partial \mathbf{u}^i}\right]^{\mathrm{T}} (\mathbf{u} - \mathbf{u}^i)$$
$$+ \boldsymbol{\lambda}^{\mathrm{T}} \frac{\partial \mathbf{f}(\mathbf{x}^i, \mathbf{u}^i)}{\partial \mathbf{x}^i} (\mathbf{x} - \mathbf{x}^i) + \boldsymbol{\lambda}^{\mathrm{T}} \left[\frac{\partial \mathbf{f}^{\mathrm{T}}(\mathbf{x}^i, \mathbf{u}^i)}{\partial \mathbf{u}^i}\right]^{\mathrm{T}} (\mathbf{u} - \mathbf{u}^i) \tag{4.2-16}$$

In order to reduce the complexity of this expression, it is convenient to satisfy Eq. (4.2-14) such that

$$\begin{aligned} \Delta H^i &= H(\mathbf{x}, \mathbf{u}, \lambda) - H(\mathbf{x}^i, \mathbf{u}^i, \boldsymbol{\lambda}^i) \\ &= \left\{\frac{\partial \theta(\mathbf{x}^i, \mathbf{u}^i)}{\partial \mathbf{u}^i} + \boldsymbol{\lambda}^{i\mathrm{T}} \left[\frac{\partial \mathbf{f}^{\mathrm{T}}(\mathbf{x}^i, \mathbf{u}^i)}{\partial \mathbf{u}^i}\right]^{\mathrm{T}}\right\}^{\mathrm{T}} \Delta \mathbf{u}^i \\ &= \left[\frac{\partial H(\mathbf{x}^i, \mathbf{u}^i, \boldsymbol{\lambda}^i)}{\partial \mathbf{u}^i}\right]^{\mathrm{T}} \Delta \mathbf{u}^i \end{aligned} \tag{4.2-17}$$

where

$$\Delta \mathbf{u}^i = \mathbf{u} - \mathbf{u}^i, \qquad \Delta \mathbf{x}^i = \mathbf{x} - \mathbf{x}^i \tag{4.2-18}$$

In order to make the steepest descent to a minimum, we wish to make ΔH^i (or ΔJ^i) as negative as possible. Thus we set

$$\Delta \mathbf{u}^i = -K^i \frac{\partial H(\mathbf{x}^i, \mathbf{u}^i, \boldsymbol{\lambda}^i)}{\partial \mathbf{u}^i} \tag{4.2-19}$$

where K^i is a positive scalar which is chosen with criteria such as convergence and descent rate in mind.

The steps in this "first-order" gradient procedure are therefore:

1. determine $\mathbf{u}^i$;
2. obtain $\mathbf{x}^i$ from $\mathbf{f}(\mathbf{x}^i, \mathbf{u}^i) = \mathbf{0}$;
3. evaluate $\boldsymbol{\lambda}^i$ from

$$\boldsymbol{\lambda}^i = -\left[\frac{\partial \mathbf{f}^{\mathrm{T}}(\mathbf{x}^i, \mathbf{u}^i)}{\partial \mathbf{x}^i}\right]^{-1} \frac{\partial \theta(\mathbf{x}^i, \mathbf{u}^i)}{\partial \mathbf{x}^i}$$

4. determine

$$\frac{\partial H(\mathbf{x}^i, \mathbf{u}^i, \boldsymbol{\lambda}^i)}{\partial \mathbf{u}^i} = \frac{\partial \theta(\mathbf{x}^i, \mathbf{u}^i)}{\partial \mathbf{u}^i} + \frac{\partial \mathbf{f}^{\mathrm{T}}(\mathbf{x}^i, \mathbf{u}^i)}{\partial \mathbf{x}^i} \boldsymbol{\lambda}^i$$

5. compute

$$\mathbf{u}^{i+1} = \mathbf{u}^i - K^i \left[\frac{\partial H(\mathbf{x}^i, \mathbf{u}^i, \boldsymbol{\lambda}^i)}{\partial \mathbf{u}^i}\right]$$

6. repeat procedure until there is no change in control from iteration to iteration.

We note that there are many ways in which this procedure could be derived and stated. We could obtain

$$\Delta J^i = J - J^i \cong \left[\frac{\partial \theta(\mathbf{x}^i, \mathbf{u}^i)}{\partial \mathbf{x}^i}\right]^{\mathrm{T}} \Delta\mathbf{x}^i + \left[\frac{\partial \theta(\mathbf{x}^i, \mathbf{u}^i)}{\partial \mathbf{u}^i}\right]^{\mathrm{T}} \Delta\mathbf{u}^i = 0 \quad (4.2\text{-}20)$$

by taking the first-order terms in the derivative of the cost function, Eq. (4.2-9). If we expand $\mathbf{f}(\mathbf{x}, \mathbf{u}) = \mathbf{0}$ in a Taylor series about $\mathbf{x}^i$, $\mathbf{u}^i$ and retain first-order terms, we have

$$\left[\frac{\partial \mathbf{f}(\mathbf{x}^i, \mathbf{u}^i)}{\partial \mathbf{x}^i}\right] \Delta\mathbf{x}^i + \left[\frac{\partial \mathbf{f}^{\mathrm{T}}(\mathbf{x}^i, \mathbf{u}^i)}{\partial \mathbf{u}^i}\right]^{\mathrm{T}} \Delta\mathbf{u}^i = \mathbf{0} \quad (4.2\text{-}21)$$

Substituting Eq. (4.2-21) into Eq. (4.2-20) results in

$$\Delta J^i = \left\{-\left[\frac{\partial \theta(\mathbf{x}^i, \mathbf{u}^i)}{\partial \mathbf{x}^i}\right]^{\mathrm{T}} \left[\frac{\partial \mathbf{f}(\mathbf{x}^i, \mathbf{u}^i)}{\partial \mathbf{x}^i}\right]^{-1} \left[\frac{\partial \mathbf{f}^{\mathrm{T}}(\mathbf{x}^i, \mathbf{u}^i)}{\partial \mathbf{u}^i}\right]^{\mathrm{T}} + \left[\frac{\partial \theta(\mathbf{x}^i, \mathbf{u}^i)}{\partial \mathbf{u}^i}\right]^{\mathrm{T}}\right\} \Delta\mathbf{u}^i \quad (4.2\text{-}22)$$

The term within { } is just the expression $[\partial H/\partial \mathbf{u}]^{\mathrm{T}}$, and thus equivalence of these two procedures is established, since to make the steepest descent we again select $\Delta\mathbf{u}^i$ as in Eq. (4.2-19).

We may naturally question whether there is any merit inherent in incorporating second-order terms in development of these gradient algorithms. If we retain terms of first *and* second order in our expansion of the cost function, we obtain, since along $\mathbf{f}(\mathbf{x}, \mathbf{u}) = \mathbf{0}$ we have $J = H$,

$$\Delta J^i = \left[\frac{\partial H(\mathbf{x}^i, \mathbf{u}^i, \boldsymbol{\lambda}^i)}{\partial \mathbf{x}^i}\right]^{\mathrm{T}} \Delta\mathbf{x}^i + \left[\frac{\partial H(\mathbf{x}^i, \mathbf{u}^i, \boldsymbol{\lambda}^i)}{\partial \mathbf{u}^i}\right]^{\mathrm{T}} \Delta\mathbf{u}^i + \tfrac{1}{2}[\Delta\mathbf{x}^{i\mathrm{T}} \Delta\mathbf{u}^{i\mathrm{T}}]$$
$$\times \begin{bmatrix} \dfrac{\partial^2 H(\mathbf{x}^i, \mathbf{u}^i, \boldsymbol{\lambda}^i)}{(\partial \mathbf{x}^i)^2} & \dfrac{\partial}{\partial \mathbf{u}^i}\dfrac{\partial H(\mathbf{x}^i, \mathbf{u}^i, \boldsymbol{\lambda}^i)}{\partial \mathbf{x}^i} \\ \left[\dfrac{\partial}{\partial \mathbf{u}^i}\dfrac{\partial H(\mathbf{x}^i, \mathbf{u}^i, \boldsymbol{\lambda}^i)}{\partial \mathbf{x}^i}\right]^{\mathrm{T}} & \dfrac{\partial^2 H(\mathbf{x}^i, \mathbf{u}^i, \boldsymbol{\lambda}^i)}{(\partial \mathbf{u}^i)^2} \end{bmatrix} \begin{bmatrix} \Delta\mathbf{x}^i \\ \Delta\mathbf{u}^i \end{bmatrix} \tag{4.2-23}$$

We will adjust $\boldsymbol{\lambda}^i$ such that Eq. (4.2-14) is satisfied and $\partial H/\partial \mathbf{x} = \mathbf{0}$. Also we will require that Eq. (4.2-10) hold for first-order changes. Thus we will again require Eq. (4.2-21) to be valid. By substituting Eqs. (4.2-14) and (4.2-21) in the foregoing, we obtain

$$\Delta J^i = \left[\frac{\partial H(\mathbf{x}^i, \mathbf{u}^i, \boldsymbol{\lambda}^i)}{\partial \mathbf{u}^i}\right]^{\mathrm{T}} \Delta\mathbf{u}^i + \frac{1}{2}\,\Delta\mathbf{u}^{i\mathrm{T}} \left[-\left[\frac{\partial}{\partial \mathbf{u}}\frac{\partial H}{\partial \boldsymbol{\lambda}}\right]^{\mathrm{T}} \left[\frac{\partial}{\partial \mathbf{x}}\frac{\partial H}{\partial \boldsymbol{\lambda}}\right]^{\mathrm{T}} \mathbf{I}\right]$$
$$\times \begin{bmatrix} \dfrac{\partial^2 H}{\partial \mathbf{x}^2} & \dfrac{\partial}{\partial \mathbf{u}}\dfrac{\partial H}{\partial \mathbf{x}} \\ \left[\dfrac{\partial}{\partial \mathbf{u}}\dfrac{\partial H}{\partial \mathbf{x}}\right]^{\mathrm{T}} & \dfrac{\partial^2 H}{\partial \mathbf{u}^2} \end{bmatrix} \begin{bmatrix} -\left[\dfrac{\partial}{\partial \mathbf{x}}\dfrac{\partial H}{\partial \boldsymbol{\lambda}}\right]\left[\dfrac{\partial}{\partial \mathbf{u}}\dfrac{\partial H}{\partial \boldsymbol{\lambda}}\right] \\ \mathbf{I} \end{bmatrix} \Delta\mathbf{u}^i \tag{4.2-24}$$

where H, $\mathbf{x}$, $\mathbf{u}$, and $\boldsymbol{\lambda}$ are evaluated on the ith iteration. We see that we now have the possibility of minimizing ΔJ with respect to choice of $\Delta\mathbf{u}^i$. This is a standard problem in matrix calculus. We obtain

$$\Delta\mathbf{u}^i = -\left\{\left[\frac{\partial}{\partial \mathbf{u}}\frac{\partial H}{\partial \boldsymbol{\lambda}}\right]^{\mathrm{T}}\left[\frac{\partial}{\partial \mathbf{x}}\frac{\partial H}{\partial \boldsymbol{\lambda}}\right]^{\mathrm{T}}\left[\frac{\partial^2 H}{\partial \mathbf{x}^2}\right]\left[\frac{\partial}{\partial \mathbf{x}}\frac{\partial H}{\partial \boldsymbol{\lambda}}\right]\left[\frac{\partial}{\partial \mathbf{u}}\frac{\partial H}{\partial \boldsymbol{\lambda}}\right]\right.$$
$$+\left[\frac{\partial^2 H}{\partial \mathbf{u}^2}\right] - \left[\frac{\partial}{\partial \mathbf{u}}\frac{\partial H}{\partial \boldsymbol{\lambda}}\right]^{\mathrm{T}}\left[\frac{\partial}{\partial \mathbf{x}}\frac{\partial H}{\partial \boldsymbol{\lambda}}\right]^{\mathrm{T}}\left[\frac{\partial}{\partial \mathbf{u}}\frac{\partial H}{\partial \mathbf{x}}\right]$$
$$\left. - \left[\frac{\partial}{\partial \mathbf{u}}\frac{\partial H}{\partial \mathbf{x}}\right]\left[\frac{\partial}{\partial \mathbf{x}}\frac{\partial H}{\partial \boldsymbol{\lambda}}\right]\left[\frac{\partial}{\partial \mathbf{u}}\frac{\partial H}{\partial \boldsymbol{\lambda}}\right]\right\}^{-1} \frac{\partial H}{\partial \mathbf{u}} \tag{4.2-25}$$

and see that we do indeed have a procedure which specifies the change in control from iteration to iteration. Since we have minimized ΔJ to quadratic terms in $\Delta\mathbf{u}$ and $\Delta\mathbf{x}$, we expect faster convergence in this second-order gradient or second variation method than we obtain in the first-order gradient method. This is obtained at the expense of greater computational complexity and increased convergence difficulties however.

The steps in the computation are relatively straightforward and are quite similar to those in the first order gradient method:

1. Select a $\mathbf{u}^i$;
2. determine $\mathbf{x}^i$ from $\mathbf{f}(\mathbf{x}^i, \mathbf{u}^i) = \mathbf{0}$;
3. determine

$$H(\mathbf{x}^i, \mathbf{u}^i, \boldsymbol{\lambda}^i) = \theta(\mathbf{x}^i, \mathbf{u}^i) + \boldsymbol{\lambda}^{iT}\mathbf{f}(\mathbf{x}^i, \mathbf{u}^i)$$

4. evaluate $\boldsymbol{\lambda}^i$ from

$$\boldsymbol{\lambda}^i = -\left[\frac{\partial \mathbf{f}(\mathbf{x}^i, \mathbf{u}^i)}{\partial \mathbf{x}^i}\right]^{-1} \frac{\partial \theta(\mathbf{x}^i, \mathbf{u}^i)}{\partial \mathbf{x}^i} \qquad \text{or} \qquad \frac{\partial H(\mathbf{x}^i, \mathbf{u}^i, \boldsymbol{\lambda}^i)}{\partial \mathbf{x}^i} = \mathbf{0}$$

5. obtain necessary derivatives so as to compute $\Delta\hat{\mathbf{u}}^i$ from Eq. (4.2-25);
6. determine $\mathbf{u}^{i+1} = \mathbf{u}^i + \Delta\hat{\mathbf{u}}^i$;
7. repeat computation until $\Delta\mathbf{u}^i$ changes little from iteration to iteration.

The computational algorithm is considerably simplified if there is no equality constraint relation and $\theta(\mathbf{x}, \mathbf{u}) = \theta(\mathbf{u})$. Equation (4.2-25) becomes in this case

$$\Delta\hat{\mathbf{u}}^i = -\left[\frac{d^2\theta(\mathbf{u}^i)}{(d\mathbf{u}^i)^2}\right]^{-1} \frac{d\theta(\mathbf{u}^i)}{d\mathbf{u}^i} \tag{4.2-26}$$

Such that the iteration algorithm is

$$\mathbf{u}^{i+1} = \mathbf{u}^i - \left[\frac{d^2\theta(\mathbf{u}^i)}{(d\mathbf{u}^i)^2}\right]^{-1} \frac{d\theta(\mathbf{u}^i)}{d\mathbf{u}^i} \tag{4.2-27}$$

Example 4.2-2. Again we consider the problem of solving $\mathbf{Au} = \mathbf{b}$ by minimizing the cost function

$$J = \theta(\mathbf{u}) = \tfrac{1}{2}(\mathbf{Au} - \mathbf{b})^T\mathbf{R}(\mathbf{Au} - \mathbf{b})$$

We obtain from Eq. (4.2-27)

$$\mathbf{u}^{i+1} = \mathbf{u}^i - [\mathbf{A}^T\mathbf{R}\mathbf{A}]^{-1}\mathbf{A}^T\mathbf{R}(\mathbf{A}\mathbf{u}^i - \mathbf{b}) = \mathbf{A}^{-1}\mathbf{b}$$

We see that convergence is obtained in one step *regardless* of $\mathbf{u}^i$. We obtain $\mathbf{u}^{i+1} = \mathbf{A}^{-1}\mathbf{b}$, which is the known solution. This feature of one-step convergence is inherent in linear problems. However, the computational algorithm involves $\mathbf{A}^{-1}$ and, in this simple case, the reason for using an iterative technique was to avoid problems inherent in taking the matrix inverse. In any case, the greatest use of the gradient algorithm is to solve sets of nonlinear algebraic equations, and for this the second variation method sometimes has great advantages.

Thus far, we have optimized the function $J = \theta(\mathbf{u})$ by a first-order gradient method and a second-order gradient or second variation method. For the gradient method we obtained

$$\Delta\mathbf{u}^i = -K^i \frac{d\theta(\mathbf{u}^i)}{d\mathbf{u}^i} \tag{4.2-28}$$

whereas for the second variation method we obtained

$$\Delta\mathbf{u}^i = -\left[\frac{d^2\theta(\mathbf{u}^i)}{(d\mathbf{u}^i)^2}\right]^{-1} \frac{d\theta(\mathbf{u}^i)}{d\mathbf{u}^i} \tag{4.2-29}$$

The first-order method possesses the advantage of computational simplicity and the disadvantage of slow convergence near the optimum [unless the optimum first-order gradient method (Sage, 1968) is employed, which is generally very difficult to accomplish]. The second variation method possesses the advantage of rapid convergence near the optimum solution and the disadvantages of needing a closer to optimum initial value of the initial $\mathbf{u}$ for convergence and considerably greater computational complexity.

The conjugate gradient method (Fletcher and Powell, 1963) attempts to combine the best features of both of the foregoing methods while lessening their disadvantages. Rather than attempting to compute $[d^2\theta(\mathbf{u})/(d\mathbf{u})^2]^{-1}$, a sequence of direction vectors $\gamma^1, \gamma^2, \ldots, \gamma^m$ is generated that are conjugate or orthogonal with respect to $d^2\theta(\mathbf{u})/(d\mathbf{u})^2$ in that

$$\gamma^{iT}\left[\frac{d^2\theta(\mathbf{u}^i)}{(d\mathbf{u}^i)^2}\right]\gamma^j = 0, \qquad i \neq j \tag{4.2-30}$$

Then a series of searches is made along each of the conjugate direction vectors γ^i to find the optimum length in that direction in which to proceed. Thus we use

$$\mathbf{u}^{i+1} = \mathbf{u}^i - K^i\gamma^i \tag{4.2-31}$$

where K^i is a positive scalar chosen in an optimum way such that

$$K^i = \min_{K_i} \theta(\mathbf{u}^i - K^i \gamma^i) \tag{4.2-32}$$

and thus we implement the optimum gradient method.

It is straightforward to show that the algorithm

$$\gamma^i = -\frac{d\theta(\mathbf{u}^i)}{d\mathbf{u}^i} + \frac{\| d\theta(\mathbf{u}^i)/d\mathbf{u}^i \|^2}{\| d\theta(\mathbf{u}^{i-1})/d\mathbf{u}^{i-1} \|^2} \gamma^{i-1} \tag{4.2-33}$$

generates a set of conjugate direction vectors as defined by Eq. (4.2-30). We start by assuming that $\gamma^1 = d\theta(\mathbf{u}^1)/d\mathbf{u}^1$. In practice, the determination of an optimum K^i is all but impossible. It is generally straightforward to try several candidate values of K^i in close proximity to the value used on the previous iteration and select the one which yields a minimum of Eq. (4.2-32). The method proceeds as follows:

1. choose $\mathbf{u}^i$;
2. determine $\gamma^i = d\theta(\mathbf{u}^i)/d\mathbf{u}^i$;
3. determine K^i so as to minimize $\theta(\mathbf{u}^i - K^i\gamma^i)$;
4. compute $\mathbf{u}^{i+1} = \mathbf{u}^i - K^i\gamma^i$;
5. determine

$$\gamma^{i+1} = -\frac{d\theta(\mathbf{u}^{i+1})}{d\mathbf{u}^{i+1}} + \frac{\| d\theta(\mathbf{u}^{i+1})/d\mathbf{u}^{i+1} \|^2}{\| d\theta(\mathbf{u}^i)/d\mathbf{u}^i \|^2} \gamma^i$$

6. repeat computation for new iterate starting at Step 3. Continue until $\mathbf{u}$ does not change appreciably from iteration to iteration.

Example 4.2-3. Again we consider the determination of $\mathbf{u}$ from the relation $\mathbf{Au} = \mathbf{b}$. The quadratic cost function $J = \theta(\mathbf{u}) = \frac{1}{2}(\mathbf{Au} - \mathbf{b})^T\mathbf{R}(\mathbf{Au} - \mathbf{b})$ is used. We assume a solution $\mathbf{u}^1$ and obtain $\gamma^i = \mathbf{A}^T\mathbf{R}(\mathbf{Au}^1 - \mathbf{b})$. Then we note that the computation proceeds as

$$\mathbf{u}^2 = \mathbf{u}^1 - k^1\mathbf{A}^T\mathbf{R}(\mathbf{Au}^1 - \mathbf{b})$$

where k^1 is determined so as to minimize

$$\begin{aligned} J^2 = \theta(\mathbf{u}^2) &= \tfrac{1}{2}(\mathbf{Au}^1 - \mathbf{AA}^T\mathbf{RAu}^1k^1 + \mathbf{AA}^T\mathbf{Rb}k^1 - \mathbf{b})^T \\ &\quad \times \mathbf{R}(\mathbf{Au}^1 - \mathbf{AA}^T\mathbf{RAu}^1k^1 + \mathbf{AA}^T\mathbf{Rb}k^1 - \mathbf{b}) \\ &= [(\mathbf{I} - \mathbf{AA}^T\mathbf{R}k^1)(\mathbf{Au}^1 - \mathbf{b})]^T\mathbf{R}[(\mathbf{I} - \mathbf{AA}^T\mathbf{R}k^1)(\mathbf{Au}^1 - \mathbf{b})] \end{aligned}$$

This value of k is easily obtained as

$$k^1 = \frac{(\mathbf{Au}^1 - \mathbf{b})^T\mathbf{RAA}^T\mathbf{R}(\mathbf{Au}^1 - \mathbf{b})}{(\mathbf{Au}^1 - \mathbf{b})^T\mathbf{RAA}^T\mathbf{RAA}^T\mathbf{R}(\mathbf{Au}^1 - \mathbf{b})}$$

Thus $\mathbf{u}^2$ is determined. From this γ^2 is found, k^2 is found, and then $\mathbf{u}^3$ ... . It is possible to show that this problem will converge to the correct solution in M, steps where $\mathbf{u}$ is an M vector. This statement applies to any quadratic cost linear constraint problem when solved with the conjugate gradient method. Unfortunately, it does not apply to nonquadratic cost or nonlinear equality constraint problems.

It is a relatively simple matter to show that the minimization problem for the cost function and equality constraint

$$J = \theta(\mathbf{x}, \mathbf{u}), \qquad \mathbf{f}(\mathbf{x}, \mathbf{u}) = \mathbf{0} \tag{4.2-34}$$

proceeds in a fashion analogous to the procedure just outlined for the no constraint case. All that is necessary is to substitute $H(\mathbf{x}^i, \mathbf{u}^i, \boldsymbol{\lambda}^i)$ for $\theta(\mathbf{u}^i)$ in the just outlined steps of the conjugate gradient procedure. Of course, it is necessary to solve the additional relations

$$\frac{\partial H(\mathbf{x}^i, \mathbf{u}^i, \boldsymbol{\lambda}^i)}{\partial \boldsymbol{\lambda}^i} = \frac{\partial H(\mathbf{x}^i, \mathbf{u}^i, \boldsymbol{\lambda}^i)}{\partial \mathbf{x}^i} = \mathbf{0} \tag{4.2-35}$$

at each iteration stage. We have introduced the static gradient or single-stage decision process here primarily because of the simplicity of the presentation, rather than because of any direct use of these procedures in system identification. As we shall see, however, these methods carry over to the multistage decision process or continuous decision process problem with only minor modifications.

4.3. Gradient Methods for Multiple Stage and Continuous System Identification

We may easily extend our work of the previous section to include discrete multistage processes. The cost function

$$J = \theta[\mathbf{x}(k_f)] + \sum_{k=k_0}^{k_f-1} \varphi[\mathbf{x}(k), \mathbf{u}(k), k] \tag{4.3-1}$$

is first considered. It is desired to minimize this cost function subject to the equality constraint

$$\mathbf{x}(k+1) = \boldsymbol{\phi}[\mathbf{x}(k), \mathbf{u}(k), k], \qquad \mathbf{x}(k_0) = \mathbf{x}_0 \tag{4.3-2}$$

We may formally obtain the two-point boundary value problem, the solution of which determines the optimal control $\mathbf{u}(k)$ and trajectory $\mathbf{x}(k)$ by defining the Hamiltonian

$$H = \varphi[\mathbf{x}(k), \mathbf{u}(k), k] + \lambda^{\mathrm{T}}(k+1)\, \boldsymbol{\phi}[\mathbf{x}(k), \mathbf{u}(k), k] \tag{4.3-3}$$

and then obtain the solution of the TPBVP

$$\frac{\partial H}{\partial \boldsymbol{\lambda}(k+1)} = \mathbf{x}(k+1), \qquad \mathbf{x}(k_0) = \mathbf{x}_0 \tag{4.3-4}$$

$$\frac{\partial H}{\partial \mathbf{x}(k)} = \boldsymbol{\lambda}(k), \qquad \boldsymbol{\lambda}(k_{\mathrm{f}}) = \frac{\partial \theta[\mathbf{x}(k_{\mathrm{f}})]}{\partial \mathbf{x}(k_{\mathrm{f}})} \tag{4.3-5}$$

$$\frac{\partial H}{\partial \mathbf{u}(k)} = \mathbf{0} \tag{4.3-6}$$

In order to overcome the analytical difficulties associated with solving this two-point boundary value problem, we examine the first-order gradient method. To insure that the equality constraints of Eq. (4.3-2) are satisfied, Lagrange multipliers are used to adjoin the equality constraints to the cost function of Eq. (4.3-1), such that there results after using Eq. (4.3-3)

$$\begin{aligned} J &= \theta[\mathbf{x}(k_{\mathrm{f}})] + \boldsymbol{\Gamma}^{\mathrm{T}}\mathbf{x}(k_0) + \sum_{k=k_0}^{k_{\mathrm{f}}-1} H - \boldsymbol{\lambda}^{\mathrm{T}}(k+1)\,\mathbf{x}(k+1) \\ &= \theta[\mathbf{x}(k_{\mathrm{f}})] - \boldsymbol{\lambda}^{\mathrm{T}}(k_{\mathrm{f}})\,\mathbf{x}(k_{\mathrm{f}}) + [\boldsymbol{\Gamma}^{\mathrm{T}} + \boldsymbol{\lambda}^{\mathrm{T}}(k_0)]\,\mathbf{x}(k_0) + \sum_{k=k_0}^{k_{\mathrm{f}}-1} [H - \boldsymbol{\lambda}^{\mathrm{T}}(k)\,\mathbf{x}(k)] \end{aligned} \tag{4.3-7}$$

We take the first variation, or first-order difference, and obtain for the first-order terms

$$\begin{aligned} \Delta J &= [\boldsymbol{\Gamma}^{\mathrm{T}} + \boldsymbol{\lambda}^{\mathrm{T}}(k_0)]\,\boldsymbol{\Delta}\mathbf{x}(k_0) + \left[\frac{\partial \theta[\mathbf{x}(k_{\mathrm{f}})]}{\partial \mathbf{x}(k_{\mathrm{f}})} - \boldsymbol{\lambda}(k_{\mathrm{f}})\right]^{\mathrm{T}} \boldsymbol{\Delta}\mathbf{x}(k_{\mathrm{f}}) \\ &\quad + \sum_{k=k_0}^{k_{\mathrm{f}}-1} \left\{ \left[\frac{\partial H}{\partial \mathbf{x}(k)} - \boldsymbol{\lambda}(k)\right]^{\mathrm{T}} \boldsymbol{\Delta}\mathbf{x}(k) + \left[\frac{\partial H}{\partial \mathbf{u}(k)}\right]^{\mathrm{T}} \boldsymbol{\Delta}\mathbf{u}(k) \right\} \end{aligned}$$

We fix $\boldsymbol{\Delta}\mathbf{x}(k_0) = \mathbf{0}$, since $\mathbf{x}(k_0)$ is given. For simplicity, we will also let

$$\frac{\partial H}{\partial \mathbf{x}(k)} = \boldsymbol{\lambda}(k), \qquad \boldsymbol{\lambda}(k_{\mathrm{f}}) = \frac{\partial \theta[\mathbf{x}(k_{\mathrm{f}})]}{\partial \mathbf{x}(k_{\mathrm{f}})} \tag{4.3-8}$$

such that the first-order change in the cost function becomes

$$\Delta J = \sum_{k=k_0}^{k_f-1} \left[\frac{\partial H}{\partial \mathbf{u}(k)}\right]^{\mathrm{T}} \Delta\mathbf{u}(k) \tag{4.3-9}$$

In order to make the steepest descent to a minimum, we choose

$$\Delta\mathbf{u}(k) = -K \frac{\partial H}{\partial \mathbf{u}(k)} \tag{4.3-10}$$

where K is a scalar which is chosen with consideration given to such factors as convergence. The computational steps for this first-order gradient for multistage decision process are:

1. determine or guess $\mathbf{u}^i(k)$;
2. compute $\mathbf{x}^i(k)$ from Eq. (4.3-2) for $k_0 \leqslant k \leqslant k_f$;
3. solve the adjoint equation [Eq. (4.3-5)] backward in stage from $k_f \geqslant k \geqslant k_0$ to determine $\boldsymbol{\lambda}^i(k)$;
4. determine

$$\frac{\partial H}{\partial \mathbf{u}^i(k)} = \frac{\partial \varphi[\mathbf{x}^i(k), \mathbf{u}^i(k), k]}{\partial \mathbf{u}^i(k)} + \frac{\partial \boldsymbol{\phi}^{\mathrm{T}}[\mathbf{x}^i(k), \mathbf{u}^i(k), k]}{\partial \mathbf{u}^i(k)} \boldsymbol{\lambda}(k+1) \tag{4.3-11}$$

5. obtain

$$\Delta\mathbf{u}^i(k) = -K^i \frac{\partial H}{\partial \mathbf{u}^i(k)}$$

6. compute the new control iterate

$$\mathbf{u}^{i+1}(k) = \mathbf{u}^i(k) + \Delta\mathbf{u}^i(k) \tag{4.3-12}$$

7. repeat the computation starting at Step 2. Continue until the state and control do not change significantly from iteration to iteration.

There will be many circumstances which occur in system identification which cannot be readily cast into the form of the first-order gradient method which we have just developed. The most important of these occur when the initial state is unspecified and there are constant parameters $\mathbf{p}(k)$ to be identified. In this case the cost function becomes

$$J = \theta_f[\mathbf{x}(k_f)] + \theta_0[\mathbf{x}(k_0)] + \sum_{k=k_0}^{k_f-1} \varphi[\mathbf{x}(k), \mathbf{p}(k), \mathbf{u}(k), k] \tag{4.3-13}$$

where the system model is

$$\mathbf{x}(k+1) = \boldsymbol{\phi}[\mathbf{x}(k), \mathbf{p}(k), \mathbf{u}(k), k] \tag{4.3-14}$$

and the constant but unknown parameters are presented by

$$\mathbf{p}(k+1) = \mathbf{p}(k) \tag{4.3-15}$$

Certainly it is possible to form an adjoined state vector consisting of the state and parameter vector and eliminate the need to consider the parameter vector explicitly. For reasons which will soon become apparent, we will not take this seemingly simpler (in terms of notation) approach but will explicitly retain the parameter vector $\mathbf{p}(k)$ in our development of a first-order gradient algorithm.

We define the Hamiltonian

$$H = \varphi[\mathbf{x}(k), \mathbf{p}(k), \mathbf{u}(k), k] + \boldsymbol{\lambda}^{\mathrm{T}}(k+1)\, \boldsymbol{\phi}[\mathbf{x}(k), \mathbf{p}(k), \mathbf{u}(k), k] + \boldsymbol{\Gamma}^{\mathrm{T}}(k+1)\, \mathbf{p}(k) \tag{4.3-16}$$

and then adjoin the equality constraints to the cost function to obtain the first-order perturbation:

$$\begin{aligned} \Delta J = {} & \left[\frac{\partial \theta_0[\mathbf{x}(k_0)]}{\partial \mathbf{x}(k_0)} + \boldsymbol{\lambda}(k_0)\right]^{\mathrm{T}} \Delta\mathbf{x}(k_0) + \left[\frac{\partial \theta_{\mathrm{f}}[\mathbf{x}(k_{\mathrm{f}})]}{\partial \mathbf{x}(k_{\mathrm{f}})} - \boldsymbol{\lambda}(k_{\mathrm{f}})\right] \Delta\mathbf{x}(k_{\mathrm{f}}) \\ & + \boldsymbol{\Gamma}^{\mathrm{T}}(k_0)\Delta\mathbf{p}(k_0) - \boldsymbol{\Gamma}^{\mathrm{T}}(k_{\mathrm{f}})\Delta\mathbf{p}(t_{\mathrm{f}}) + \sum_{k=k_0}^{k_{\mathrm{f}}-1} \left\{\left[\frac{\partial H}{\partial \mathbf{x}(k)} - \boldsymbol{\lambda}(k)\right]^{\mathrm{T}} \Delta\mathbf{x}(k) \right. \\ & \left. + \left[\frac{\partial H}{\partial \mathbf{p}(k)} - \boldsymbol{\Gamma}(k)\right]^{\mathrm{T}} \Delta\mathbf{p}(k) + \left[\frac{\partial H}{\partial \mathbf{u}(k)}\right]^{\mathrm{T}} \Delta\mathbf{u}(k)\right\} \end{aligned} \tag{4.3-17}$$

As before, we will define the adjoint variables in a certain way such that ΔJ is simplified. Specifically, we will let

$$\boldsymbol{\lambda}(k) = \frac{\partial H}{\partial \mathbf{x}(k)}, \qquad \boldsymbol{\lambda}(k_{\mathrm{f}}) = \frac{\partial \theta_{\mathrm{f}}[\mathbf{x}(k_{\mathrm{f}})]}{\partial \mathbf{x}(k_{\mathrm{f}})} \tag{4.3-18}$$

$$\boldsymbol{\Gamma}(k) = \frac{\partial H}{\partial \mathbf{p}(k)}, \qquad \boldsymbol{\Gamma}(k_{\mathrm{f}}) = \mathbf{0} \tag{4.3-19}$$

such that the first-order change in the cost function becomes

$$\Delta J = \left[\frac{\partial \theta_0[\mathbf{x}(k_0)]}{\partial \mathbf{x}(k_0)} + \boldsymbol{\lambda}(k_0)\right]^{\mathrm{T}} \Delta\mathbf{x}(k_0) + \boldsymbol{\Gamma}^{\mathrm{T}}(k_0)\Delta\mathbf{p}(k_0) + \sum_{k=k_0}^{k_{\mathrm{f}}-1} \left[\frac{\partial H}{\partial \mathbf{u}(k)}\right]^{\mathrm{T}} \Delta\mathbf{u}(k) \tag{4.3-20}$$

We now select $\Delta\mathbf{x}(k_0)$, $\Delta\mathbf{p}(k_0)$, and $\Delta\mathbf{u}(k)$ such that we obtain the steepest descent to a minimum. Thus we let

$$\Delta\mathbf{x}(k_0) = -K_{\Delta\mathbf{x}}\left[\frac{\partial\theta_0[\mathbf{x}(k_0)]}{\partial\mathbf{x}(k_0)} + \boldsymbol{\lambda}(k_0)\right] \tag{4.3-21}$$

$$\Delta\mathbf{p}(k_0) = -K_{\Delta\mathbf{p}}\boldsymbol{\Gamma}(k_0) \tag{4.3-22}$$

$$\Delta\mathbf{u}(k) = -K_{\Delta\mathbf{u}}\frac{\partial H}{\partial\mathbf{u}(k)} \tag{4.3-23}$$

We see that at each iteration we now update the initial conditions $\Delta\mathbf{x}(k_0)$. $\Delta\mathbf{p}(k_0)$, and $\Delta\mathbf{u}(k)$. In the previous formulation, $\mathbf{x}(k_0)$ was fixed, so $\Delta\mathbf{x}(k_0)$ was of necessity equal to $\mathbf{0}$.

In many problems $\mathbf{x}(k_0)$ will be given. If this is the case, we simply use the given $\mathbf{x}(k_0)$ to start the solution of the system model equation. Also, it will often occur that there is no control $\mathbf{u}(k)$. In this case, the algorithm for updating (and of course calculating $\partial H/\partial\mathbf{u}$) $\mathbf{u}(k)$ is eliminated. The steps in the computation are:

1. Determine or guess $\mathbf{x}^i(k_0)$, $\mathbf{p}^i(k_0)$, and $\mathbf{u}^i(k)$.
2. Solve the difference Eqs. (4.3-14) and (4.3-15) with the given values of $\mathbf{x}^i(k_0)$ and $\mathbf{p}^i(k_0)[\mathbf{p}^i(k+1) = \mathbf{p}^i(k)]$ for the state vector $\mathbf{x}^i(k)$, $k_0 \leqslant k \leqslant k_f$.
3. Solve the adjoint Eqs. (4.3.18) and (4.3-19) to determine $\boldsymbol{\lambda}^i(k)$ and $\boldsymbol{\Gamma}^i(k)$ for $k_f \geqslant k \geqslant k_0$.
4. Determine the incremental changes

$$\Delta\mathbf{x}^i(k_0) = -K^i_{\Delta\mathbf{x}}\left[\frac{\partial\theta_0[\mathbf{x}^i(k_0)]}{\partial\mathbf{x}^i(k_0)} + \boldsymbol{\lambda}^i(k_0)\right] \tag{4.3-24}$$

$$\Delta\mathbf{p}^i(k_0) = -K^i_{\Delta\mathbf{p}}\boldsymbol{\Gamma}^i(k_0) \tag{4.3-25}$$

$$\Delta\mathbf{u}^i(k) = -K^i_{\Delta\mathbf{u}}\frac{\partial H}{\partial\mathbf{u}^i(k)} \tag{4.3-26}$$

5. Compute the control and initial conditions for the next iteration:

$$\mathbf{x}^{i+1}(k_0) = \mathbf{x}^i(k_0) + \Delta\mathbf{x}^i(k_0) \tag{4.3-27}$$

$$\mathbf{p}^{i+1}(k_0) = \mathbf{p}^i(k_0) + \Delta\mathbf{p}^i(k_0) \tag{4.3-28}$$

$$\mathbf{u}^{i+1}(k) = \mathbf{u}^i(k) + \Delta\mathbf{u}^i(k) \tag{4.3-29}$$

6. Repeat the iterations starting at Step 2. Continue until there is negligible change in $\mathbf{x}(k)$, $\mathbf{p}(k)$, and $\mathbf{u}(k)$ from iteration to iteration.

Extension of these algorithms to the continuous case is a simple matter. The first-order gradient algorithms for minimization of the cost function

$$J = \theta_f[\mathbf{x}(t_f)] + \theta_0[\mathbf{x}(t_0)] + \int_{t_0}^{t_f} \boldsymbol{\mu}[\mathbf{x}(t), \mathbf{p}(t), \mathbf{u}(t), t]\, dt \qquad (4.3\text{-}30)$$

for the differential equality constraints relating the state variable $\mathbf{x}(t)$, the constant unknown parameter $\mathbf{p}(t)$, and the control $\mathbf{u}(t)$

$$\dot{\mathbf{x}} = \mathbf{f}[\mathbf{x}(t), \mathbf{p}(t), \mathbf{u}(t), t] \qquad (4.3\text{-}31)$$

$$\dot{\mathbf{p}} = \mathbf{0} \qquad (4.3\text{-}32)$$

are:

1. Determine the Hamiltonian

$$H = \varphi[\mathbf{x}(t), \mathbf{p}(t), \mathbf{u}(t), t] + \boldsymbol{\lambda}^T(t)\, \mathbf{f}[\mathbf{x}(t), \mathbf{p}(t), \mathbf{u}(t), t] \qquad (4.3\text{-}33)$$

2. Guess initial values of $\mathbf{u}(t)$, $\mathbf{x}(t_0)$, and $\mathbf{p}(t_0)$.
3. With these $\mathbf{u}^i(t)$, $\mathbf{x}^i(t_0)$, and $\mathbf{p}^i(t_0)$, solve Eqs. (4.3-31) and (4.3-32) for $\mathbf{x}^i(t)$ and $\mathbf{p}^i(t) = \mathbf{p}^i(t_0)$.
4. Solve the adjoint equations

$$\dot{\boldsymbol{\lambda}}^i = -\frac{\partial H}{\partial \mathbf{x}^n(t)}, \qquad \boldsymbol{\lambda}^i(t_f) = \frac{\partial \theta_f[\mathbf{x}(t_f)]}{\partial \mathbf{x}(t_f)} \qquad (4.3\text{-}34)$$

$$\dot{\boldsymbol{\Gamma}}^i = -\frac{\partial H}{\partial \mathbf{p}^n(t)}, \qquad \boldsymbol{\Gamma}^i(t_f) = \mathbf{0} \qquad (4.3\text{-}35)$$

backward in time.

5. Determine the incremental changes

$$\Delta\mathbf{u}^i(t) = -K^i_{\Delta\mathbf{u}} \frac{\partial H}{\partial \mathbf{u}^i(k)} \qquad (4.3\text{-}36)$$

$$\Delta\mathbf{x}^i(t_0) = -K^i_{\Delta\mathbf{x}} \left[\frac{\partial \theta_0[\mathbf{x}^i(t_0)]}{\partial \mathbf{x}^i(t_0)} + \boldsymbol{\lambda}^i(t_0)\right] \qquad (4.3\text{-}37)$$

$$\Delta\mathbf{p}^i(t_0) = -K_{\Delta}\mathbf{p}^i\boldsymbol{\Gamma}^i(t_0) \qquad (4.3\text{-}38)$$

6. Compute new trial values

$$\mathbf{u}^{i+1}(t) = \mathbf{u}^i(t) + \Delta\mathbf{u}^i(t) \qquad (4.3\text{-}39)$$

$$\mathbf{x}^{i+1}(t_0) = \mathbf{x}^i(t_0) + \Delta\mathbf{x}^i(t_0) \qquad (4.3\text{-}40)$$

$$\mathbf{p}^{i+1}(t_0) = \mathbf{p}^i(t_0) + \Delta\mathbf{p}^i(t_0) \qquad (4.3\text{-}41)$$

7. Repeat the computation starting at Step 3. Continue until the parameter change is "slight" from iteration to iteration.

It is possible to add any of several different equality and inequality constraints to the system identification problem. Since those such as state and control variable inequality constraints and terminal manifold equality constraints do not often enter system identification problems, we will not discuss these topics here (Sage, 1968). After a brief diversion for two examples, we will turn our attention to the second-order gradient or second variation and conjugate gradient method for multistage or continuous system identification problems.

Example 4.3-1. We consider the problem of system identification, illustrated in Fig. 4.3-1, by tuning an adjustable model. The unknown system is assumed to be described by the transfer function relation

$$\frac{Z(s)}{U(s)} = H(s) = \frac{A(s)}{B(s)} = \frac{a_0 + a_1 + \cdots + a_{n-1}s^{n-1}}{b_0 + b_1 s + \cdots + s^n}$$

where the parameters

$$\mathbf{a}^{\mathrm{T}} = [a_0 \quad a_1 \cdots a_{n-1}]^{\mathrm{T}}, \qquad \mathbf{b}^{\mathrm{T}} = [b_0 \quad b_1 \cdots b_{n-1}]^{\mathrm{T}}$$

are assured to be constant but unknown. The scalar input $u(t)$ with Laplace transform $U(s)$, and scalar output $z(t)$ with Laplace transform $Z(s)$ are assumed to be known. The system is known to be initially unexcited.

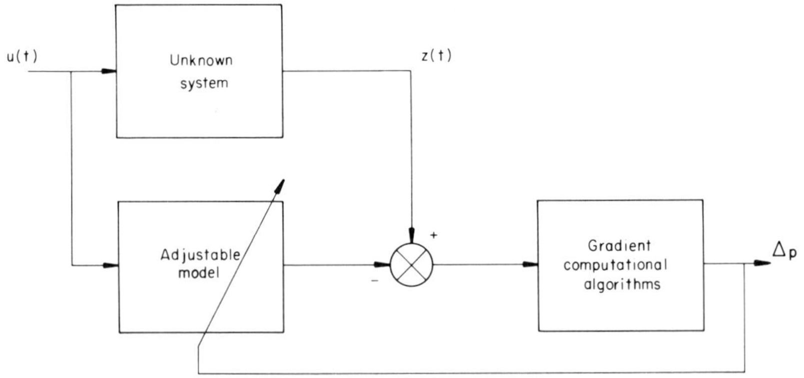

FIG. 4.3-1. Adjustable model for Example 4.3-1.

The unknown system may be conveniently represented in the state variable format by the representation

$$\dot{\mathbf{x}} = \mathbf{F}(\mathbf{b})\,\mathbf{x}(t) + \mathbf{g}(\mathbf{a})\,u(t)$$
$$y(t) = \mathbf{h}^{\mathrm{T}}\mathbf{x}(t)$$

where

$$\mathbf{F}(\mathbf{b}) = \begin{bmatrix} -b_{n-1} & 1 & 0 & 0 & \cdots \\ -b_{n-2} & 0 & 1 & 0 & \\ \vdots & 0 & 0 & 1 & \vdots \\ & & & & 1 \\ b_0 & 0 & 0 & & 0 \end{bmatrix}, \qquad \mathbf{g}(\mathbf{a}) = \begin{bmatrix} a_{n-1} \\ a_{n-2} \\ \vdots \\ a_1 \\ a_0 \end{bmatrix}, \qquad \mathbf{h} = \begin{bmatrix} 1 \\ 0 \\ 0 \\ \vdots \\ 0 \end{bmatrix}$$

We will adjust the parameter vector **p**, consisting of the adjoined vectors **a** and **b**, until the cost function

$$J = \frac{1}{2}\int_0^{t_f} [z(t) - y(t)]^2\, dt$$

is minimum. We thus see that this problem is a particular case of the continuous system identification problem, solution of which was posed by the first-order gradient method. It is instructive to present the computational algorithms in reasonable detail. We define

$$H = \tfrac{1}{2}[z(t) - \mathbf{h}^{\mathrm{T}}\mathbf{x}(t)]^2 + \boldsymbol{\lambda}^{\mathrm{T}}(t)[\mathbf{F}(\mathbf{b})\mathbf{x}(t) + \mathbf{g}(\mathbf{a})u(t)] + \boldsymbol{\Gamma}_{\mathbf{a}}^{\mathrm{T}}[\mathbf{0}] + \boldsymbol{\Gamma}_{\mathbf{b}}^{\mathrm{T}}[\mathbf{0}]$$

$u(t)$ is a *known* input such that it is not negotiable for purposes of optimization. The computation proceeds as follows:

1. We guess an initial value of **a** and **b**, $\mathbf{a}^i$ and $\mathbf{b}^i$.
2. We solve the differential equation

$$\dot{\mathbf{x}}^i = \mathbf{F}(\mathbf{b}^i)\,\mathbf{x}^i(t) + \mathbf{g}(\mathbf{a}^i)\,\mathbf{u}(t), \qquad \mathbf{x}^i(0) = \mathbf{0}$$

3. The adjoint equations are solved backward in time

$$\dot{\boldsymbol{\lambda}}^i = \mathbf{h}[\mathbf{z}(t) - \mathbf{h}^{\mathrm{T}}\mathbf{x}^i(\dot{t})] - \mathbf{F}^{\mathrm{T}}(\mathbf{b}^i)\,\boldsymbol{\lambda}^i(t), \qquad \boldsymbol{\lambda}^i(t_f) = \mathbf{0}$$
$$\dot{\boldsymbol{\Gamma}}_{\mathbf{a}} = \boldsymbol{\lambda}^i(t)\,u(t), \qquad \boldsymbol{\Gamma}_{\mathbf{a}}^i(t_f) = \mathbf{0}$$
$$\dot{\boldsymbol{\Gamma}}_{\mathbf{b}}^i = x_1^i(t)\,\boldsymbol{\lambda}^i(t), \qquad \boldsymbol{\Gamma}_{\mathbf{b}}^i(t_f) = \mathbf{0}$$

4. The incremental changes

$$\Delta\mathbf{a}^i = -K_{\mathbf{a}}^i\boldsymbol{\Gamma}_{\mathbf{a}}^i(0)$$
$$\Delta\mathbf{b}^i = -K_{\mathbf{b}}^i\boldsymbol{\Gamma}_{\mathbf{b}}^i(0)$$

are determined.

5. New iterates

$$\mathbf{a}^{i+1} = \mathbf{a}^i + \Delta\mathbf{a}^i$$
$$\mathbf{b}^{i+1} = \mathbf{b}^i + \mathbf{b}\Delta^i$$

are computed and the procedure repeated starting at Step 2 until convergence is reached.

It is important to note that it is necessary to have an input $u(t)$ to identify $\mathbf{p}$ and that the "speed" of identification will depend upon the input.

For this problem we have assumed that the unknown system and the model can be tuned perfectly. Of course $z(t)$ could have come from a noise-corrupted version of the system output, or the system and model may be different in order. In this case, it is a sounder procedure to assume that $\mathbf{x}(0)$ is not specified and to determine $\Delta\mathbf{x}^i(0)$ from that portion of the first-order change which contains

$$\Delta J = \boldsymbol{\lambda}^{\mathrm{T}}(0)\, \Delta\mathbf{x}(0)$$

such that the algorithms

$$\Delta\mathbf{x}^i(0) = -K_{\Delta\mathbf{x}}\boldsymbol{\lambda}^i(0)$$
$$\mathbf{x}^{i+1}(0) = \mathbf{x}^i(0) + \Delta\mathbf{x}^i(0)$$

are used to iterate the initial condition vector which is needed to solve the differential equation in Step 2. The disadvantage of this procedure is that nonzero initial conditions for the identified model will generally result.

The equivalent discrete version of this problem

$$\frac{Z(z)}{U(z)} = \mathscr{H}(z) = \frac{\alpha_0 + \alpha_1 z + \cdots + \alpha_{n-1} z^n}{\beta_0 + \beta_1 z + \cdots + \beta_{n-1} z^n}$$

may be formulated and solved by the discrete gradient algorithms. Also, the computations may be simplified and approximated somewhat in many instances. The reader should notice, for example, the very close relationship which exists between the algorithms of this example and those for model reference identification in Chap. 2.

One of the biggest possibilities for simplification consists in assuming that the identification of the constant parameters $\mathbf{a}$ and $\mathbf{b}$ may be solved by a static gradient procedure. Strictly speaking, this is not a correct procedure, although the simplicity of the computational algorithms may justify the approximations involved. For the general problem of minimizing

$$J = \int_{t_0}^{t_f} \phi[\mathbf{x}(t), \mathbf{p}(t), t]\, dt$$

subject to the equality constraints

$$\dot{\mathbf{x}} = \mathbf{f}[\mathbf{x}(t), \mathbf{p}(t), t], \qquad \mathbf{x}(t_0) = \mathbf{x}_0$$

$$\dot{\mathbf{p}} = \mathbf{0}$$

we assume an initial $\mathbf{p}^i$ and solve the system differential equation and evaluate the cost J^i. A slightly different $\mathbf{p}^i$, $\mathbf{p}^i + \epsilon_j$ is then assumed and the cost $J^i + \eta_j$ determined. The jth component of the cost gradient is then approximated by

$$\frac{dJ}{dp_j^{\,i}} \approx \frac{(J^i + \eta_j) - J^i}{(\mathbf{p}^i + \epsilon_j) - \mathbf{p}^i} = \frac{\eta_j}{\epsilon_j}$$

By repeating this procedure for perturbations of different components of the parameter vector, an approximation to the gradient vector $dJ/d\mathbf{p}^i$ is determined. The first-order change in the parameter vector in the direction of steepest descent to a minimum of the cost function is then

$$\Delta\mathbf{p}^i = -K_{\Delta\mathbf{p}} \frac{dJ}{d\mathbf{p}^i}$$

and the new trial parameter vector is

$$\mathbf{p}^{i+1} = \mathbf{p}^i + \Delta\mathbf{p}^i$$

While this method is simple and is the basis for several of the iteration schemes in Chap. 2, the errors due to the approximate evaluation of $dJ/d\mathbf{p}^i$ (exact evaluation consists of treating the problem as a dynamic problem which yields the algorithms previously presented) are difficult to evaluate, and the procedure is subject to considerable error, particularly when the cost function is not very sensitive to changes in the parameter vector. Unfortunately, this insensitivity is often the case, particularly where measurement or observation noise and unknown inputs are involved.

Just as in the single-stage case, we may consider the series expansion of the cost function for the multistage case to include terms of second order. For the cost function of Eq. (4.3-17), the first- and second-order terms are

$$
\begin{aligned}
\Delta J = {} & \left[\frac{\partial \theta_0}{\partial \mathbf{x}_0} + \boldsymbol{\lambda}_0\right]^{\mathrm{T}} \Delta\mathbf{x}_0 + \frac{1}{2}\Delta\mathbf{x}_0^{\mathrm{T}} \frac{\partial^2 \theta_0}{\partial \mathbf{x}_0^2} \Delta\mathbf{x}_0 \\
& + \left[\frac{\partial \theta_{\mathrm{f}}}{\partial \mathbf{x}_{\mathrm{f}}} - \boldsymbol{\lambda}_{\mathrm{f}}\right]^{\mathrm{T}} \Delta\mathbf{x}_{\mathrm{f}} + \frac{1}{2}\Delta\mathbf{x}_{\mathrm{f}}^{\mathrm{T}} \frac{\partial^2 \theta_{\mathrm{f}}}{\partial \mathbf{x}_{\mathrm{f}}^2} \Delta\mathbf{x}_{\mathrm{f}} \\
& + \boldsymbol{\Gamma}_0^{\mathrm{T}}\Delta\mathbf{p}_0 - \boldsymbol{\Gamma}_{\mathrm{f}}^{\mathrm{T}}\Delta\mathbf{p}_{\mathrm{f}} \\
& + \sum_{k=k_0}^{k_{\mathrm{f}}-1} \Bigg\{ \left[\frac{\partial H}{\partial \mathbf{x}} - \boldsymbol{\lambda}\right]^{\mathrm{T}} \Delta\mathbf{x} + \left[\frac{\partial H}{\partial \mathbf{u}}\right]^{\mathrm{T}} \Delta\mathbf{u} + \left[\frac{\partial H}{\partial \mathbf{p}} - \boldsymbol{\Gamma}\right]^{\mathrm{T}} \Delta\mathbf{p} \\
& + \frac{1}{2}\begin{bmatrix}\Delta\mathbf{x}\\ \Delta\mathbf{u}\\ \Delta\mathbf{p}\end{bmatrix}^{\mathrm{T}}
\begin{bmatrix}
\dfrac{\partial^2 H}{\partial \mathbf{x}^2} & \dfrac{\partial}{\partial \mathbf{u}}\dfrac{\partial H}{\partial \mathbf{x}} & \dfrac{\partial}{\partial \mathbf{p}}\dfrac{\partial H}{\partial \mathbf{x}} \\
\left[\dfrac{\partial}{\partial \mathbf{u}}\dfrac{\partial H}{\partial \mathbf{x}}\right]^{\mathrm{T}} & \dfrac{\partial^2 H}{\partial \mathbf{u}^2} & \dfrac{\partial}{\partial \mathbf{u}}\dfrac{\partial H}{\partial \mathbf{p}} \\
\left[\dfrac{\partial}{\partial \mathbf{p}}\dfrac{\partial H}{\partial \mathbf{x}}\right]^{\mathrm{T}} & \left[\dfrac{\partial}{\partial \mathbf{u}}\dfrac{\partial H}{\partial \mathbf{p}}\right]^{\mathrm{T}} & \dfrac{\partial^2 H}{\partial \mathbf{p}^2}
\end{bmatrix}
\begin{bmatrix}\Delta\mathbf{x}\\ \Delta\mathbf{u}\\ \Delta\mathbf{p}\end{bmatrix} \Bigg\}
\end{aligned}
\tag{4.3-42}
$$

where the stage index k has been eliminated to simplify the notation. To simplify this expression, we will again require Eqs. (4.3-18) and (4.3-19) to be true such that ΔJ simplifies to

$$
\begin{aligned}
\Delta J = {} & \left[\frac{\partial \theta_0}{\partial \mathbf{x}_0} + \boldsymbol{\lambda}_0\right] \Delta\mathbf{x}_0 + \tfrac{1}{2}\Delta\mathbf{x}_0^{\mathrm{T}} \frac{\partial^2 \theta_0}{\partial \mathbf{x}_0^2} \Delta\mathbf{x}_0 + \tfrac{1}{2}\Delta\mathbf{x}_{\mathrm{f}}^{\mathrm{T}} \frac{\partial^2 \theta_{\mathrm{f}}}{\partial \mathbf{x}_{\mathrm{f}}^2} \Delta\mathbf{x}_{\mathrm{f}} \\
& + \boldsymbol{\Gamma}_0^{\mathrm{T}}\Delta\mathbf{p}_0 + \sum_{k=k_0}^{k_{\mathrm{f}}-1} \Bigg\{ \left[\frac{\partial H}{\partial \mathbf{u}}\right]^{\mathrm{T}} \Delta\mathbf{u} \\
& + \frac{1}{2}\begin{bmatrix}\Delta\mathbf{x}\\ \Delta\mathbf{u}\\ \Delta\mathbf{p}\end{bmatrix}^{\mathrm{T}}
\begin{bmatrix}
\dfrac{\partial^2 H}{\partial \mathbf{x}^2} & \dfrac{\partial}{\partial \mathbf{u}}\dfrac{\partial H}{\partial \mathbf{x}} & \dfrac{\partial}{\partial \mathbf{p}}\dfrac{\partial H}{\partial \mathbf{x}} \\
\left[\dfrac{\partial}{\partial \mathbf{u}}\dfrac{\partial H}{\partial \mathbf{x}}\right]^{\mathrm{T}} & \dfrac{\partial^2 H}{\partial \mathbf{u}^2} & \dfrac{\partial}{\partial \mathbf{u}}\dfrac{\partial H}{\partial \mathbf{p}} \\
\left[\dfrac{\partial}{\partial \mathbf{p}}\dfrac{\partial H}{\partial \mathbf{x}}\right]^{\mathrm{T}} & \left[\dfrac{\partial}{\partial \mathbf{u}}\dfrac{\partial H}{\partial \mathbf{p}}\right]^{\mathrm{T}} & \dfrac{\partial^2 H}{\partial \mathbf{p}^2}
\end{bmatrix}
\begin{bmatrix}\Delta\mathbf{x}\\ \Delta\mathbf{u}\\ \Delta\mathbf{p}\end{bmatrix} \Bigg\}
\end{aligned}
\tag{4.3-43}
$$

We now note that we may conduct an ancillary minimization of Eq. (4.3-43) with respect to the perturbations $\Delta\mathbf{x}$, $\Delta\mathbf{p}$, and $\Delta\mathbf{u}$. This minimization will be conducted such that first-order perturbations of the system equations (4.3-14) and (4.3-15) are valid. These are

$$\begin{aligned}\Delta\mathbf{x}(k+1) &= \frac{\partial\boldsymbol{\phi}}{\partial\mathbf{x}(k)}\Delta\mathbf{x}(k) + \frac{\partial\boldsymbol{\phi}}{\partial\mathbf{u}(k)}\Delta\mathbf{u}(k) + \frac{\partial\boldsymbol{\phi}}{\partial\mathbf{p}(k)}\Delta\mathbf{p}(k)\\ &= \left[\frac{\partial}{\partial\mathbf{x}(k)}\frac{\partial H}{\partial\boldsymbol{\lambda}(k)}\right]\Delta\mathbf{x}(k) + \left[\frac{\partial}{\partial\mathbf{u}(k)}\frac{\partial H}{\partial\boldsymbol{\lambda}(k)}\right]\Delta\mathbf{u}(k)\\ &\quad + \left[\frac{\partial}{\partial\mathbf{p}(k)}\frac{\partial H}{\partial\boldsymbol{\lambda}(k)}\right]\Delta\mathbf{x}(k)\end{aligned} \tag{4.3-44}$$

$$\Delta\mathbf{p}(k+1) = \Delta\mathbf{p}(k) \tag{4.3-45}$$

Routine application of the discrete maximum principle leads to the TPBVP of Eqs. (4.3-44), (4.3-45), and

$$\begin{aligned}\Delta\mathbf{u}(k) = -\left[\frac{\partial^2 H}{\partial u^2(k)}\right]^{-1}\Bigg\{&\left[\frac{\partial}{\partial\mathbf{u}(k)}\frac{\partial H}{\partial\mathbf{u}(k)}\right]^{\mathrm{T}}\Delta\mathbf{x}(k) + \left[\frac{\partial}{\partial\mathbf{u}(k)}\frac{\partial}{\partial\boldsymbol{\lambda}(k)}\right]^{\mathrm{T}}\Delta\boldsymbol{\lambda}(k+1)\\ &+ \left[\frac{\partial}{\partial\mathbf{u}(k)}\frac{\partial H}{\partial\mathbf{p}(k)}\right]^{\mathrm{T}}\Delta\mathbf{p}(k) + \frac{\partial H}{\partial\mathbf{u}(k)}\Bigg\}\end{aligned} \tag{4.3-46}$$

$$\begin{aligned}\Delta\boldsymbol{\lambda}(k) = &\left[\frac{\partial^2 H}{\partial\mathbf{x}(k)^2}\right]\Delta\mathbf{x}(k) + \left[\frac{\partial}{\partial\mathbf{u}(k)}\frac{\partial H}{\partial\mathbf{x}(k)}\right]\Delta\mathbf{u}(k) + \left[\frac{\partial}{\partial\mathbf{p}(k)}\frac{\partial H}{\partial\mathbf{x}(k)}\right]\Delta\mathbf{p}(k)\\ &\left[\frac{\partial}{\partial\boldsymbol{\lambda}(k)}\frac{\partial H}{\partial\mathbf{x}(k)}\right]\Delta\boldsymbol{\lambda}(k+1)\end{aligned} \tag{4.3-47}$$

$$\begin{aligned}\Delta\boldsymbol{\Gamma}(k) = \Delta\boldsymbol{\Gamma}(k+1) &+ \left[\frac{\partial}{\partial\boldsymbol{\lambda}(k)}\frac{\partial H}{\partial\mathbf{p}(k)}\right]\Delta\boldsymbol{\lambda}(k+1)\\ &+ \left[\frac{\partial^2 H}{\partial\mathbf{p}(k)^2}\right]\Delta\mathbf{p}(k) + \left[\frac{\partial}{\partial\mathbf{u}(k)}\frac{\partial H}{\partial\mathbf{p}(k)}\right]\Delta\mathbf{u}(k) + \left[\frac{\partial}{\partial\mathbf{x}(k)}\frac{\partial H}{\partial\mathbf{p}(k)}\right]\Delta\mathbf{x}(k)\end{aligned} \tag{4.3-48}$$

where $\Delta\boldsymbol{\lambda}$ and $\Delta\boldsymbol{\Gamma}$ are Lagrange multipliers which satisfy the two-point boundary conditions

$$\Delta\boldsymbol{\lambda}(k_0) = -\frac{\partial\theta_0(k_0)}{\partial\mathbf{x}(k_0)} - \boldsymbol{\lambda}(k_0) - \frac{\partial^2\theta_0(k_0)}{\partial\mathbf{x}(k_0)^2}\Delta\mathbf{x}(k_0) \tag{4.3-49}$$

$$\Delta\boldsymbol{\Gamma}(k_0) = -\boldsymbol{\Gamma}(k_0) \tag{4.3-50}$$

$$\Delta\boldsymbol{\lambda}(k_f) = \frac{\partial^2\theta_f(k_f)}{\partial\mathbf{x}(k_f)^2}\Delta\mathbf{x}(k_f) \tag{4.3-51}$$

$$\Delta\boldsymbol{\Gamma}(k_f) = \mathbf{0} \tag{4.3-52}$$

The use of the second variation technique has specified the change in control from iteration to iteration as it did in the single-stage case. The computational algorithms for this second variation method are considerably more complex than those for the first-order gradient method. Convergence is quadratic and much faster than with the first-order algorithms. The region of initial conditions and initial assumed controls for which the method converges are much more restricted with the second-order gradient method however.

The steps in the computation are:

1. Assume an initial control $\mathbf{u}^i(k)$, state $\mathbf{x}^i(k_0)$, and parameter value $\mathbf{p}^i(k_0)$.
2. Solve the state and parameter equations (4.3-14) and (4.3-15) to determine $\mathbf{x}^i(k)$ and $\mathbf{p}^i(k)$. Clearly the equation for $\mathbf{p}^i(k)$ does not have to be implemented on the computer since $\mathbf{p}^i(k)=\mathbf{p}^i(k_0)$.
3 Solve the adjoint equations (4.3-18) and (4.3-19) backward from k_f to k_0.
4. Solve the linear two-point boundary value problem consisting of Eqs. (4.3-44)–(4.3-52). Since this TPBVP is linear, a variety of methods can be used to solve it (Sage, 1968). One particularly successful method is to use a Riccati transformation. We will gain considerable experience with solving linear TPBVP's in Chap. 6. Solution of this TBPVP will yield the needed quantities $\Delta\mathbf{x}^i(k_0)$ and $\Delta\mathbf{p}^i(k_0)$.
5. Determine $\Delta\mathbf{u}^i(k)$ from Eq. (4.3-46).
6. Update to obtain $\mathbf{x}^{i+1}(k_0)$, $\mathbf{p}^{i+1}(k_0)$, and $\mathbf{u}^{i+1}(k)$ using Eqs. (4.3-27)–(4.3-29).
7. Repeat computation starting at Step 2 until there is only very slight change in the $\Delta\mathbf{x}(k_0)$, $\Delta\mathbf{p}(k_0)$, and $\Delta\mathbf{u}(k)$ from iteration to iteration.

As we indicated, a Riccati transformation may be used to simplify computation of the $\Delta\mathbf{x}(k)$, $\Delta\mathbf{u}(k)$, and $\Delta\mathbf{p}(k)$. Rather than simply make this transformation on the relations just presented for the discrete second variation method, let us approach the problem in a slightly different way.

We will again assume that an initial control $\mathbf{u}^i(k)$ and state and parameter $\mathbf{x}^i(k_0)$ and $\mathbf{p}^i(k_0)$ have been chosen. The state and parameter equations (4.3-14) and (4.3-15) are solved in forward time and the adjoint equations (4.3-18) and (4.3-19) are solved in backward time.

It is then further assumed that first-order perturbations of Eqs. (4.3-14), (4.3-15), (4.3-18), and (4.3-19) are satisfied. These are

$$\Delta\mathbf{x}(k+1) = \left[\frac{\partial}{\partial\mathbf{x}(k)}\frac{\partial H}{\partial\boldsymbol{\lambda}(k+1)}\right]\Delta\mathbf{x}(k) + \left[\frac{\partial}{\partial\mathbf{u}(k)}\frac{\partial H}{\partial\boldsymbol{\lambda}(k+1)}\right]\Delta\mathbf{u}(k)$$
$$+ \left[\frac{\partial}{\partial\mathbf{p}(k)}\frac{\partial H}{\partial\boldsymbol{\lambda}(k+1)}\right]\Delta\mathbf{p}(k) = \frac{\partial\boldsymbol{\phi}}{\partial\mathbf{x}(k)}\Delta\mathbf{x}(k)$$
$$+ \frac{\partial\boldsymbol{\phi}}{\partial\mathbf{u}(k)}\Delta\mathbf{u}(k) + \frac{\partial\boldsymbol{\phi}}{\partial\mathbf{p}(k)}\Delta\mathbf{p}(k) \tag{4.3-53}$$

$$\Delta\mathbf{p}(k+1) = \Delta\mathbf{p}(k) \tag{4.3-54}$$

$$\Delta\boldsymbol{\lambda}(k) = \left[\frac{\partial^2 H}{\partial\mathbf{x}(k)^2}\right]\Delta\mathbf{x}(k) + \left[\frac{\partial}{\partial\mathbf{u}(k)}\frac{\partial H}{\partial\mathbf{x}(k)}\right]\Delta\mathbf{u}(k)$$
$$+ \left[\frac{\partial}{\partial\mathbf{p}(k)}\frac{\partial H}{\partial\mathbf{x}(k)}\right]\Delta\mathbf{p}(k) + \left[\frac{\partial}{\partial\boldsymbol{\lambda}(k+1)}\frac{\partial H}{\partial\mathbf{x}(k)}\right]\Delta\boldsymbol{\lambda}(k+1) \tag{4.3-55}$$

$$\Delta\boldsymbol{\Gamma}(k) = \left[\frac{\partial}{\partial\mathbf{x}(k)}\frac{\partial H}{\partial\mathbf{p}(k)}\right]\Delta\mathbf{x}(k) + \left[\frac{\partial}{\partial\mathbf{u}(k)}\frac{\partial H}{\partial\mathbf{p}(k)}\right]\Delta\mathbf{u}(k)$$
$$+ \left[\frac{\partial^2 H}{\partial\mathbf{p}(k)^2}\right]\Delta\mathbf{p}(k) + \left[\frac{\partial}{\partial\boldsymbol{\lambda}(k+1)}\frac{\partial H}{\partial\mathbf{p}(k)}\right]\Delta\boldsymbol{\lambda}(k+1)$$
$$+ \Delta\boldsymbol{\Gamma}(k+1) \tag{4.3-56}$$

where H is defined by Eq. (4.3-16). The control $\Delta\mathbf{u}$ is determined by making a *known* linear perturbation in $\partial H/\partial\mathbf{u}$ such that

$$\Delta\left[\frac{\partial H}{\partial\mathbf{u}(k)}\right] = \left[\frac{\partial}{\partial\mathbf{u}(k)}\frac{\partial H}{\partial\mathbf{x}(k)}\right]^{\mathrm{T}}\Delta\mathbf{x}(k) + \left[\frac{\partial^2 H}{\partial\mathbf{u}(k)^2}\right]\Delta\mathbf{u}(k)$$
$$+ \left[\frac{\partial}{\partial\mathbf{u}(k)}\frac{\partial H}{\partial\boldsymbol{\lambda}(k+1)}\right]^{\mathrm{T}}\Delta\boldsymbol{\lambda}(k+1) + \left[\frac{\partial}{\partial\mathbf{u}(k)}\frac{\partial H}{\partial\mathbf{p}(k)}\right]^{\mathrm{T}}\Delta\mathbf{p}(k) \tag{4.3-57}$$

The linear perturbed two-point boundary conditions are

$$\Delta\boldsymbol{\lambda}(k_0) = -\frac{\partial\theta_0[\mathbf{x}(k_0)]}{\partial\mathbf{x}(k_0)} - \boldsymbol{\lambda}(k_0) - \frac{\partial^2\theta_0[\mathbf{x}(k_0)]}{\partial\mathbf{x}(k_0)^2}\Delta\mathbf{x}(k_0) \tag{4.3-58}$$

$$\Delta\boldsymbol{\lambda}(k_f) = \frac{\partial^2\theta_f[\mathbf{x}(k_f)]}{\partial\mathbf{x}(k_f)^2}\Delta\mathbf{x}(k_f)^2 \tag{4.3-59}$$

$$\Delta\boldsymbol{\Gamma}(k_0) = -\boldsymbol{\Gamma}(k_0), \qquad \Delta\boldsymbol{\Gamma}(k_f) = \mathbf{0} \tag{4.3-60}$$

These relations may be arranged into the simpler form

$$\begin{bmatrix} \Delta\mathbf{x}(k+1) \\ \Delta\boldsymbol{\lambda}(k) \\ \Delta\mathbf{p}(k+1) \\ \Delta\boldsymbol{\Gamma}(k) \end{bmatrix} = \begin{bmatrix} \mathbf{C}_{11}(k) & \mathbf{C}_{12}(k) & \mathbf{C}_{13}(k) & \mathbf{0} \\ \mathbf{C}_{21}(k) & \mathbf{C}_{11}(k) & \mathbf{C}_{23}(k) & \mathbf{0} \\ \mathbf{0} & \mathbf{0} & \mathbf{I} & \mathbf{0} \\ \mathbf{C}_{23}^{T}(k) & \mathbf{C}_{13}^{T}(k) & \mathbf{C}_{33}(k) & \mathbf{I} \end{bmatrix} \begin{bmatrix} \Delta\mathbf{x}(k) \\ \Delta\boldsymbol{\lambda}(k+1) \\ \Delta\mathbf{p}(k) \\ \Delta\boldsymbol{\Gamma}(k+1) \end{bmatrix} + \begin{bmatrix} \Delta\boldsymbol{\omega}_1(k) \\ \Delta\boldsymbol{\omega}_2(k) \\ \mathbf{0} \\ \Delta\boldsymbol{\omega}_3(k) \end{bmatrix} \tag{4.3-61}$$

which are subject to the two-point boundary conditions of Eqs. (4.3-58)–(4.3-60).

Due to the nature of the boundary conditions, it seems reasonable to assume solutions to the TPBVP of the form

$$\Delta\boldsymbol{\lambda}(k) = \boldsymbol{\Xi}_{\lambda\mathbf{x}}(k)\,\Delta\mathbf{x}(k) + \boldsymbol{\Xi}_{\lambda\mathbf{p}}\Delta\mathbf{p}(k) + \boldsymbol{\omega}_{\lambda}(k) \tag{4.3-62}$$

$$\Delta\boldsymbol{\Gamma}(k) = \boldsymbol{\Xi}_{\Gamma\mathbf{x}}(k)\,\Delta\mathbf{x}(k) + \boldsymbol{\Xi}_{\Gamma\mathbf{p}}\Delta\mathbf{p}(k) + \boldsymbol{\omega}_{\Gamma}(k) \tag{4.3-63}$$

These are substituted into Eq. (4.3-61) and the expressions manipulated such that we obtain, for nonzero $\Delta\mathbf{x}$ and $\Delta\mathbf{p}$, the requirements that a set of matrix Riccati equations hold. The perturbation equations are thus similar to those of the closed loop regulator problem. Due to the great complexity of the final algorithms, they will not be presented here. Obtaining the explicit algorithms for a given identification problem is a straightforward but often tedious exercise.

The steps in the application of this *successive sweep* second variation method, which is originally due in continuous from McReynolds and Bryson (1965), is essentially the same as those outlined previously for the basic second variation technique. The essential difference is that a particular technique, the backward sweep integration of the nonhomogeneous Riccati equations is obtained. Equations (4.3-53) and (4.3-54) are then propagated forward in time with initial conditions determined by combining Eqs. (4.3-58), (4.3-60), (4.3-62), and (4.3-63):

$$\Delta\mathbf{x}(k_0) = \left\{\boldsymbol{\Xi}_{\lambda\mathbf{p}}^{-1}(k_0)\left[\boldsymbol{\Xi}_{\lambda\mathbf{x}}(k_0) + \frac{\partial^2\theta_0[\mathbf{x}(k_0)]}{\partial\mathbf{x}(k_0)^2}\right] + \boldsymbol{\Xi}_{\Gamma\mathbf{p}}^{-1}(k_0)\,\boldsymbol{\Xi}_{\Gamma\mathbf{x}}(k_0)\right\}^{-1}$$
$$\times \{\boldsymbol{\Xi}_{\Gamma\mathbf{p}}^{-1}(k_0)\,\boldsymbol{\omega}_{\Gamma}(k_0) - \boldsymbol{\Xi}_{\lambda\mathbf{p}}(k_0)\,\boldsymbol{\omega}_{\lambda}(k_0)\} \tag{4.3-64}$$

$$\Delta\mathbf{p}(k_0) = \left\{\left[\boldsymbol{\Xi}_{\lambda\mathbf{x}}(k_0) + \frac{\partial^2\theta_0[\mathbf{x}(k_0)]}{\partial\mathbf{x}(k_0)^2}\right]^{-1}\boldsymbol{\Xi}_{\lambda\mathbf{p}}(k_0) - \boldsymbol{\Xi}_{\Gamma\mathbf{x}}(k_0)\,\boldsymbol{\Xi}_{\Gamma\mathbf{p}}(k_0)\right\}^{-1}$$
$$\times \left\{\boldsymbol{\Xi}_{\Gamma\mathbf{x}}(k_0)\,\boldsymbol{\omega}_{\Gamma}(k_0) - \left[\boldsymbol{\Xi}_{\lambda\mathbf{x}}(k_0) + \frac{\partial^2\theta_0[\mathbf{x}(k_0)]}{\partial\mathbf{x}(k_0)^2}\right]^{-1}\boldsymbol{\omega}_{\lambda}(k_0)\right\} \tag{4.3-65}$$

It is sometimes desirable to propagate Eqs. (4.3-53) and (4.3-54) forward in time together with Eqs. (4.3-14) and (4.3-15). The ability to choose $\Delta(\partial H/\partial \mathbf{u})$ from iteration can be a significant beneficial feature to this method as opposed to the basic second variation method.

The conjugate gradient algorithms for solution of the dynamic system identification problem is a relatively straightforward (conceptually) extension of the conjugate gradient algorithms for static systems. It is necessary to determine initiating $\mathbf{x}_0{}^i$, $\mathbf{p}_0{}^i$, and $\mathbf{u}^i$. The state equations are then solved forward in time and the adjoint equations solved in reverse time. This allows determination of the various gradients of Eqs. (4.3-21)–(4.3-23) for the discrete case or Eqs. (4.3-36)–(4.3-38) for the continuous case. The steps outlined on p. 91 for the static conjugate gradient method are then used for each of the three gradient vectors of the dynamic problem. It is convenient to define, for either the discrete or continuous gradient vectors,

$$c_{\Delta \mathbf{u}}^{i} = \frac{\partial H}{\partial \mathbf{u}^i}$$

$$c_{\Delta \mathbf{x}_0}^{i} = \left[\frac{\partial \theta_0{}^i}{\partial \mathbf{x}_0{}^i} + \boldsymbol{\lambda}_0{}^i\right]$$

$$c_{\Delta \mathbf{p}}^{i} = \boldsymbol{\Gamma}_0{}^i$$

and define the integral inner product notation as

$$\langle \mathbf{f}(t), \mathbf{g}(t) \rangle = \int_{t_0}^{t_f} \mathbf{f}(t)\, \mathbf{g}^{\mathrm{T}}(t)\, dt$$

for the continuous case and

$$\langle \mathbf{f}(k), \mathbf{g}(k) \rangle = \sum_{k=k_0}^{k_f-1} \mathbf{f}(k)\, \mathbf{g}^{\mathrm{T}}(k)$$

for the discrete case. The steps of the conjugate method are:

1. Choose $\mathbf{x}_0{}^i$, $\mathbf{p}^i$, $\mathbf{u}^i$.
2. Solve the system equations forward in time and the adjoint equations backward in time.
3. Determine the gradients $\mathbf{c}_{\Delta u}^i$, $\mathbf{c}_{\Delta x_0}^i$, and $\mathbf{c}_{\Delta p}^i$.
4. Determine $K_{\Delta \mathbf{u}}$, $K_{\Delta \mathbf{x}_0}$, and $K_{\Delta \mathbf{p}}^i$ such as to minimize the original cost function

$$J[\mathbf{u}^i - K_{\Delta \mathbf{u}}^i c_{\Delta \mathbf{u}}^i,\ \mathbf{x}_0{}^i - K_{\Delta \mathbf{x}_0}^i c_{\Delta \mathbf{x}}^i,\ \mathbf{p}^i + K_{\Delta \mathbf{p}}^i c_{\Delta \mathbf{p}}^i]$$

This steps represents the optimum gradient method and will normally be very difficult to implement. Interpolation with several values for each $K_{\Delta\mathbf{u}}$, $K_{\Delta\mathbf{x}_0}$, and $K_{\Delta\mathbf{p}}$ will normally yield acceptable values.

5. Choose, using the K's determined in Step 4,

$$\mathbf{u}^{i+1} = \mathbf{u}^{i} - K^{i}_{\Delta\mathbf{u}} c^{i}_{\Delta\mathbf{u}}$$

$$\mathbf{x}_0^{i+1} = \mathbf{x}_0^{i} - K^{i}_{\Delta\mathbf{x}_0} c^{i}_{\Delta\mathbf{x}_0}$$

$$\mathbf{p}^{i+1} = \mathbf{p} - K^{i}_{\Delta\mathbf{p}} c^{i}_{\Delta\mathbf{p}}$$

6. Solve the state equation forward in time and the adjoint equation backward in time using the values in Step 5.
7. Determine the *conjugate gradient* directions

$$c^{i+1}_{\Delta\mathbf{u}} = -\frac{\partial H}{\partial \mathbf{u}^{i+1}} + \frac{\langle \partial H/\partial \mathbf{u}^{i+1}, \partial H/\partial \mathbf{u}^{i+1} \rangle}{\langle c^{i}_{\Delta\mathbf{u}}, c^{i}_{\Delta\mathbf{u}} \rangle} c^{i}_{\Delta\mathbf{u}}$$

$$c^{i+1}_{\Delta\mathbf{x}_0} = -\left[\frac{\partial \theta_0^{i+1}}{\partial \mathbf{x}_0^{i+1}} + \boldsymbol{\lambda}_0^{i+1}\right] + \frac{\|(\partial \theta_0^{i+1}/\partial \mathbf{x}_0^{i+1} + \boldsymbol{\lambda}_0^{i+1}\|^2}{\| c^{i}_{\Delta\mathbf{x}_0} \|^2} c^{i}_{\Delta\mathbf{x}_0}$$

$$c^{i+1}_{\Delta\mathbf{p}} = -\boldsymbol{\Gamma}_0^{i+1} + \frac{\| \boldsymbol{\Gamma}_0^{i+1} \|^2}{\| c^{i}_{\Delta\mathbf{p}} \|^2} c^{i}_{\Delta\mathbf{p}}$$

8. Repeat the iterations starting at Step 4. Thus we see that after the initial stage the conjugate directions are used as the directions of search. In order to illustrate the specific computational steps, let us reconsider Example 4.3-1 for a first-order system.

Example 4.3-2. It is instructive to consider the conjugate gradient algorithms for identification of the parameter b in the linear system

$$\dot{x} = -bx(t) + u(t), \qquad u(t) = 1, \quad x(0) = 0$$

such as to minimize

$$J = \frac{1}{2} \int_0^{t_f} [z(t) - x(t)]^2 \, dt$$

The basic gradient procedure is very simple and is a special case of the general result developed in Example 4.3-1:

1. We guess an initial b^i.
2. We solve the equation

$$\dot{x}^i = -b^i x^i(t) + 1, \qquad x^i(0) = 0$$

3. The adjoint equations

$$\dot{\lambda} = z(t) - x^i(t) - b^i \lambda^i(t), \qquad \lambda^i(t_f) = 0$$

$$\dot{\Gamma}_b{}^i = x^i(t)\,\lambda^i(t), \qquad \Gamma_b{}^i(t_f) = 0$$

are solved.
4. A new parameter iterate,

$$b^{i+1} = b^i - K^i \Gamma_b{}^i(0)$$

is determined.
5. The computation is repeated starting at Step 2.

In order to apply the conjugate gradient algorithms, it is necessary to replace the foregoing Step 5 by the following:

5.′ Determine $\hat{K}^i$ such as to minimize the cost function

$$J = \frac{1}{2} \int_0^{t_f} [z(t) - x(t)]^2 \, dt$$

and then compute the resulting $\hat{b}^{i+1}$.
6. Solve the state and adjoint equations with the just obtained $\hat{b}^{i+1}$.
7. Determine the conjugate gradient direction

$$c_{\Delta b}^{i+1} = -\Gamma^{i+1}(0) + \frac{[\Gamma^{i+1}(0)]^2}{c_{\Delta b}^{i}}$$

where the initial iterate $c_{\Delta b}^0$ is just $\Gamma^0(0)$.
8. Use, on the next iteration, the parameter

$$b^{i+1} = b^i - \hat{K}^i c_{\Delta b}^i$$

9. Repeat the computations starting at Step 2.

4.4. Summary

We have developed several sets of algorithms to solve the system identification problem. System identification by minimization of several of the cost functions of Chap. 3 has been considered in the several examples considered here. We will now consider some features of what may be considered a statistical gradient method—the method of stochastic approximation.

5

SYSTEM IDENTIFICATION USING STOCHASTIC APPROXIMATION

5.1. Introduction

In this chapter, we will present a survey of the methods of stochastic approximation with application to system identification. Our discussions will be such as to give physical and heuristic meaning to stochastic approximation rather than to reproduce the rigorous proofs which can be found in the literature cited in the bibliography. The origins of stochastic approximation are the works of Robbins and Monro (1951), Kiefer and Wolfowitz (1952), Dvoretzky (1956), and Blum (1954). A readable engineering survey of stochastic approximation has been presented by Sakrison (1966).

The Robbins–Monro algorithm is the stochastic analog of the simple gradient algorithm for finding the unique root of the equation

$$\mathbf{h}(\mathbf{x}) = \mathbf{0} \tag{5.1-1}$$

which is

$$\mathbf{x}^{i+1} = \mathbf{x}^{i} - K^{i}\mathbf{h}(\mathbf{x}^{i}) \tag{5.1-2}$$

where K^i is a sequence of real numbers which must satisfy certain conditions in order to insure convergence of the algorithms.

When there is measurement noise present such that $\mathbf{h}(\mathbf{x})$ cannot be measured but only a noisy observation of finite variance

$$\mathbf{z} = \mathbf{h}(\mathbf{x}) + \mathbf{v} \tag{5.1-3}$$

where $\mathbf{v}$ is zero mean observation noise, then $\mathbf{h}(\mathbf{x})$ is spoken of as the regression function of $\mathbf{z}$ on $\mathbf{x}$, since, for independent $\mathbf{x}$ and $\mathbf{z}$,

$$\mathscr{E}\{\mathbf{z} \mid \mathbf{x}\} = \int_{-\infty}^{\infty} \mathbf{z} p(\mathbf{z} \mid \mathbf{x})\, d\mathbf{z} = \mathbf{h}(\mathbf{x}) \tag{5.1-4}$$

Now, Eq. (5.1-2) is no longer an acceptable algorithm since $\mathbf{h}(\mathbf{x})$ cannot be observed. However the expectation of Eq. (5.1-3) conditioned upon $\mathbf{x}$ is just Eq. (5.1-4), and we see that a stochastic algorithm to find the roots of the regression Eq. (5.1-4) is

$$\mathbf{x}^{i+1} = \mathbf{x}^i - K^i \mathbf{z}(\mathbf{x}^i) \tag{5.1-5}$$

where we note that $\mathbf{z}$ is a function of $\mathbf{x}$ to indicate explicitly the iterative nature of the algorithm. The $\mathbf{x}^i$ determined from the solution to Eq. (5.1-5) are a sequence of random variables which hopefully converge to the solution of Eq. (5.1-1). The contribution of Robbins and Monro was to show that this convergence occurs if the three requirements

$$\lim_{i \to \infty} K^i = 0, \qquad \sum_{i=1}^{\infty} K^i = \infty, \qquad \sum_{i=1}^{\infty} (K^i)^2 < \infty \tag{5.1-6}$$

are met. A simple K^i which meets these requirements is

$$K^i = \frac{\alpha}{\beta + i} \tag{5.1-7}$$

Also, it is required that the regression function $\mathbf{h}(\mathbf{x})$ be bounded on either side of a true solution by straight lines, such that it is not possible to overshoot the solution $\mathbf{x}$ which cannot be corrected by a K^i satisfying Eqs. (5.1-6). Thus for the scalar case

$$| h(x) | \leqslant a \mid x - \hat{x} \mid + b, \qquad a, b > 0 \tag{5.1-8}$$

This latter requirement is not especially severe. We will give heuristic justification to Eq. (5.1-6) in the next section, where we give a dynamic interpretation to the stochastic approximation method.

Kiefer and Wolfowitz extended the stochastic approximation method to include finding the extremum of an unknown unimodal regression function $\theta(\mathbf{u})$. The approach is the exact analog of the deterministic gradient procedure, which, as we know, yields for an optimization algorithm

$$\mathbf{u}^{i+1} = \mathbf{u}^i - K^i \frac{d\theta(\mathbf{u}^i)}{d\mathbf{u}^i} \tag{5.1-9}$$

In the noisy case, we observe

$$l = \theta(\mathbf{u}) + \xi \tag{5.1-10}$$

such that we replace the deterministic algorithm (5.1-9) by the stochastic algorithm

$$\mathbf{u}^{i+1} = \mathbf{u}^i - K^i \frac{dl(\mathbf{u}^i)}{d\mathbf{u}^i} \tag{5.1-11}$$

which corresponds to Eq. (5.1-9) in the sense that the conditioned expectation of Eq. (5.1-11) yields Eq. (5.1-9). In some instances, direct differentiation to obtain $dl(\mathbf{u}^i)/d\mathbf{u}^i$ is not possible, and in this case the approximation

$$\frac{dl(\mathbf{u}^i)}{d\mathbf{u}^i} = \frac{l(\mathbf{u}^i + \Delta\mathbf{u}^i) - l(\mathbf{u}^i - \Delta\mathbf{u}^i)}{2\Delta\mathbf{u}^i} \tag{5.1-12}$$

is used, such that the Kiefer–Wolfowitz algorithm becomes

$$\mathbf{u}^{i+1} = \mathbf{u}^i - K^i \left[\frac{l(\mathbf{u}^i + \Delta\mathbf{u}^i) - l(\mathbf{u}^i - \Delta\mathbf{u}^i)}{2\Delta\mathbf{u}^i}\right] \tag{5.1-13}$$

Convergence requirements are now

$$\begin{gathered} \lim_{i\to\infty} K^i = 0, \qquad \lim_{i\to\infty} \Delta\mathbf{u}^i = \mathbf{0}, \\ \sum_{i=1}^{\infty} K^i = \infty, \qquad \sum_{i=1}^{\infty} (K^i)^2 = \infty, \qquad \sum_{i=1}^{\infty} \left[\frac{K^i}{\Delta u_j{}^i}\right]^2 < \infty \end{gathered} \tag{5.1-14}$$

as well as a requirement corresponding to Eq. (5.1-8).

A basic idea in stochastic approximation is that a stochastic counterpart for any deterministic algorithm exists. Following this notion, Dvoretzky (1956) formulated a generalized stochastic approximation method consisting of a deterministic algorithm $\mathscr{D}$ with an additive component $\mathscr{n}$

$$x^{i+1} = \mathscr{D}(x^1, x^2, \ldots, x^i) + \mathscr{n}^i \tag{5.1-15}$$

It is possible to show that the Robbins–Monro and the Kiefer–Wolfowitz algorithms are special cases of the Dvoretzky algorithm.

In the methods of stochastic approximation, which we see are very similar to the gradient methods of the previous chapter, there are various methods of increasing the rate of convergence of the estimates. Perhaps the simplest is to keep the value of K^i constant until the sign of the observation $[\mathbf{z}(\mathbf{x}^i)$ or $l(\mathbf{u}^i)]$ changes and then change K^i such that it satisfies the aforementioned relations. The motivation behind this scheme is that when the zero of $\mathbf{h}(\mathbf{x})$ or $d\theta(\mathbf{u})/d\mathbf{u}$ is far away from $\mathbf{x}$ or $\mathbf{u}$, the observations will be likely to be of the same sign; but when we are near to the zero, the sign of the observation will often change.

Another method would be to use

$$\mathbf{x}^{i+1} = \mathbf{x}^i - K^i \operatorname{sign}[\mathbf{z}(\mathbf{x}^i)] \tag{5.1-16}$$

or

$$\mathbf{u}^{i+1} = \mathbf{u}^i - K^i \operatorname{sign}\left[\frac{dl(\mathbf{u}^i)}{d\mathbf{u}^i}\right] \tag{5.1-17}$$

This approach greatly accelerates convergence for regression functions where $\mathbf{h}(\mathbf{x})$ decreases rapidly away from $\mathbf{x} = \mathbf{0}$ [such as $h(x) = x \exp(-x)$].

Dvoretzky (1956) has proven that if

$$\operatorname{var}\{z(x) \mid x\} < V_v < \infty$$

and if the regression function $h(x) = \mathscr{E}\{z(x) \mid x\}$ is bounded such that

$$0 < A \mid x - \hat{x} \mid \leqslant h(x) \leqslant B \mid x - \hat{x} \mid < \infty \tag{5.1-18}$$

and also

$$\mid x^i - \hat{x} \mid \leqslant C = \left(\frac{2\sigma^2}{A(B - A)}\right)^{1/2} \tag{5.1-19}$$

then the sequence

$$K^i = \frac{AC^2}{V_v + iA^2C^2} \tag{5.1-20}$$

yields the upper bound

$$\operatorname{var}\{(x^i - \hat{x})^2\} \leqslant \frac{V_v C^2}{V_v + (i - 1) A^2C^2} \tag{5.1-21}$$

In addition to the close connection to the gradient method, there exists a close relationship between stochastic approximation and optimal filter theory. It is well known, for instance, that the problem of obtaining the best linear conditional mean estimate of $\mathbf{x}$ given the observations

$$\mathbf{z}(k) = \mathbf{Hx} + \mathbf{v}(k) \tag{5.1-22}$$

where $\mathbf{v}(k)$ is zero mean white noise such that $\text{var}\{\mathbf{v}(k)\} = \mathbf{I}$, is given by

$$\begin{aligned}\hat{\mathbf{x}}(k+1) &= \hat{\mathbf{x}}(k) + \mathbf{V}_{\tilde{\mathbf{x}}}(k+1)\,\mathbf{H}^{\mathrm{T}}[\mathbf{z}(k+1) - \mathbf{H}\hat{\mathbf{x}}(k)] \\ &= \hat{\mathbf{x}}(k) + \mathbf{V}_{\tilde{\mathbf{x}}}(k)\,\mathbf{H}^{\mathrm{T}}[\mathbf{H}\mathbf{V}_{\tilde{\mathbf{x}}}(k)\,\mathbf{H}^{\mathrm{T}} + \mathbf{I}]^{-1}[\mathbf{z}(k+1) - \mathbf{H}\hat{\mathbf{x}}(k)]\end{aligned} \tag{5.1-23}$$

where

$$\mathbf{V}_{\tilde{\mathbf{x}}}(k+1) = \mathbf{V}_{\tilde{\mathbf{x}}}(k) - \mathbf{V}_{\tilde{\mathbf{x}}}(k)\,\mathbf{H}^{\mathrm{T}}[\mathbf{H}\mathbf{V}_{\tilde{\mathbf{x}}}(k)\,\mathbf{H}^{\mathrm{T}} + \mathbf{I}]^{-1}\,\mathbf{H}\mathbf{V}_{\tilde{\mathbf{x}}}(k) \tag{5.1-24}$$

or

$$\mathbf{V}_{\tilde{\mathbf{x}}}(k+1)\,\mathbf{H}^{\mathrm{T}} = \mathbf{V}_{\tilde{\mathbf{x}}}(k)\,\mathbf{H}^{\mathrm{T}}[\mathbf{H}\mathbf{V}_{\tilde{\mathbf{x}}}(k)\,\mathbf{H}^{\mathrm{T}} + \mathbf{I}]^{-1} \tag{5.1-25}$$

Repeated application of Eq. (5.1-25) leads immediately to

$$\mathbf{V}_{\tilde{\mathbf{x}}}(k)\,\mathbf{H}^{\mathrm{T}} = \mathbf{V}_{\tilde{\mathbf{x}}}(0)\,\mathbf{H}^{\mathrm{T}}[k\mathbf{H}\mathbf{V}_{\tilde{\mathbf{x}}}(0)\,\mathbf{H}^{\mathrm{T}} + \mathbf{I}]^{-1}$$

such that

$$\lim_{k\to\infty} \mathbf{V}_{\tilde{\mathbf{x}}}(k)\,\mathbf{H}^{\mathrm{T}} = \frac{1}{k}\,\mathbf{V}_{\tilde{\mathbf{x}}}(0)\,\mathbf{H}^{\mathrm{T}}[\mathbf{H}\mathbf{V}_{\tilde{\mathbf{x}}}(0)\,\mathbf{H}^{\mathrm{T}}]^{-1} \tag{5.1-26}$$

and the error variance decreases to zero as we expect, since we are estimating a constant. The filter algorithm (5.1-23) thus becomes asymptotically (for large k)

$$\hat{\mathbf{x}}(k+1) = \hat{\mathbf{x}}(k) + \frac{1}{k+1}\,\mathbf{V}_{\tilde{\mathbf{x}}}(0)\,\mathbf{H}^{\mathrm{T}}[\mathbf{H}\mathbf{V}_{\tilde{\mathbf{x}}}(0)\,\mathbf{H}^{\mathrm{T}}]^{-1}[\mathbf{z}(k+1) - \mathbf{H}\hat{\mathbf{x}}(k)] \tag{5.1-27}$$

We see that Eq. (5.1-27) is a multidimensional stochastic approximation algorithm. If we assume further that $\mathbf{H}$ is square and nonsingular, we obtain from the foregoing

$$\hat{\mathbf{x}}(k+1) = \hat{\mathbf{x}}(k) + \frac{1}{k+1}\,[\mathbf{x} + \mathbf{H}^{-1}\mathbf{v}(k) - \hat{\mathbf{x}}(k)]$$

which is easily solved to give the weak law of large numbers

$$\hat{\mathbf{x}}(k) = \frac{1}{k} \sum_{i=1}^{k} \mathbf{H}^{-1}\mathbf{z}(i)$$

This brief treatment (Ho, 1962) of the interconnection between stochastic approximation and optimum linear filter theory shows that the two methods are closely related. There are however very significant differences. The stochastic approximation method does not effectively use the prior statistics, and the optimal filtering method does. In other words, the *optimum matrix* K^i is selected by the optimum filtering methods but not by the stochastic approximation methods. Also, the optimum filtering methods can effectively and easily treat problems in which plant noise is present. This is not easily accomplished via the stochastic approximation methods.

We will now turn our attention to a brief look at a dynamic formulation of the stochastic approximation method and its application to system identification.

5.2. Stochastic Approximation for Stochastic Dynamic Systems

We have indicated in Chap. 4 that determination of the values of $\mathbf{u}$ which product the extremum of $J = \theta(\mathbf{u})$ may often be obtained by solving in an iterative fashion.

$$\mathbf{u}^{i+1} = \mathbf{u}^i - K^i \frac{d\theta(\mathbf{u}^i)}{d\mathbf{u}^i}$$

and have also indicated in the previous section that, if we observe $l = \theta(\mathbf{u}) + \epsilon$ rather than $\theta(\mathbf{u})$, the algorithm

$$\mathbf{u}^{i+1} = \mathbf{u}^i - K^i \frac{dl(\mathbf{u}^i)}{d\mathbf{u}^i}$$

will, under appropriate restrictions on K^i, extremize $\theta(\mathbf{u})$. In this section, we wish to consider this latter problem in more fundamental fashion. In addition, we wish to relate the stochastic approximation method to problems of system identification, so we will study a dynamic version of the stochastic approximation formulation of the last section.

We wish to find methods to determine a control input $\mathbf{u}(k)$ or $\mathbf{u}(t)$ and a parameter vector $\mathbf{p}$ such that we minimize

$$J = \mathscr{E}\left\{\theta_f[\mathbf{x}(k_f)] + \theta_0[\mathbf{x}(k_0)] + \sum_{k=k_0}^{k_0-1} \varphi[\mathbf{x}(k), \mathbf{p}(k), \mathbf{u}(k), k]\right\} \tag{5.2-1}$$

subject to the constraints

$$\mathbf{x}(k+1) = \boldsymbol{\phi}[\mathbf{x}(k), \mathbf{p}(k), \mathbf{u}(k), \boldsymbol{\zeta}(k), k], \qquad \mathbf{p}(k+1) = \mathbf{p}(k) \tag{5.2-2}$$

for the discrete case. For the continuous case, we desire to minimize

$$J = \mathscr{E}\left\{\theta_f[\mathbf{x}(t_f)] + \theta_0[\mathbf{x}(t_0)] + \int_{t_0}^{t_f} \varphi[\mathbf{x}(t), \mathbf{p}(t), \mathbf{u}(t), t]\, dt\right\} \tag{5.2-3}$$

subject to the differential equality constraints

$$\dot{\mathbf{x}} = \mathbf{f}[\mathbf{x}(t), \mathbf{u}(t), \mathbf{p}(t), \boldsymbol{\zeta}(t), t], \qquad \dot{\mathbf{p}} = \mathbf{0} \tag{5.2-4}$$

Here $\boldsymbol{\zeta}(k)$ and $\boldsymbol{\zeta}(t)$ represent stochastic processes. These are the same cost functions and constraint equations which we used in Sect. 4.3 [Eqs. (4.3-13)–(4.3-15) for the discrete case and Eqs. (4.3-30)–(4.3.32) for the continuous case] except that we take the expectation of the cost function and include the stochastic process $\boldsymbol{\zeta}(t)$ (which represents the plant and measurement noise) since the expectation must be taken with respect to the random variables in the process.

The problem which we have just formulated is a rather formidable open-loop optimal control and estimation-identification problem. Analytical solutions to this problem will, in general, be most difficult to obtain. Often, open-loop control solutions are not as acceptable as closed-loop solutions. In one particular case—when the system dynamics are completely linear, when all noise terms enter additively, when the cost function is quadratic in the control and states, and when there are no parameters to identify—the separation theorem or certainty equivalence principle [first presented in the optimal control literature by Kalman (Sage, 1968)] applies, and the optimal closed-loop solution consists of an optimum linear regulator-type controller which uses the output of an optimum linear filter as the observed state variables which with to implement the controller. We will explore solution possibilities to the stochastic control-estimation-identification problem by first considering some simpler problems and developing the necessary restrictions for the simpler problems, which include those of the previous section, and then examine the more complex cases.

First, we consider the problem of extremizing (most often minimizing) the cost function

$$J = \mathscr{E}\{\theta(\mathbf{u}, \boldsymbol{\zeta})\} \tag{5.2-5}$$

where $\boldsymbol{\zeta}$ is a random variable with known probability density $p(\boldsymbol{\zeta})$. To minimize Eq. (5.2-5), we set $\partial J/\partial \mathbf{u} = \mathbf{0}$ to obtain from

$$J = \int_{-\infty}^{\infty} \theta(\mathbf{u}, \boldsymbol{\zeta})\, p(\boldsymbol{\zeta})\, d\boldsymbol{\zeta} \tag{5.2-6}$$

$$\frac{\partial J}{\partial \mathbf{u}} = \int_{-\infty}^{\infty} \frac{\partial \theta(\mathbf{u}, \boldsymbol{\zeta})}{\partial \mathbf{u}}\, p(\boldsymbol{\zeta})\, d\boldsymbol{\zeta} = \mathscr{E}\left\{\frac{\partial \theta(\mathbf{u}, \boldsymbol{\zeta})}{\partial \mathbf{u}}\right\} = \mathbf{0} \tag{5.2-7}$$

We recognize that analytical solution of Eq. (5.2-7) is often impossible and so attempt solution by means of the iterative algorithm

$$\mathbf{u}^{i+1} = \mathbf{u}^i - K^i \left[\frac{\partial \theta(\mathbf{u}^i, \boldsymbol{\zeta}^i)}{\partial \mathbf{u}^i}\right] \tag{5.2-8}$$

where K^i is a positive scalar sequence. The gradient $\partial \boldsymbol{\theta}/\partial \mathbf{u}$ is random and consists of two components: the component representing the effect of $\mathbf{u}$ on θ, and that due to the particular value of the noise $\boldsymbol{\zeta}^i$. It is convenient to let

$$\frac{\partial \theta(\mathbf{u}^i, \boldsymbol{\zeta}^i)}{\partial \mathbf{u}^i} = \mathscr{E}\left\{\frac{\partial \theta(\mathbf{u}^i, \boldsymbol{\zeta}^i)}{\partial \mathbf{u}^i}\right\} + \mathbf{v}^i = \mathscr{g}\{\theta^i\} + \mathbf{v}^i \tag{5.2-9}$$

where $\mathbf{v}^i$ represents the random component of the gradient and, by definition, has zero mean as is easily seen by taking the expectation of Eq. (5.2-9) and comparing this result with Eq. (5.2-9). By combining the foregoing two equations, we have

$$\mathbf{u}^{i+1} = \mathbf{u}^i - K^i[\mathscr{g}\{\theta^i\} + \mathbf{v}^i] \tag{5.2-10}$$

We should remember that actual computation proceeds using Eq. (5.2-8). Equation (5.2-9) is, however, convenient for the analytical work to follow. We have a sequence $\partial \theta/\partial \mathbf{u}^i$, for $i = 1, 2,\ldots$, and hope that for sufficiently large i, $\mathbf{u}^{i+1}$ approaches $\mathbf{u}^i$. In the limit, we require that

$$\lim_{i\to\infty} \mathbf{u}^{i+1} = \lim_{i\to\infty} \mathbf{u}^i$$

By taking the limit of both sides of Eq. (5.2-9) we see that we must require

$$\lim_{i\to\infty} K^i = 0 \tag{5.2-11}$$

or the iterative sequence represented by Eq. (5.2-9) will not in general terminate on a constant value. This is true even if $\mathscr{g}\{\theta^i\}$ is zero, which indicates, at least on the average, that $\partial\theta/\partial\mathbf{u}$ is in fact zero. The noise term $\mathbf{v}^i$ is not zero, and this drives the $\mathbf{u}^i$ to a nonoptimum value. Of course the K^i sequence should not approach zero too rapidly for increasing i, or $\mathscr{g}\{\theta^i\}$ will not be able to drive $\mathbf{u}^i$ to the correct value. Also, the average effect of each additional noise term $\mathbf{v}^i$, $i = 1, 2,...,$ must be less than the previous noise term such that this noise term cannot continue to cause computational error.

It has been shown in the fundamental stochastic approximation literature that the requirement that $\mathscr{g}\{\theta^i\}$ be able to drive $\mathbf{u}^i$ to the correct value requires that

$$\sum_{i=1}^{\infty} K^i = \infty \tag{5.2-12}$$

Since $\mathbf{v}^i$ has zero mean, we know that

$$\sum_{i=1}^{\infty} K^i \mathbf{v}^i = \mathbf{0}$$

In order to not let the effect of the noise $\mathbf{v}^i$ be cumulative, we also require that

$$\sum_{i=1}^{\infty} (K^i)^2 (v_j{}^i)^2 < \infty, \qquad \text{for all} \quad j$$

It can be shown that this requirement is satisfied if the requirement

$$\sum_{i=1}^{\infty} (K^i)^2 < \infty \tag{5.2-13}$$

is imposed on the K^i sequence, and if the noise $\mathbf{v}^i$ has finite variance

$$\mathbf{V}_{\mathbf{v}^i} = \operatorname{var}\{\mathbf{v}^i\} \leqslant \mathscr{V} < \infty \tag{5.2-14}$$

A particular value of K^i, but certainly not the only value, which satisfies requirements (5.2-11)–(5.2-13) is

$$K^i = k/i \tag{5.2-15}$$

Stochastic approximation unfortunately does not tell us anything about the proper value of k except that it is nonnegative. Optimum filter theory, which we explored in Chaps. 2 and 3 and which we will consider again in Chap. 7, tells us that k is determined by the relative magnitudes of the plant and measurement noises, and that this gain should really be a matrix gain.

From our remarks concerning the relationships between stochastic approximation and optimal filter theory in the last section, indeed we see that these relations and restrictions on K^i are to be expected. Formal proofs are presented by Kiefer and Wolfowitz (1952), Blum (1954), and Kushner (1965). A very readable engineering proof is presented by Ho and Newbold (1967).

We now wish to extend our efforts to include stochastic extremization where equality constraints are present. Thus we desire to extremize (minimize)

$$J = \mathscr{E}\{\theta(\mathbf{x}, \mathbf{u})\} \tag{5.2-16}$$

subject to the equality constraint

$$\mathbf{f}(\mathbf{x}, \mathbf{u}, \boldsymbol{\zeta}) = \mathbf{0} \tag{5.2-17}$$

Let us assume that we can determine the probability that $\boldsymbol{\zeta} = \boldsymbol{\zeta}^i$ is P_i, for $i = 1, 2,..., M$.[1] Now we consider the problem with the cost function and equality constraint:

$$J^i = \theta(\mathbf{x}, \mathbf{u}), \qquad \mathbf{f}(\mathbf{x}, \mathbf{u}, \boldsymbol{\zeta}^i) = \mathbf{0} \tag{5.2-18}$$

This is a simple single-stage decision problem, since $\boldsymbol{\zeta}^i$ is assumed to be known. We determine the optimum solution (Chap. 3, Sage, 1968; Bryson and Ho, 1969) by defining the Hamiltonian

$$H^i = \theta(\mathbf{x}, \mathbf{u}) + \boldsymbol{\lambda}^T(\mathbf{x}, \mathbf{u}, \boldsymbol{\zeta}^i) \tag{5.2-19}$$

[1] Difficulties arise with this interpretation in the often encountered case in which $\boldsymbol{\zeta}$ is a continuous random variable. The difficulties are purely formal, and our reasoning here is easily extended to the continuous case as we shall see.

and then solving simultaneously the relationships

$$\frac{\partial H^i}{\partial \boldsymbol{\lambda}} = \mathbf{0}, \qquad \frac{\partial H^i}{\partial \mathbf{u}} = \mathbf{0}, \qquad \frac{\partial H^i}{\partial \mathbf{x}} = \mathbf{0} \tag{5.2-20}$$

We expect this particular problem, where $\boldsymbol{\zeta} = \boldsymbol{\zeta}^i$, to occur with probability P_i. Solving the original problem is therefore equivalent to solving a sequence of the deterministic problem where $\boldsymbol{\zeta} = \boldsymbol{\zeta}^i$ and weighting the sum of these solutions by the probability P_i. The original problem of extremizing the cost function of Eq. (5.2-16) subject to the constraint of Eq. (5.2-17) is solved by determining the solution to the necessary conditions

$$\mathbf{f}(\mathbf{x}, \mathbf{u}, \boldsymbol{\zeta}) = \mathbf{0}, \qquad \sum_{i=1}^{M} P_i \frac{\partial H^i}{\partial \mathbf{u}} = \mathbf{0}, \qquad \sum_{i=1}^{M} P_i \frac{\partial H^i}{\partial \mathbf{x}} = \mathbf{0} \tag{5.2-21}$$

If the distribution of $\boldsymbol{\zeta}$ is continuous, we see that Eq. (5.2-21) becomes

$$\mathbf{f}(\mathbf{x}, \mathbf{u}, \boldsymbol{\zeta}) = \mathbf{0}, \qquad \int_{-\infty}^{\infty} p(\boldsymbol{\zeta}^i) \frac{\partial H^i}{\partial \mathbf{u}} \, d\boldsymbol{\zeta}^i = \mathbf{0}, \qquad \int_{-\infty}^{\infty} p(\boldsymbol{\zeta}^i) \frac{\partial H^i}{\partial \mathbf{x}} \, d\boldsymbol{\zeta}^i = \mathbf{0}$$

or

$$\mathbf{f}(\mathbf{x}, \mathbf{u}, \boldsymbol{\zeta}) = \mathbf{0}, \qquad \mathscr{E}\left\{\frac{\partial H}{\partial \mathbf{u}}\right\} = \mathbf{0}, \qquad \mathscr{E}\left\{\frac{\partial H}{\partial \mathbf{x}}\right\} = \mathbf{0} \tag{5.2-22}$$

Unfortunately, we will not often be able to solve Eq. (5.2-22), due to the nonlinearities and expectations involved. Instead, we look for a set of gradient algorithms to solve Eqs. (5.2-22) iteratively.

The stochastic approximation method is thus very similar again to the basic gradient method for single-stage decision processes. We wish to minimize the cost function of Eq. (5.2-16) and equality constraint of Eq. (5.2-17):

$$J = \mathscr{E}\{\theta(\mathbf{x}, \mathbf{u})\}, \qquad \mathbf{f}(\mathbf{x}, \mathbf{u}, \boldsymbol{\zeta}) = \mathbf{0}$$

We choose an initial control $\mathbf{u}^i$ and a $\boldsymbol{\zeta}^i$, a random sample determined in accordance with the probability density function $\mathbf{p}(\boldsymbol{\zeta})$. The state $\mathbf{x}^i$ is then determined from $\mathbf{f}(\mathbf{x}^i, \mathbf{u}^i, \boldsymbol{\zeta}^i) = \mathbf{0}$. The adjoint equation $\partial H/\partial \mathbf{x}^i = \mathbf{0}$ is then used to update the control. A new random sample for $\boldsymbol{\zeta}$ is chosen and the iteration repeated. The steps in the computation are:

1. Choose $\mathbf{u}^i$.
2. Choose $\boldsymbol{\zeta}^i$ according to the density $p(\boldsymbol{\zeta})$.
3. Solve $\mathbf{f}(\mathbf{x}^i, \mathbf{u}^i, \boldsymbol{\zeta}^i) = \mathbf{0}$ for $\mathbf{x}^i$.
4. Solve

$$\frac{\partial H^i}{\partial \mathbf{x}^i} = \frac{\partial \theta(\mathbf{x}^i, \mathbf{u}^i)}{\partial \mathbf{x}^i} + \frac{\partial \mathbf{f}^{\mathrm{T}}(\mathbf{x}^i, \mathbf{u}^i, \boldsymbol{\zeta}^i)}{\partial \mathbf{x}^i} \boldsymbol{\lambda}^i = \mathbf{0}$$

for $\boldsymbol{\lambda}^i$.

5. Determine the gradient

$$\frac{\partial H^i}{\partial \mathbf{u}^i} = \frac{\partial \theta(\mathbf{x}^i, \mathbf{u}^i)}{\partial \mathbf{u}^i} + \frac{\partial \mathbf{f}^{\mathrm{T}}(\mathbf{x}^i, \mathbf{u}^i, \boldsymbol{\zeta}^i)}{\partial \mathbf{u}^i} \boldsymbol{\lambda}^i$$

6. Update the control using the stochastic approximation algorithm

$$\mathbf{u}^{i+1} = \mathbf{u}^i - K^i \frac{\partial H^i}{\partial \mathbf{u}^i} = \mathbf{u}^i - K^i \left[\frac{\partial \theta^i}{\partial \mathbf{u}^i} + \frac{\partial \mathbf{f}^{i\mathrm{T}}}{\partial \mathbf{u}^i} \boldsymbol{\lambda}^i \right]$$

where K^i satisfies all aforementioned convergence requirements. Repeat the iteration starting with Step 2.

Example 5.2-1. It is well known (Sage, 1968; Ho and Bryson, 1969; Sage and Melsa, 1971) that the optimum linear filtering problem of finding the best linear filter to minimize

$$J = \mathscr{E}\{\| \mathbf{x}(t) - \hat{\mathbf{x}}(t) \|_{\mathbf{S}}^2 \mid Z(t)\}$$

for the linear stationary system with uncorrelated plant and measurement noise

$$\dot{\mathbf{x}} = \mathbf{F}\mathbf{x}(t) + \mathbf{w}(t), \qquad \mathscr{E}\{\mathbf{w}(t)\} = \mathbf{0}, \qquad \operatorname{cov}\{\mathbf{w}(t), \mathbf{w}(\tau)\} = \boldsymbol{\Psi}_{\mathrm{w}} \delta_{\mathrm{D}}(t - \tau)$$

$$\mathbf{z} = \mathbf{H}\mathbf{x}(t) + \mathbf{v}(t), \qquad \mathscr{E}\{\mathbf{v}(t)\} = \mathbf{0}, \qquad \operatorname{cov}\{\mathbf{v}(t), \mathbf{v}(\tau)\} = \boldsymbol{\Psi}_{\mathrm{v}} \delta_{\mathrm{D}}(t - \tau)$$

is given by

$$\dot{\hat{\mathbf{x}}} = \mathbf{F}\hat{\mathbf{x}}(t) + \mathscr{K}(t)[\mathbf{z}(t) - \mathbf{H}\hat{\mathbf{x}}(t)], \qquad \hat{\mathbf{x}}(t_0) = \mathscr{E}\{\mathbf{x}(t_0)\}$$

$$\mathscr{K}(t) = \mathbf{V}_{\tilde{\mathbf{x}}}(t)\, \mathbf{H}^{\mathrm{T}} \boldsymbol{\Psi}_{\mathrm{v}}^{-1}$$

$$\dot{\mathbf{V}}_{\tilde{\mathbf{x}}} = \mathbf{F}\mathbf{V}_{\tilde{\mathbf{x}}}(t) + \mathbf{V}_{\tilde{\mathbf{x}}}(t)\, \mathbf{F}^{\mathrm{T}} - \mathbf{V}_{\tilde{\mathbf{x}}}(t)\, \mathbf{H}^{\mathrm{T}} \boldsymbol{\Psi}_{\mathrm{v}}^{-1} \mathbf{H} \mathbf{V}_{\tilde{\mathbf{x}}}(t) + \boldsymbol{\Psi}_{\mathrm{w}}(t)$$

$$\mathbf{V}_{\tilde{\mathbf{x}}}(t_0) = \operatorname{var}\{\mathbf{x}(t_0)\}$$

We wish to assume that the filter has been in operation for a sufficient length of time such that steady state conditions have been reached. Further, we assume that we do not know the plant and measurement noise coefficients $\boldsymbol{\Psi}_{\mathrm{w}}$ and $\boldsymbol{\Psi}_{\mathrm{v}}$, which are random over an ensemble, but only their average (nominal) values

$$\mathscr{E}\{\boldsymbol{\Psi}_{\mathrm{w}}\} = \bar{\boldsymbol{\Psi}}_{\mathrm{w}}, \qquad \mathscr{E}\{\boldsymbol{\Psi}_{\mathrm{v}}\} = \bar{\boldsymbol{\Psi}}_{\mathrm{v}}$$

Three methods of filter design are apparent:

1. We may implement the suboptimum filter

$$\dot{\hat{\mathbf{x}}}_1 = \mathbf{F}\hat{\mathbf{x}}(t) + \mathscr{K}_1[\mathbf{z}(t) - \mathbf{H}\hat{\mathbf{x}}(t)]$$

$$\mathscr{K}_1 = \boldsymbol{\Xi}_1\mathbf{H}^{\mathrm{T}}\bar{\boldsymbol{\Psi}}_{\mathrm{v}}^{-1}$$

$$\mathbf{0} = \mathbf{F}\boldsymbol{\Xi}_1 + \boldsymbol{\Xi}_1\mathbf{F}^{\mathrm{T}} - \boldsymbol{\Xi}_1\mathbf{H}^{\mathrm{T}}\bar{\boldsymbol{\Psi}}_{\mathrm{v}}^{-1}\mathbf{H}\boldsymbol{\Xi}_1 + \bar{\boldsymbol{\Psi}}_{\mathrm{w}}$$

   which is the stationary Kalman filter implemented with the nominal or average plant and measurement noise variance coefficients.

2. We may implement the suboptimum adaptive filter

$$\dot{\hat{\mathbf{x}}}_2 = \mathbf{F}\hat{\mathbf{x}}_2(t) + \mathscr{K}_2(t)[\mathbf{z}(t) - \mathbf{H}\hat{\mathbf{x}}(t)]$$

$$\mathscr{K}_2(t) = \boldsymbol{\Xi}_2\mathbf{H}^{\mathrm{T}}\hat{\boldsymbol{\Psi}}_{\mathrm{v}}^{-1}(t)$$

$$\dot{\boldsymbol{\Xi}}_2 = \mathbf{F}\boldsymbol{\Xi}_2(t) + \boldsymbol{\Xi}_2(t)\,\mathbf{F}^{\mathrm{T}} - \dot{\boldsymbol{\Xi}}_2(t)\,\mathbf{H}^{\mathrm{T}}\hat{\boldsymbol{\Psi}}_{\mathrm{v}}^{-1}(t)\,\mathbf{H}\boldsymbol{\Xi}_1(t) + \hat{\boldsymbol{\Psi}}_{\mathrm{w}}(t)$$

   where some suitable adaptive estimation algorithms (Sage and Husa, 1969a) are used to estimate the random variance coefficients. For many problems this solution may be unacceptable from a practical standpoint as being too complex computationally.

3. Rather than implement the filter of method 1, we may implement the no more complex filter

$$\dot{\hat{\mathbf{x}}} = \mathbf{F}\hat{\mathbf{x}}(t) + \mathscr{K}_3[\mathbf{z}(t) - \mathbf{H}\hat{\mathbf{x}}(t)]$$

   where $\mathscr{K}_3$ is a constant chosen to minimize the estimation error and considering the random (over an ensemble) nature of $\boldsymbol{\Psi}_{\mathrm{w}}$ and $\boldsymbol{\Psi}_{\mathrm{v}}$.

It is a simple matter to show (Sage, 1968; Sage and Melsa, 1971) that the true error variance for the suboptimal filter

$$\dot{\hat{\mathbf{x}}}_{\text{so}} = \mathbf{F}\hat{\mathbf{x}}_{\text{so}}(t) + \mathscr{K}_{\text{so}}(t)[\mathbf{z}(t) - \mathbf{H}(t)\,\hat{\mathbf{x}}_{\text{so}}(t)]$$

which is defined by $E = \text{var}\{\mathbf{x}(t) - \mathbf{x}_{\text{so}}(t)\}$, is given by

$$\dot{\mathbf{E}} = \mathbf{D}(t)\,\mathbf{E}(t) + \mathbf{E}(t)\,\mathbf{D}^{\text{T}}(t) + \mathscr{K}_{\text{so}}(t)\,\boldsymbol{\Psi}_{\text{v}}(t)\,\mathscr{K}_{\text{so}}^{\text{T}}(t) + \boldsymbol{\Psi}_{\text{w}}(t),$$

$$\mathbf{E}(t_0) = \text{var}\{\mathbf{x}(t_0)\}$$

where $\mathbf{D}(t) = \mathbf{F} - \mathscr{K}_{\text{so}}(t)\mathbf{H}$. This error variance algorithm clearly applies to either methods 1 or 3. We are particularly interested in the steady state error variance algorithm for constant $\mathscr{K}_{\text{so}}$. This is the positive definite solution of

$$\mathbf{0} = \mathbf{DE} + \mathbf{ED}^{\text{T}} + \mathscr{K}_{\text{so}}\boldsymbol{\Psi}_{\text{v}}\mathscr{K}_{\text{so}}^{\text{T}} + \boldsymbol{\Psi}_{\text{w}}$$

where $\mathbf{D} = \mathbf{F} - \mathscr{K}_{\text{so}}\mathbf{H}$. The cost function which we will use is

$$J = \mathscr{E}\{\text{trace}\,\mathbf{E}\}$$

in which we imply the nonnegative definite solution for $\mathbf{E}$. For the filter of method 1, the gain $\mathscr{K}_{\text{so}}$ is determined by the nominal values of $\bar{\boldsymbol{\Psi}}_{\text{w}}$ and $\bar{\boldsymbol{\Psi}}_{\text{v}}$. For method 3, we wish to determine $\mathscr{K}_{\text{so}}$ such that J is a minimum. Because of the random nature of $\boldsymbol{\Psi}_{\text{w}}$ and $\boldsymbol{\Psi}_{\text{v}}$, we will use the stochastic approximation method which we have just developed. We wish to minimize, by choice of $\mathscr{K}_{\text{so}}$, the cost function

$$J = \mathscr{E}\{\text{trace}\,\mathbf{E}\}$$

subject to the equality constraint

$$\mathbf{0} = (\mathbf{F} - \mathscr{K}_{\text{so}}\mathbf{H})\mathbf{E} + \mathbf{E}(\mathbf{F}^{\text{T}} - \mathbf{H}^{\text{T}}\mathscr{K}_{\text{so}}^{\text{T}}) + \mathscr{K}_{\text{so}}\boldsymbol{\Psi}_{\text{v}}\mathscr{K}_{\text{so}}^{\text{T}} + \boldsymbol{\Psi}_{\text{w}}$$

where $\boldsymbol{\Psi}_{\text{w}}$ and $\boldsymbol{\Psi}_{\text{v}}$ are random variables with known probability densities.

We define the Hamiltonian (a random variable)

$$H = \text{trace}(\mathbf{E}) + \text{trace}\{\boldsymbol{\Lambda}[(\mathbf{F} - \mathscr{K}_{\text{so}}\mathbf{H})\mathbf{E} + \mathbf{E}(\mathbf{F}^{\text{T}} - \mathbf{H}^{\text{T}}\mathscr{K}_{\text{so}}^{\text{T}}) + \mathscr{K}_{\text{so}}\boldsymbol{\Psi}_{\text{v}}\mathscr{K}_{\text{so}}^{\text{T}} + \boldsymbol{\Psi}_{\text{w}}]\}$$

where $\mathbf{\Lambda}$ is a symmetric matrix Lagrange multiplier. The optimum value of $\mathscr{K}_{so}$ is determined from solution of

$$\frac{\partial H}{\partial \mathbf{\Lambda}} = 0, \qquad \mathscr{E}\left\{\frac{\partial H}{\partial \mathbf{E}}\right\} = 0, \qquad \mathscr{E}\left\{\frac{\partial H}{\partial \mathscr{K}_{so}}\right\} = 0$$

These relations yield

$$0 = (\mathbf{F} - \mathscr{K}_{so}\mathbf{H})\mathbf{E} + \mathbf{E}(\mathbf{F}^T - \mathbf{H}^T\mathscr{K}_{so}^T) + \mathscr{K}_{so}\mathbf{\Psi}_v\mathscr{K}_{so}^T + \mathbf{\Psi}_w$$

$$\mathbf{E}\{\mathbf{I} + \mathbf{\Lambda}(\mathbf{F} - \mathscr{K}_{so}\mathbf{H}) + (\mathbf{F}^T - \mathbf{H}^T\mathscr{K}_{so}^T)\mathbf{\Lambda}\} = 0$$

$$\mathscr{E}\{\mathbf{\Lambda}\mathbf{E}\mathbf{H}^T + \mathbf{\Lambda}\mathscr{K}_{so}\mathbf{\Psi}_v\} = 0$$

In this particular, the expectations are easily taken, with the result that

$$\mathscr{K}_{so} = -\mathscr{E}\{\mathbf{E}\}\,\mathbf{H}^T\mathbf{\Psi}_v^{-1}$$

where $\mathscr{E}\{\mathbf{E}\}$ is defined by the equality constraint. By substituting this relation for $\mathscr{K}_{so}$ in the equality constraint, we see that a simpler expression for the equality constraint with the optimum $\mathbf{E}$ is

$$0 = \mathbf{F}\mathscr{E}\{\mathbf{E}\} + \mathscr{E}\{\mathbf{E}\}\,\mathbf{F}^T - \mathscr{E}\{\mathbf{E}\}\,\mathbf{H}^T\bar{\mathbf{\Psi}}_v^{-1}\mathbf{H}\mathscr{E}\{\mathbf{E}\} + \bar{\mathbf{\Psi}}_w$$

We reach the perhaps not unsurprising conclusion that the "best" Kalman gain here is that obtained by using $\bar{\mathbf{\Psi}}_w$ and $\bar{\mathbf{\Psi}}_v$ in implementing the Kalman filter.

It is of interest to examine this result further for a simple case. The message and observation models and average variances are

$$\dot{x} = w(t), \qquad \bar{\Psi}_w = 1$$

$$z = x(t) + v(t), \qquad \bar{\Psi}_v = 1 = \Psi_v$$

The random deviation of Ψ_w from its average value of 1 is uniform with a maximum deviation $\pm\alpha$ such that the variance of Ψ_w is $\alpha^2/3$.

The actual error variance is determined from

$$E = \frac{\mathscr{K}_{so}^2 + \Psi_w}{2\mathscr{K}_{so}}$$

so we see that E also has a uniform distribution with an average value

$$\mathscr{E}\{E\} = \frac{\mathscr{K}_{so}}{2} + \frac{1}{2\mathscr{K}_{so}}$$

and a maximum deviation $\pm\alpha/(2\mathscr{K}_{so})$. In this particular case, we see that the "best" $\mathscr{K}_{so}$ to minimize $\mathscr{E}\{E\}$ is just $\mathscr{K}_{so} = 1$, which is the value obtained from the Kalman filter algorithms if we use $\overline{\Psi}_w = \overline{\Psi}_v = 1$. Best here means in the sense of minimizing the average error variance.

Since Ψ_w may be actually anywhere in the range $1 - \alpha < \Psi_w < 1 + \alpha$ we see that the *maximum* and *minimum* errors resulting from the strategy of using $\mathscr{K}_{so} = 1$ are $E_{max} = [1 + (1 + \alpha)]/2$, and $E_{min} = [1 + (1 - \alpha)]/2$, where α is of course less that 1, since Ψ_w cannot be negative. An alternate strategy is to minimize the maximum error. In this case, we see that maximum error occurs at $\Psi_w = 1 + \alpha$, the best gain for this value of Ψ_w is $\mathscr{K}_{so} = \sqrt{\Psi_w} = (1 + \alpha)^{1/2}$, and the minimum value of the maximum error is then $E_{min-max} = (1 + \alpha)^{1/2}$. The actual error is $E = [(1 + \alpha) + \Psi_w]/[2(1 + \alpha)]^{1/2}$ and varies from $1/(1 + \alpha)^{1/2}$ to $(1 + \alpha)^{1/2}$. The average error with this min–max criterion is $(2 + \alpha)/[2(1 + \alpha)]^{1/2}$, whereas the average error with the original $\mathscr{E}\{E\}$ criterion is 1.

Figure 5.2-1 indicates that manner in which the error variances

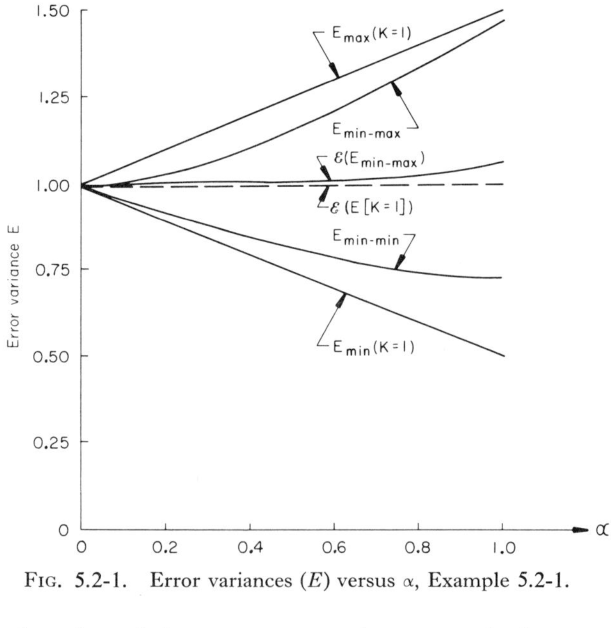

FIG. 5.2-1. Error variances (E) versus α, Example 5.2-1.

vary as a function of the parameter α. As expected, the average error variance is less (but not by much) for the case in which the criterion is to minimize the average error variance. The minimum value of the maximum error variance criterion leads to a smaller maximum error variance (but again not by much). Thus we might conclude that, for this case, the two criteria lead to somewhat similar results. This will often, but certainly not always, be the case. This is fortuitous, since minimizing the expected value of a cost function is easier (but still often quite difficult) than minimizing the maximum error.

Example 5.2-2. We reconsider Example 5.2-1, except that we now assume that a modeling error exists in determination of **F**, **G**, and **H**. The actual process is therefore

$$\dot{\mathbf{x}} = \mathbf{F}\mathbf{x}(t) + \mathbf{G}\mathbf{w}(t), \qquad \operatorname{cov}\{\mathbf{w}(t), \mathbf{w}(\tau)\} = \boldsymbol{\Psi}_{\mathrm{w}}(t)\,\delta_{\mathrm{D}}(t-\tau)$$

$$\mathbf{z}(t) = \mathbf{H}\mathbf{x}(t) + \mathbf{v}(t), \qquad \operatorname{cov}\{\mathbf{v}(t), \mathbf{v}(\tau)\} = \boldsymbol{\Psi}_{\mathrm{v}}(t)\,\delta_{\mathrm{D}}(t-\tau)$$

where the variance coefficients of the zero mean uncorrelated random processes are assumed to be known. **F**, **G**, and **H** are not known with precision, so nominal or average values, $\bar{\mathbf{F}}$, $\bar{\mathbf{G}}$, and $\bar{\mathbf{H}}$ are used for them.

One possibility in constructing a filter for the process is to implement a standard Kalman filter assuming that the actual parameters are $\bar{\mathbf{F}}$, $\bar{\mathbf{G}}$, and $\bar{\mathbf{H}}$. In this case, the algorithms

$$\dot{\hat{\mathbf{x}}} = \bar{\mathbf{F}}\hat{\mathbf{x}}(t) + \mathscr{K}_1[\mathbf{z}(t) - \bar{\mathbf{H}}\hat{\mathbf{x}}(t)]$$

$$\mathscr{K}_1 = \boldsymbol{\Xi}_1\bar{\mathbf{H}}^{\mathrm{T}}\boldsymbol{\Psi}_{\mathrm{v}}^{-1}$$

$$\mathbf{0} = \bar{\mathbf{F}}\boldsymbol{\Xi}_1 + \boldsymbol{\Xi}_1\bar{\mathbf{F}}^{\mathrm{T}} - \boldsymbol{\Xi}_1\bar{\mathbf{H}}^{\mathrm{T}}\boldsymbol{\Psi}_{\mathrm{v}}^{-1}\bar{\mathbf{H}}\boldsymbol{\Xi}_1 + \mathbf{G}\boldsymbol{\Psi}_{\mathrm{w}}\mathbf{G}^{\mathrm{T}}$$

are implemented for the stationary case. An alternate and better procedure is to implement the filter

$$\dot{\hat{\mathbf{x}}} = \bar{\mathbf{F}}\hat{\mathbf{x}}(t) + \mathscr{K}_{\mathrm{so}}[\mathbf{z}(t) - \bar{\mathbf{H}}\hat{\mathbf{x}}(t)]$$

and adjust $\mathscr{K}_{\mathrm{so}}$ such that the expectation of the actual error variance $\mathscr{E}\{\mathbf{V}_{\tilde{\mathbf{x}}}\} = \mathscr{E}\{\operatorname{var}\{\mathbf{x}(t) - \hat{\mathbf{x}}(t)\}\}$ is a minimum.

By subtracting the equation for the actual process and the suboptimal filter, we obtain the error expression

$$\dot{\tilde{\mathbf{x}}} = (\Delta\mathbf{F} - \mathscr{K}_{\mathrm{so}}\Delta\mathbf{H})\,\mathbf{x}(t) + (\bar{\mathbf{F}} - \mathscr{K}_{\mathrm{so}}\bar{\mathbf{H}})\,\tilde{\mathbf{x}}(t) + \mathbf{G}\mathbf{w}(t) - \mathscr{K}_{\mathrm{so}}\mathbf{v}(t)$$

where

$$\Delta\mathbf{F} = \mathbf{F} - \bar{\mathbf{F}}, \qquad \Delta\mathbf{H} = \mathbf{H} - \bar{\mathbf{H}}$$

are used to represent the modeling errors. It is convenient to define an augmented state vector and input vector

$$\mathbf{X}(t) = \begin{bmatrix} \tilde{\mathbf{x}}(t) \\ \mathbf{x}(t) \end{bmatrix}, \qquad \mathbf{W}(t) = \begin{bmatrix} \mathbf{w}(t) \\ \mathbf{v}(t) \end{bmatrix}$$

such that we have a new state equation

$$\dot{\mathbf{X}} = \mathbf{A}(t)\,\mathbf{X}(t) + \mathbf{B}(t)\,\mathbf{W}(t)$$

where

$$\mathbf{A}(t) = \begin{bmatrix} \bar{\mathbf{F}} - \mathscr{K}_{\text{so}}\bar{\mathbf{H}} & \Delta\mathbf{F} - \mathscr{K}_{\text{so}}\Delta\mathbf{H} \\ \mathbf{0} & \mathbf{F} \end{bmatrix}, \qquad \mathbf{B}(t) = \begin{bmatrix} \mathbf{G} & -\mathscr{K}_{\text{so}} \\ \mathbf{G} & \mathbf{0} \end{bmatrix}$$

It is well known that the differential equation for the variance of $\mathbf{X}$ is given by

$$\dot{\mathbf{V}}_{\mathbf{X}} = \mathbf{A}(t)\,\mathbf{V}_{\mathbf{X}}(t) + \mathbf{V}_{\mathbf{X}}(t)\,\mathbf{A}^{\mathrm{T}}(t) + \mathbf{B}(t)\,\mathbf{\Psi}_{\mathbf{W}}\mathbf{B}^{\mathrm{T}}(t)$$

We are interested in the stationary solution of this equation, so we set $\dot{\mathbf{V}}_{\mathbf{X}} = \mathbf{0}$. The best suboptimum gain will be chosen such that $\mathscr{E}\{\text{trace}\,\mathbf{V}_{\tilde{\mathbf{x}}}\}$ is a minimum. Since $\mathbf{V}_{\mathbf{x}}$ is not influenced by the random parameter $\mathbf{F}$, $\mathbf{G}$, and $\mathbf{H}$, minimization of $\mathscr{E}\{\text{trace}\,\mathbf{V}_{\mathbf{X}}\}$ is entirely equivalent to minimization of $\mathscr{E}\{\text{trace}\,\mathbf{V}_{\tilde{\mathbf{x}}}\}$.

Thus we desire to minimize

$$J = \mathscr{E}\{\text{trace}\,\mathbf{V}_{\mathbf{X}}\}$$

where the equality constraint

$$\mathbf{0} = \mathbf{A}(\mathbf{p}, \mathscr{K}_{\text{so}})\,\mathbf{V}_{\mathbf{X}} + \mathbf{V}_{\mathbf{X}}\mathbf{A}^{\mathrm{T}}(\mathbf{p}, \mathscr{K}_{\text{so}}) + \mathbf{B}(\mathbf{p}, \mathscr{K}_{\text{so}})\,\mathbf{\Psi}_{\mathbf{W}}\mathbf{B}^{\mathrm{T}}(\mathbf{p}, \mathscr{K}_{\text{so}})$$

is satisfied. $\mathbf{p}$ is used to indicate the random parameters, $\mathbf{F}$, $\mathbf{G}$, and $\mathbf{H}$.

We define the (random variable) Hamiltonian

$$H = \text{trace}\{\mathbf{V}_{\mathbf{X}}\} + \text{trace}\{\mathbf{\Lambda}(\mathbf{A}\mathbf{V}_{\mathbf{X}} + \mathbf{V}_{\mathbf{X}}\mathbf{A}^{\mathrm{T}} + \mathbf{B}\mathbf{\Psi}_{\mathbf{W}}\mathbf{B}^{\mathrm{T}})\}$$

The conditions

$$\frac{\partial H}{\partial \mathbf{\Lambda}} = \mathbf{0}, \qquad \mathscr{E}\left\{\frac{\partial H}{\partial \mathbf{V}_{\mathbf{X}}}\right\} = \mathbf{0}, \qquad \mathscr{E}\left\{\frac{\partial H}{\partial \mathscr{K}_{\text{so}}}\right\} = \mathbf{0}$$

are necessary for a minimum of J. The stochastic approximation algorithm is:

1. Determine a $\mathscr{K}_{so}^{i}$.
2. Determine a sample $\mathbf{p}^i(\mathbf{F}^i$, $\mathbf{G}^i$, and $\mathbf{H}^i)$ in accordance with the known probability density of $\mathbf{p}$.
3. Choose the symmetric Lagrange multiplier matrix such that

$$\frac{\partial H}{\partial \mathbf{V_X}^i} = \mathbf{I} + \mathbf{\Lambda}^i\mathbf{A}^i + \mathbf{A}^{i\mathrm{T}}\mathbf{\Lambda}^i = \mathbf{0}$$

4. Calculate $\partial H^i/\partial \mathscr{K}_{so}^{i}$ using the now known values of $\mathscr{K}_{so}^{i}$, $\mathbf{F}^i$, $\mathbf{G}^i$, $\mathbf{H}^i$, and $\mathbf{\Lambda}^i$.
5. Determine a new control iterate according to the stochastic approximation algorithm

$$\mathscr{K}_{so}^{i+1} = \mathscr{K}_{so}^{i} - K^i \frac{\partial H^i}{\partial \mathscr{K}_{so}^{i}}$$

6. Repeat the iterations starting with Step 2.

The specific nominal system

$$\dot{x} = -x(t) + w(t), \qquad \Psi_w(t) = 1$$

$$z = x(t) + v(t), \qquad \Psi_v(t) = 1$$

is now considered. Here $\bar{F} = -1$, $G = \bar{G} = 1$, and $H = \bar{H} = 1$. The nominal Kalman gain is $\mathscr{K}_1 = 0.414$. The suboptimal Kalman gain is determined by minimization of

$$J = \mathscr{E}\{\text{trace}\, \mathbf{V_X}\} = \mathscr{E}\{V_{\tilde{x}} + V_x\} = V_x + \mathscr{E}\{V_{\tilde{x}}\}$$

subject to the equality constraints

$$0 = -2(1 + \mathscr{K}_{so})\, V_{\tilde{x}} + 2(F + 1)\, V_{x\tilde{x}} + 1 + \mathscr{K}_{so}^2$$

$$0 = (F - 1 - \mathscr{K}_{so})\, V_{x\tilde{x}} + (F + 1)\, V_x + 1$$

$$0 = 2FV_x + 1$$

We may now use directly the just obtained algorithms or easily combine the foregoing three equations so as to obtain

$$V_{\tilde{x}} = \frac{1 + \mathscr{K}_{so}^2}{2(1 + \mathscr{K}_{so})} - \frac{(1 + F)(1 - F)}{2F(\mathscr{K}_{so} + 1 - F)(1 + \mathscr{K}_{so})}$$

where the first term indicates the error variance if F is in fact equal to -1. The computation for this specific problem is:

1. Select $\mathscr{K}_{\mathrm{so}}^i$.
2. Choose F^i in accordance with the known probability density of F.
3. Determine $\partial V_{\mathrm{x}}/\partial \mathscr{K}_{\mathrm{so}}^i$.
4. Update the control by using

$$\mathscr{K}_{\mathrm{so}}^{i+1} = \mathscr{K}_{\mathrm{so}}^i - K^i \frac{\partial V_x}{\partial \mathscr{K}_{\mathrm{so}}^i}$$

5. Repeat the computations starting at Step 2.

Figure 5.2-2 shows typical illustrations of the convergence of $\mathscr{K}_{\mathrm{so}}$.

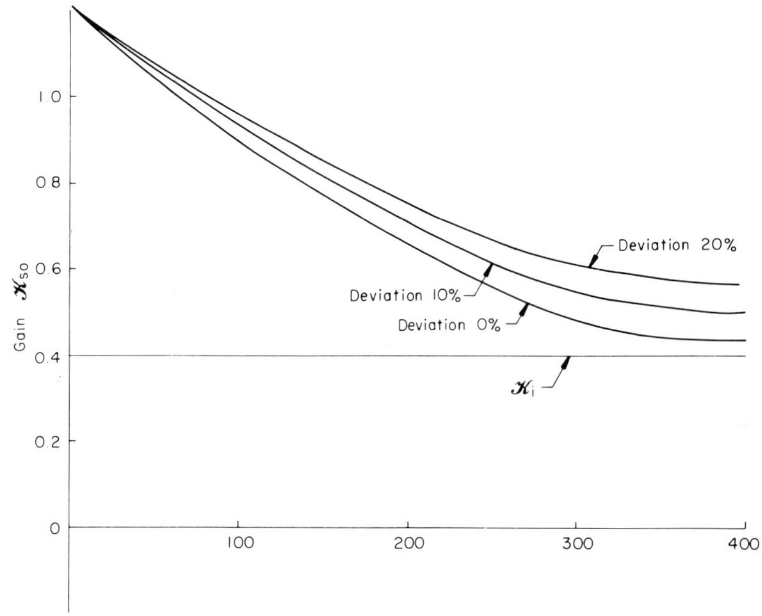

FIG. 5.2-2. Kalman gain values, Example 5.2-2.

The parameter F was assumed to have a probability density uniform about -1 with maximum deviation $\pm\alpha$. The initial gain $\mathscr{K}_{\mathrm{so}}$ was chosen to be 1.2 and the K^i for the stochastic approximation algorithm chosen to be $1/i$. Several results from this figure are worth noting.

In particular, convergence is much slower than might ordinarily be expected with a gradient algorithm. The suboptimal Kalman gain is always greater than the optimal gain for the case where $F = -1$, and this gain increases as the uncertainty in F increases. No attempt at optimization of the algorithms to insure rapid convergence was made in this example. It is conceivable that more rapid convergence could be obtained if this were accomplished.

Extension of our results to dynamic situations is relatively straightforward. We consider minimization of the cost function

$$J = \mathscr{E}\left\{\theta_f[\mathbf{x}(k_f)] + \theta_0[\mathbf{x}(k_0)] + \sum_{k=k_0}^{k_0-1} \varphi[\mathbf{x}(k), \mathbf{u}(k), \boldsymbol{\zeta}(k), k]\right\} \tag{5.2-23}$$

subject to the difference equality constraint

$$\mathbf{x}(k+1) = \boldsymbol{\phi}[\mathbf{x}(k), \mathbf{u}(k), \boldsymbol{\zeta}(k), k] \tag{5.2-24}$$

where $\mathbf{x}(k)$ is a generalized state vector which includes all unknown parameters, and $\boldsymbol{\zeta}(k)$ is a vector stochastic process with presumed known probability density function $p[\boldsymbol{\zeta}(k)]$. Just as in the single-stage case, we may take a particular sample of the stochastic process $\boldsymbol{\zeta}^i(k)$ and formulate a deterministic optimization problem. We define the Hamiltonian

$$H^i = \varphi[\mathbf{x}(k), \mathbf{u}(k), \boldsymbol{\zeta}^i(k), k] + \boldsymbol{\lambda}^T(k+1)\, \boldsymbol{\phi}[\mathbf{x}(k), \mathbf{u}(k), \boldsymbol{\zeta}^i(k), k] \tag{5.2-25}$$

and obtain in a straightforward fashion the canonic equations and associated two-point boundary conditions

$$\frac{\partial H^i}{\partial \boldsymbol{\lambda}(k+1)} = \mathbf{x}(k+1), \qquad \frac{\partial H^i}{\partial \mathbf{u}(k)} = \mathbf{0}, \qquad \frac{\partial H^i}{\partial \mathbf{x}(k)} = \boldsymbol{\lambda}(k)$$
$$\boldsymbol{\lambda}(k_0) = -\frac{\partial \theta_0[\mathbf{x}(k_0)]}{\partial \mathbf{x}(k_0)}, \qquad \boldsymbol{\lambda}(k_f) = \frac{\partial \theta_f[\mathbf{x}(k_f)]}{\partial \mathbf{x}(k_f)} \tag{5.2-26}$$

Now this problem occurs with probability P_i, for $i = 1, 2,\ldots, M$. Thus solving the original stochastic problem is equivalent to solving a weighted sum of problems with appropriate weighting coefficients

P_i. The original problem of Eqs. (5.2-23) and (5.2-24) results in the necessary conditions

$$\frac{\partial H}{\partial \boldsymbol{\lambda}(k+1)} = \mathbf{x}(k+1)$$

$$\sum_{i=1}^{M} P_i \frac{\partial H^i}{\partial \mathbf{u}(k)} = \mathbf{0}$$

$$\sum_{i=1}^{M} P_i \left[\frac{\partial H^i}{\partial \mathbf{x}(k)} - \boldsymbol{\lambda}(k)\right] = \mathbf{0}$$

$$\sum_{i=1}^{M} P_i \left[\boldsymbol{\lambda}(k_0) + \frac{\partial \theta_0[\mathbf{x}(k_0)]}{\partial \mathbf{x}(k_0)}\right] = \mathbf{0}$$

$$\sum_{i=1}^{M} P_i \left[\boldsymbol{\lambda}(k_f) - \frac{\partial \theta_f[\mathbf{x}(k_f)]}{\partial \mathbf{x}(k_f)}\right] = \mathbf{0}$$

We pass formally from the discrete distribution of P_i to a continuous distribution and then have the relations for the discrete time stochastic maximum principle

$$H = \varphi[\mathbf{x}(k), \mathbf{u}(k), \boldsymbol{\zeta}(k), k] + \boldsymbol{\lambda}^{\mathrm{T}}(k+1)\, \boldsymbol{\phi}[\mathbf{x}(k), \mathbf{u}(k), \boldsymbol{\zeta}(k) k]$$

$$\frac{\partial H}{\partial \boldsymbol{\lambda}(k+1)} = \mathbf{x}(k+1), \qquad \mathscr{E}\left\{\frac{\partial H}{\partial \mathbf{u}(k)}\right\} = \mathbf{0}, \qquad \mathscr{E}\left\{\frac{\partial H}{\partial \mathbf{x}(k)} - \boldsymbol{\lambda}(k)\right\} = \mathbf{0} \tag{5.2-27}$$

$$\mathscr{E}\left\{\boldsymbol{\lambda}(k_0) + \frac{\partial \theta_0[\mathbf{x}(k_0)]}{\partial \mathbf{x}(k_0)}\right\} = \mathscr{E}\left\{\boldsymbol{\lambda}(k_f) - \frac{\partial \theta_f[\mathbf{x}(k_f)]}{\partial \mathbf{x}(k_f)}\right\} = \mathbf{0}$$

Use of the stochastic approximation algorithms is directly patterned after our previous results. The steps are:

1. Choose a $\mathbf{u}^i(k)$.
2. Choose an $\mathbf{x}^i(k_0)$.
3. Determine a sample sequence $\boldsymbol{\zeta}^i(k)$ consistent with the probability density $p[\boldsymbol{\zeta}(k)]$.
4. Solve the difference equation

$$\mathbf{x}^i(k+1) = \boldsymbol{\phi}[\mathbf{x}^i(k), \mathbf{u}^i(k), \boldsymbol{\zeta}^i(k), k], \qquad \mathbf{x}^i(k_0) \text{ given}$$

forward in stage from k_0 to k_f.

5. Solve the adjoint equation

$$\boldsymbol{\lambda}^i(k) = \frac{\partial H}{\partial \mathbf{x}^i(k)} = \frac{\partial \varphi^i}{\partial \mathbf{x}^i(k)} + \frac{\partial \boldsymbol{\phi}^{i\mathrm{T}}}{\partial \mathbf{x}^i(k)} \boldsymbol{\lambda}^i(k+1)$$

backward in stage with the terminal condition

$$\boldsymbol{\lambda}(k_\mathrm{f}) = \frac{\partial \theta_\mathrm{f}[\mathbf{x}(k_\mathrm{f})]}{\partial \mathbf{x}(k_\mathrm{f})}$$

6. Determine a new control iterate from the stochastic approximation algorithm

$$\mathbf{u}^{i+1} = \mathbf{u}^i - K_\mathbf{u}{}^i \frac{\partial H^i}{\partial \mathbf{u}^i}$$
$$\mathbf{u}^{i+1} = \mathbf{u}^i - K_\mathbf{u}{}^i \left[\frac{\partial \varphi^i}{\partial \mathbf{u}^i} + \frac{\partial \boldsymbol{\phi}^{i\mathrm{T}}}{\partial \mathbf{u}^i} \boldsymbol{\lambda}^i(k+1)\right] \tag{5.2-28}$$

7. Determine a new initial condition iterate from the stochastic approximation algorithm

$$\mathbf{x}^{i+1}(k_0) = \mathbf{x}^i(k_0) - K_\mathbf{x}{}^i \left[\frac{\partial \theta_0[\mathbf{x}^i(k_0)]}{\partial \mathbf{x}^i(k_0)} + \boldsymbol{\lambda}^i(k_0)\right] \tag{5.2-29}$$

8. Repeat the computations starting with Step 3.

Example 5.2-3. Let us determine some characteristics of the control $u(k)$ to minimize

$$J = \mathscr{E}\left\{\frac{1}{2} x^2(k_\mathrm{f}) + \frac{1}{2} \sum_{k=k_0}^{k_\mathrm{f}-1} u^2(k)\right\}$$

for the system

$$x(k+1) = x(k) + u(k) + w(k), \qquad x(k_0) = x_0$$

where $w(k)$ is an unknown random input with known probability density function.

The solution proceeds as follows: We define the Hamiltonian

$$H = \tfrac{1}{2}u^2(k) + \lambda(k+1)[x(k) + u(k) + w(k)]$$

The stochastic maximum principle of Eqs. (5.2-27) leads to the canonic equations

$$x(k+1) = x(k) + u(k) + w(k), \qquad x(k_0) = x_0$$

$$\mathscr{E}\{u(k) + \lambda(k+1)\} = 0, \quad \mathscr{E}\{\lambda(k+1) - \lambda(k)\} = 0, \quad \mathscr{E}\{\lambda(k_\mathrm{f}) - x(k_\mathrm{f})\} = 0$$

It is straightforward to show that the solution to these equations is

$$\hat{u}(k) = \frac{-x(k_0)}{k_f - k_0}$$

$$\hat{x}(k) = x(k_0) + \sum_{k=k_0}^{k-1} \hat{u}(k) + w(k)$$

Instead of solving the stochastic maximum principle canonic equations directly, we now use the stochastic approximation algorithm directly. The adjoint equation from Step 5 of our procedure is

$$\lambda(k) = \frac{\partial H}{\partial x(k)} = \lambda(k+1), \qquad \lambda(k_f) = x(k_f)$$

which has the solution $\lambda(k) = x(k_f)$. The stochastic approximation for the control update is

$$u^{i+1}(k) = u^i(k) - K^i \frac{\partial H}{\partial u^i(k)} = u^i(k) - K^i[u^i(k) + x(k_f)]$$

and we see that if the initial control sequence is constant, so will be all control sequences $u^i(k)$. Since

$$\frac{\partial H}{\partial u(k)} = u(k) + \lambda(k+1) = u(k+1) + x(k_f)$$

we see that

$$\mathscr{E}\left\{\frac{\partial H}{\partial u^i(k)}\right\} = u^i(k) + \mathscr{E}\{x^i(k_f)\}, \qquad \operatorname{var}\left\{\frac{\partial H}{\partial u^i(k)}\right\} = \operatorname{var}\{x^i(k_f)\}$$

such that the variance of $\partial H/\partial u^i(k)$ is a constant for all i:

$$\operatorname{var}\left\{\frac{\partial H}{\partial u^i(k)}\right\} = V_x(k_f)$$

This also follows directly from the system difference equation solution for constant u:

$$x^i(k) = x^i(k_0) + \sum_{k=k_0}^{k-1} [u^i + w^i(k)]$$

which has the moments

$$\mathscr{E}\{x^i(k_f)\} = x^i(k_0) + \sum_{k=k_0}^{k-1} u^i, \qquad \operatorname{var}\{x^i(k_f)\} = \operatorname{var}\left\{\sum_{k=k_0}^{k_f-1} w^i(k)\right\} = V_x(k_f)$$

Since $\operatorname{var}\{\partial H/\partial u^i(k)\} = \text{const}$, we see that the stochastic approximation algorithm $u^{i+1} = u^i - K^i(\partial H/\partial u^i)$ will converge in that

$$\lim_{i\to\infty} \operatorname{var}\{u^i\} = 0, \qquad \text{if} \quad K^i = 1/i$$

However, if K^i is constant with respect to iteration number as in the conventional gradient method, then we see that

$$\lim_{i\to\infty} \operatorname{var}\{u^i\} = \infty$$

and the algorithm does not converge. This may lead us to question the validity of the work in Chap. 4, in which we used the conventional first- and second-order gradient approaches. There is a central difference in the two approaches. The same random sequence was used in all iterations in the gradient methods of Chap. 4, whereas each iteration by the stochastic approximation method uses a *new* random sequence $\zeta^i(k)$ selected according to the known probability density $p[\zeta(k)]$. Thus convergence questions do not arise in the same fashion in the previous chapter as they do here. By using new random samples sequences, we expect that parameter estimates are less likely to be biased than they are if the same random sequence is used over and over again. The reason is simply that the observation record length is effectively longer if a new sequence $\zeta^i(k)$ is used for each iteration. Of course, in many practical problems only one observation record is available, and we must use this normal operating record rather than being able to subject the system to special test inputs.

The continuous case results follow directly from those for the discrete case. We desire to minimize

$$J = \mathscr{E}\left\{\theta_0[\mathbf{x}(t_0)] + \theta_f[\mathbf{x}(t_f)] + \int_{t_0}^{t_0} \varphi[\mathbf{x}(t), \mathbf{u}(t), \boldsymbol{\zeta}(t), t]\, dt\right\} \tag{5.2-30}$$

subject to the differential equality constraint

$$\dot{\mathbf{x}} = \mathbf{f}[\mathbf{x}(t), \mathbf{u}(t), \boldsymbol{\zeta}(t), t] \tag{5.2-31}$$

We define the Hamiltonian (a random variable)

$$H = \varphi[\mathbf{x}(t), \mathbf{u}(t), \boldsymbol{\zeta}(t), t] + \boldsymbol{\lambda}^{\mathrm{T}}(t)\, \mathbf{f}[\mathbf{x}(t), \mathbf{u}(t), \boldsymbol{\zeta}(t), t] \tag{5.2-32}$$

The canonic equations and associated two-point boundary value problem for the stochastic maximum principle are

$$\begin{gathered}\frac{\partial H}{\partial \boldsymbol{\lambda}(t)} = \dot{\mathbf{x}}, \qquad \mathscr{E}\left\{\frac{\partial H}{\partial \mathbf{x}(t)} + \dot{\boldsymbol{\lambda}}\right\} = \mathbf{0}, \qquad \mathscr{E}\left\{\frac{\partial H}{\partial \mathbf{u}(t)}\right\} = \mathbf{0} \\ \mathscr{E}\left\{\frac{\partial \theta_0[\mathbf{x}(t_0)]}{\partial \mathbf{x}(t_0)} + \boldsymbol{\lambda}(t_0)\right\} = \mathscr{E}\left\{\frac{\partial \theta_{\mathrm{f}}[\mathbf{x}(t_{\mathrm{f}})]}{\partial \mathbf{x}(t_{\mathrm{f}})} + \boldsymbol{\lambda}(t_{\mathrm{f}})\right\} = \mathbf{0}\end{gathered} \tag{5.2-33}$$

In cases in which the equations from the stochastic maximum principle are impractical to solve, the following stochastic approximation method may be used:

1. Choose an initial $\mathbf{x}^i(t_0)$ and $\mathbf{u}^i(t)$.
2. From the known density $p[\boldsymbol{\zeta}(t)]$, select a typical sequence $\boldsymbol{\zeta}^i(t)$.
3. Solve the system dynamic equation

$$\dot{\mathbf{x}}^i = \mathbf{f}[\mathbf{x}^i(t), \mathbf{u}^i(t), \boldsymbol{\zeta}^i(t), t]$$

forward in time $t_0 \leqslant t \leqslant t_{\mathrm{f}}$.

4. Solve the adjoint equation with its associated terminal condition

$$\dot{\boldsymbol{\lambda}}^i = -\frac{\partial H}{\partial \mathbf{x}^i(t)} = -\frac{\partial \varphi^i}{\partial \mathbf{x}^i(t)} - \frac{\partial \mathbf{f}^{i\mathrm{T}}}{\partial \mathbf{x}^i(t)} \boldsymbol{\lambda}^i(t), \qquad \boldsymbol{\lambda}^i(t_{\mathrm{f}}) = \frac{\partial \theta_{\mathrm{f}}[\mathbf{x}(t_{\mathrm{f}})]}{\partial \mathbf{x}(t_{\mathrm{f}})} \tag{5.2-34}$$

backward in time from t_{f} to t_0.

5. Determine the new control iterate

$$\begin{aligned}\mathbf{u}^{i+1}(t) &= \mathbf{u}^i(t) - K_1{}^i \frac{\partial H^i}{\partial \mathbf{u}^i(t)} \\ &= \mathbf{u}^i(t) - K_1{}^i \left[\frac{\partial \varphi^i}{\partial \mathbf{u}^i(t)} + \frac{\partial \mathbf{f}^{i\mathrm{T}}}{\partial \mathbf{u}^i(t)} \boldsymbol{\lambda}^i(t)\right]\end{aligned} \tag{5.2-35}$$

where $K_{\mathbf{u}}{}^i$ satisfies all stochastic approximation requirements.

6. Determine the new initial condition iterate

$$\mathbf{x}^{i+1}(t_0) = \mathbf{x}^i(t_0) - K_{\mathbf{x}}{}^i \left[\frac{\partial \theta_0[\mathbf{x}^i(t_0)]}{\partial \mathbf{x}^i(t_0)} - \boldsymbol{\lambda}^i(t_0)\right] \tag{5.2-36}$$

7. Repeat the computations [with a new random sequence $\zeta^i(t)$] starting at Step 2.

We have now completed our formal presentation of the stochastic approximation method. Several examples will indicate how this technique may be applied to problems of system identification.

5.3. Sequential Estimation of Linear System Dynamics Using Stochastic Approximation

In this section we will give detailed consideration to identification of the $\boldsymbol{\Phi}$ and $\boldsymbol{\Gamma}$ matrices for the constant coefficient linear discrete system

$$\mathbf{x}(k+1) = \boldsymbol{\Phi}\mathbf{x}(k) + \boldsymbol{\Gamma}w(k) \tag{5.3-1}$$

$$z(k) = \mathbf{H}\mathbf{x}(k) + v(k) = y(k) + v(k) \tag{5.3-2}$$

We will assume the process input to be a scalar such that $\boldsymbol{\Gamma}$ is a column vector $\boldsymbol{\gamma}$. Also the output will consist of only the first state variable such that

$$\mathbf{H} = [1\ 0\ 0\ \cdots\ 0] = \mathbf{h}^{\mathrm{T}}$$

Unforced systems—no measurement noise. It is convenient to assume first that the system is unforced and that there is no measurement noise. The basic reason for restricting the output to be a scalar is that there must be N measurements in the measurement noise free case to identify the system. It is convenient to define the N vector of augmented observations and use Eqs. (5.3-1) and (5.3-2) to obtain

$$\mathscr{y}(N) \triangleq \begin{bmatrix} z(1) \\ z(2) \\ \vdots \\ z(N) \end{bmatrix} = \begin{bmatrix} \mathbf{h}^{\mathrm{T}}\boldsymbol{\Phi} \\ \mathbf{h}^{\mathrm{T}}\boldsymbol{\Phi}^2 \\ \vdots \\ \mathbf{h}^{\mathrm{T}}\boldsymbol{\Phi}^N \end{bmatrix} \mathbf{x}(0) = \mathscr{A}\boldsymbol{\Phi}\mathbf{x}(0) \tag{5.3-3}$$

where

$$\mathscr{A} = \begin{bmatrix} \mathbf{h}^{\mathrm{T}} \\ \mathbf{h}^{\mathrm{T}}\boldsymbol{\Phi} \\ \vdots \\ \mathbf{h}^{\mathrm{T}}\boldsymbol{\Phi}^{N-1} \end{bmatrix} \tag{5.3-4}$$

In a similar fashion, we find

$$\mathscr{y}(N+1) \triangleq \begin{bmatrix} y(2) \\ y(3) \\ \vdots \\ y(N+1) \end{bmatrix} = \begin{bmatrix} \mathbf{h}^{\mathrm{T}}\mathbf{\Phi}^2 \\ \mathbf{h}^{\mathrm{T}}\mathbf{\Phi}^3 \\ \vdots \\ \mathbf{h}^{\mathrm{T}}\mathbf{\Phi}^{N+1} \end{bmatrix} \mathbf{x}(0) = \mathscr{A}\mathbf{\Phi}\mathbf{x}(1) = \mathscr{A}\mathbf{\Phi}^2\mathbf{x}(0)$$

and, in general,

$$\mathscr{y}(N+k) = \mathscr{A}\mathbf{\Phi}\mathbf{x}(k) \tag{5.3-5}$$

The augmented matrix

$$\mathscr{Y}(2N-1) \triangleq [\mathscr{y}(N)\ \mathscr{y}(N+1)\ \mathscr{y}(N+2)\ \cdots\ \mathscr{y}(2N-1)]$$

$$= \begin{bmatrix} y(1) & y(2) & \cdots & y(N) \\ y(2) & y(3) & & y(N+1) \\ \vdots & \vdots & & \vdots \\ y(N) & y(N+1) & & y(2N-1) \end{bmatrix} \tag{5.3-6}$$

is now formed. From Eqs. (5.3-3), (5.3-5), and the obvious extensions of these equations

$$\mathscr{Y}(2N-1) = \mathscr{A}\mathbf{\Phi}\mathscr{B} \tag{5.3-7}$$

where

$$\mathscr{B} = [\mathbf{x}(0)\ \ \mathbf{\Phi}\mathbf{x}(0)\ \ \mathbf{\Phi}^2\mathbf{x}(0)\ \cdots\ \mathbf{\Phi}^{N-1}\mathbf{x}(0)] \tag{5.3-8}$$

The matrix $\mathscr{A}$ is often spoken of as the *observability matrix*. This matrix must be nonsingular if the state vector is to be recovered from the observation sequence. The matrix $\mathscr{B}$ is called the *identifiability matrix* and must be nonsingular in order for system identification to be possible (Lee, 1964).

From Eq. (5.3-7), we can determine (identify) the system transition matrix

$$\hat{\mathbf{\Phi}} = \mathbf{\Phi} = \mathscr{A}^{-1}\mathscr{Y}(2N-1)\,\mathscr{B}^{-1} \tag{5.3-9}$$

which is *exact*, since no noise inputs were assumed. We did however use the assumption that the transition matrix was an $N \times N$ matrix. It certainly appears that we are in trouble with this result, since $\mathscr{A}$ and $\mathscr{B}$ do, in fact, depend upon $\mathbf{\Phi}$. This difficulty will soon be resolved We may repeat the argument leading to Eq. (5.3-7) using a different augmented data matrix $\mathscr{Y}(2N)$ and obtain the result

$$\mathscr{Y}(2N) = \mathscr{A}\mathbf{\Phi}^2\mathscr{B} = \mathscr{A}\mathbf{\Phi}\mathscr{A}^{-1}\mathscr{Y}(2N-1) \tag{5.3-10}$$

by use of Eq. (5.3-9). It is convenient to define

$$\mathbf{T} = \mathscr{A}\boldsymbol{\Phi}\mathscr{A}^{-1} \tag{5.3-11}$$

such that Eq. (5.3-10) becomes

$$\mathscr{Y}(2N) = \mathbf{T}\mathscr{Y}(2N-1) \tag{5.3-12}$$

From Eqs. (5.3-3), (5.3-5), and (5.3-9), we have

$$\mathscr{y}(N+1) = \boldsymbol{\Upsilon}\mathscr{y}(N) \tag{5.3-13}$$

Since $y(N) = x_1(N)$ and we also have

$$y(1) = \mathbf{h}^{\mathrm{T}}\mathscr{y}(N) \tag{5.3-14}$$

which can also be written from Eq. (5.3-3) as

$$y(1) = \mathbf{h}^{\mathrm{T}}\mathscr{y}(N) = \mathbf{h}^{\mathrm{T}}\mathscr{A}\mathbf{x}(1)$$

The relation $\mathbf{h}^{\mathrm{T}} = \mathbf{h}^{\mathrm{T}}\mathscr{A}$ is verified.

Equations (5.3-13) and (5.3-14) are linear relations equivalent to Eqs. (5.3-1) and (5.3-2). $\boldsymbol{\Upsilon}$ is related to $\boldsymbol{\Phi}$ by the transformation of Eq. (5.3-12). Thus once $\boldsymbol{\Upsilon}$ is determined from Eq. (5.3-12), $\boldsymbol{\Phi}$ follows from Eq. (5.3-11)

$$\boldsymbol{\Upsilon} = \mathscr{Y}(2N)\,\mathscr{Y}^{-1}(2N-1), \qquad \boldsymbol{\Phi} = \mathscr{A}^{-1}\boldsymbol{\Upsilon}\mathscr{A} \tag{5.3-15}$$

It is important to note that, regardless of $\boldsymbol{\Phi}$, $\boldsymbol{\Upsilon}$ is of the form

$$\boldsymbol{\Upsilon} = \left[\begin{array}{c|ccc} 0 & & & \\ 0 & & & \\ 0 & & \mathbf{I} & \\ 0 & & & \\ \hline -a_1 & -a_2 & \cdots & -a_N \end{array}\right] \tag{5.3-16}$$

This fact follows directly from the definition of $\boldsymbol{\Upsilon}$. To show this it is convenient to define the inverse of a partitioned matrix

$$\left[\begin{array}{c} \mathbf{a}^{\mathrm{T}} \\ \hline \mathbf{B} \end{array}\right]^{-1} = \left[\begin{array}{c|c} \mathbf{a}_{-1} & \mathbf{B}_{-1} \end{array}\right] \tag{5.3-17}$$

where $\mathbf{a}$ is a column vector. Since the product of a matrix by its inverse yields the identity matrix, we have

$$\begin{bmatrix} \mathbf{a}^{\mathrm{T}} \\ \hline \mathbf{B} \end{bmatrix} \left[\mathbf{a}_{-1} \;\middle|\; \mathbf{B}_{-1} \right] = \mathbf{I}$$

or

$$\mathbf{a}^{\mathrm{T}}\mathbf{a}_{-1} = 1, \qquad \mathbf{B}\mathbf{B}_{-1} = \mathbf{I}, \qquad \mathbf{a}^{\mathrm{T}}\mathbf{B}_{-1} = \mathbf{0}, \qquad \mathbf{B}\mathbf{a}_{-1} = \mathbf{0}$$

so that we see, from Eq. (5.3-4),

$$\mathscr{A}\boldsymbol{\Phi}\mathscr{A}^{-1} = \begin{bmatrix} \mathbf{h}^{\mathrm{T}} \\ \mathbf{h}^{\mathrm{T}}\boldsymbol{\Phi} \\ \mathbf{h}^{\mathrm{T}}\boldsymbol{\Phi}^{2} \\ \vdots \\ \mathbf{h}^{\mathrm{T}}\boldsymbol{\Phi}^{N-2} \\ \hline \mathbf{h}^{\mathrm{T}}\boldsymbol{\Phi}^{N-1} \end{bmatrix} \begin{bmatrix} \mathbf{h}^{\mathrm{T}}\boldsymbol{\Phi}^{-1} \\ \hline \mathbf{h}^{\mathrm{T}} \\ \mathbf{h}^{\mathrm{T}}\boldsymbol{\Phi} \\ \vdots \\ \mathbf{h}^{\mathrm{T}}\boldsymbol{\Phi}^{N-2} \end{bmatrix}^{-1} = \begin{bmatrix} \mathbf{B} \\ \hline \mathbf{c}^{\mathrm{T}} \end{bmatrix} \begin{bmatrix} \mathbf{a}^{\mathrm{T}} \\ \hline \mathbf{B} \end{bmatrix}^{-1}$$

$$= \begin{bmatrix} \mathbf{B} \\ \hline \mathbf{c}^{\mathrm{T}} \end{bmatrix} \left[\mathbf{a}_{-1} \;\middle|\; \mathbf{B}_{-1} \right] = \left[\begin{array}{c|c} 0 & \mathbf{I} \\ \hline \mathbf{c}^{\mathrm{T}}\mathbf{a}_{-1} & \mathbf{c}^{\mathrm{T}}\mathbf{B}_{-1} \end{array} \right] \tag{5.3-18}$$

which is of the form specified in Eq. (5.3-16). Since it is easier computationally to identify $\boldsymbol{\Upsilon}$ than $\boldsymbol{\Phi}$, we will hence forth consider that identification of $\boldsymbol{\Upsilon}$ is what is required. In fact, we see that there is no unique way of specifying $\boldsymbol{\Phi}$ but only a unique way of determining $\boldsymbol{\Upsilon}$. There are simply not N^2 unknown parameters associated with the unforced response of a linear system—only N parameters— and thus it was really unrealistic to expect to identify the entire $\boldsymbol{\Phi}$ matrix.

We have indicated that it was necessary to assume that the system was of order N to obtain the preceding identification algorithms. If the system is of undetermined order, we can readily use the observability requirement to determine the order of the system. If the assumed order of the system is less than or equal to the actual order of the system, the observability matrix $\mathscr{A}$ will always have an inverse. However, if the assumed order of the system is greater than the actual order, the observability matrix will be singular [an unobservable system is always reducible in order (Sage, 1968)]. Thus we may easily discern the order of the system (in this simple noise-free case) from the observability requirement.

Forced systems—no measurement noise—no numerator dynamics. We now consider the system

$$\mathbf{x}(k+1) = \boldsymbol{\Upsilon}\mathbf{x}(k) + \boldsymbol{\Gamma}w(k) \tag{5.3-19}$$

$$y(k) = \mathbf{h}^{\mathrm{T}}\mathbf{x}(k) \tag{5.3-20}$$

in which

$$\boldsymbol{\Upsilon} = \left[\begin{array}{c|c} 0 & \mathbf{I} \\ \hline \multicolumn{2}{c}{\mathbf{a}^{\mathrm{T}}} \end{array}\right], \qquad \boldsymbol{\Gamma} = \begin{bmatrix} \Gamma_1 \\ \Gamma_2 \\ \vdots \\ \Gamma_N \end{bmatrix}, \qquad \mathbf{h} = \begin{bmatrix} 1 \\ 0 \\ \vdots \\ 0 \end{bmatrix} \tag{5.3-21}$$

We shall assume that $w(k)$ is zero mean white noise

$$\mathscr{E}\{w(k)\} = 0, \qquad \operatorname{cov}\{w(k), w(j)\} = V_w \delta_{\mathrm{K}}(k-j)$$

The state variable equation may be written as the scalar difference equation

$$\begin{aligned} y(k) &+ a_1 y(k-1) + a_2 y(k-2) + \cdots \\ &= b_1 w(k-1) + b_2 w(k-2) + \cdots + b_N w(k-N) \end{aligned} \tag{5.3-22}$$

where

$$\mathbf{a} = \begin{bmatrix} a_1 \\ a_2 \\ \vdots \\ b_N \end{bmatrix}, \qquad \mathbf{b} = \begin{bmatrix} b_1 \\ b_2 \\ \vdots \\ b_N \end{bmatrix}, \qquad \mathbf{b} = \begin{bmatrix} 1 & & & \\ a_1 & 1 & & 0 \\ a_2 & a_1 & 1 & \\ \vdots & \vdots & \vdots & \\ a_{N-1} & a_{N-2} & & 1 \end{bmatrix} \boldsymbol{\Gamma} \tag{5.3-23}$$

This difference equation may then be written as

$$\mathbf{y}(k) = -\boldsymbol{y}^{\mathrm{T}}(k-1)\mathbf{a} + \boldsymbol{w}^{\mathrm{T}}(k-1)\mathbf{b} \tag{5.3-24}$$

where

$$\boldsymbol{y}(k-1) = \begin{bmatrix} y(k-N) \\ \vdots \\ y(k-2) \\ y(k-1) \end{bmatrix}, \qquad \boldsymbol{w}(k-1) = \begin{bmatrix} w(k-N) \\ \vdots \\ w(k-2) \\ w(k-1) \end{bmatrix} \tag{5.3-25}$$

For convenience let us assume that all b_i are zero except b_1, which is known to be unity. Equation (5.3-24) becomes

$$y(k) = -\mathbf{y}^{\mathrm{T}}(k-1)\mathbf{a} + w(k-1) \tag{5.3-26}$$

We thus desire to obtain the estimate $\hat{\mathbf{a}}$ which gives the root of the regression equation in Eq. (5.3-26). This yields the stochastic approximation algorithm to minimize

$$J = \mathscr{E}\{[y(k) + \mathbf{y}^{\mathrm{T}}(k-1)\mathbf{a}]^2\} \tag{5.3-27}$$

which is

$$\hat{\mathbf{a}}(k+1) = \hat{\mathbf{a}}(k) - K(k)\,\mathbf{y}(k)[y(k+1) + \mathbf{y}^{\mathrm{T}}(k)\,\hat{\mathbf{a}}(k)] \tag{5.3-28}$$

where $K(k)$ is chosen to satisfy our previously developed stochastic approximation requirements for convergence. Unfortunately, there is no insight gained into an optimum $K(k)$ by the stochastic approximation method. Let us look at the least squares identification criterion in order to obtain an alternate approach.

Motivated by Eq. (5.3-26), we seek an estimate for $\mathbf{a}$ to minimize the cost function

$$J = \sum_{i=N+1}^{k} \| w(i-1)\|^2 = \sum_{i=N+1}^{k} \| y(i) + \mathbf{y}^{\mathrm{T}}(i-1)\mathbf{a} \|^2 \tag{5.3-29}$$

where the lower index on the summation starts at stage $N+1$ because the first complete identification equation can only be obtained after the $(N+1)$th observation. If we take N additional measurements, we have

$$J = \sum_{i=N+1}^{2N} \| y(i) + \mathbf{y}^{\mathrm{T}}(i-1)\mathbf{a} \|^2 = \left\| \begin{bmatrix} y(N+1) \\ y(N+2) \\ \vdots \\ y(2N) \end{bmatrix} + \begin{bmatrix} \mathbf{y}^{\mathrm{T}}(N) \\ \mathbf{y}^{\mathrm{T}}(N+1) \\ \vdots \\ \mathbf{y}^{\mathrm{T}}(2N-1) \end{bmatrix} \mathbf{a} \right\|^2$$

$$= \| \mathbf{y}(2N) + \mathscr{Y}(2N-1)\mathbf{a} \|^2 \tag{5.3-30}$$

which is dimensionally sufficient to obtain an estimate of the parameter vector to be identified. We obtain

$$\hat{\mathbf{a}}(2N) = -[\mathscr{Y}^{\mathrm{T}}(2N-1)\,\mathscr{Y}(2N-1)]^{-1}\,\mathscr{Y}(2N-1)\,\mathbf{y}(2N) \tag{5.3-31}$$

Since $\mathscr{Y}$ is symmetric, as defined in Eq. (5.3-6), we see that we could simplify this result to

$$\hat{\mathbf{a}}(2N) = -[\mathscr{Y}(2N-1)]^{-1}\,\mathscr{y}(2N) \qquad (5.3\text{-}32)$$

where we use $\hat{\mathbf{a}}(2N)$ to indicate that $2N$ observations have been taken to yield $\hat{\mathbf{a}}$. In order to facilitate development of a sequential identification scheme,[2] this simplification will not be made. We define

$$\mathscr{P}(2N) = [\mathscr{Y}^{\mathrm{T}}(2N-1)\,\mathscr{Y}(2N-1)]^{-1} \qquad (5.3\text{-}33)$$

We now add a new measurement $y(2N+1)$ and reminimize the cost function of Eq. (5.3-30). We easily obtain

$$\hat{\mathbf{a}}(2N+1) = \mathscr{P}(2N+1)\left[\mathscr{y}(2N) \;\vdots\; \mathscr{Y}^{\mathrm{T}}(2N+1)\right]\begin{bmatrix} \mathscr{y}(2N) \\ \mathscr{y}(2N+1)\end{bmatrix} \qquad (5.3\text{-}34)$$

where

$$\mathscr{P}(2N+1) = \left\{[\mathscr{Y}^{\mathrm{T}}(2N-1) \;\vdots\; \mathscr{y}(2N)]^{-1}\begin{bmatrix} \mathscr{Y}(2N-1) \\ \hline \mathscr{y}^{\mathrm{T}}(2N)\end{bmatrix}\right\}^{-1}$$

$$= [\mathscr{P}(2N) + \mathscr{y}(2N)\,\mathscr{y}^{\mathrm{T}}(2N)]^{-1} \qquad (5.3\text{-}35)$$

To put this relation into a form more suitable for computation, we use the matrix inversion lemma, such that we have

$$\mathscr{P}(2N+1) = \mathscr{P}(2N) - \mathscr{P}(2N)\,\mathscr{y}(2N) \times [\mathscr{y}^{\mathrm{T}}(2N)\,\mathscr{P}(2N)\,\mathscr{y}(2N) + 1]^{-1}\,\mathscr{y}^{\mathrm{T}}(2N)\,\mathscr{P}(2N) \qquad (5.3\text{-}36)$$

which requires the inverse of only a scalar term. Use of this relation in Eq. (5.3-33) yields, with the definition Eq. (5.3-31),

$$\hat{\mathbf{a}}(2N+1) = \hat{\mathbf{a}}(2N) - \mathscr{P}(2N)\,\mathscr{y}(2N)[\mathscr{y}^{\mathrm{T}}(2N)\,\mathscr{P}(2N)\,\mathscr{y}(2N) + 1]^{-1} \times [y(2N+1) + \mathscr{y}^{\mathrm{T}}(2N)\,\hat{\mathbf{a}}(2N)] \qquad (5.3\text{-}37)$$

Now this procedure may be repeated for $2N+2$, $2N+3$,..., so we see that Eq. (5.3-37) may be rewritten by substituting k for $2N$. We

[2] A more systematic development utilizing invariant imbedding will be presented in Chap. 7.

see that the stochastic approximation algorithm of Eq. (5.3-28) is equivalent to the least squares curve fit result of Eq. (5.3-37) if we replace

$$K(k) = \frac{\mathscr{P}(2k)}{1 + \mathscr{y}^{\mathrm{T}}(k)\,\mathscr{P}(k)\,\mathscr{y}(k)} \tag{5.3-38}$$

We have shown in Sect. 5.1 that $\mathscr{P}(k)$ behaves as $1/k$, and we thus conclude that the least squares method and the stochastic approximation method lead to similar results.

In our development, we have assumed that $2N$ observations are taken before Eqs. (5.3-36) and (5.3-37) are implemented. This is needed such that we obtain the initial conditions

$$\hat{\mathbf{a}}(2N) = -[\mathscr{Y}(2N-1)]^{-1}\,\mathscr{y}(2N), \qquad \mathscr{P}(2N) = [\mathscr{Y}(2N-1)]^{-2} \tag{5.3-39}$$

needed to start the computations. In many practical situations, more or less arbitrary values of the initial $\hat{\mathbf{a}}$ and $\mathscr{P}$ may be used to start the computation.

Forced systems—no numerator dynamics—measurement noise. The presence of measurement noise considerably complicates the stochastic approximation identification algorithms. The model

$$\mathbf{x}(k+1) = \begin{bmatrix} 0 & \mathbf{I} \\ \hline \multicolumn{2}{c}{\mathbf{a}^{\mathrm{T}}} \end{bmatrix} \mathbf{x}(k) + \begin{bmatrix} 0 \\ \vdots \\ 1 \end{bmatrix} \mathbf{w}(k)$$

$$y(k) = [1 \;\; 0 \;\; \cdots \;\; 0]\,\mathbf{x}(k) \tag{5.3-40}$$

$$z(k) = y(k) + v(k)$$

where $v(k)$ is zero mean white noise, and where $z(k)$ represents the observation, becomes the scalar difference equation

$$\begin{aligned} z(k) &+ a_1 z(k-1) + a_2 z(k-2) + \cdots + a_N z(k-N) \\ &= b_1 w(k-1) + b_2 w(k-2) + \cdots + b_N w(k-N) + v(k) \\ &\quad + a_1 v(k-1) + \cdots + a_N v(k-N) \end{aligned} \tag{5.3-41}$$

which is rewritten as

$$z(k) = -\mathscr{z}^{\mathrm{T}}(k-1)\mathbf{a} + w(k-1) + \mathscr{v}^{\mathrm{T}}(k-1)\mathbf{a} \tag{5.3-42}$$

where

$$\mathscr{z}(k-1) = \begin{bmatrix} z(k-N) \\ \vdots \\ z(k-2) \\ z(k-1) \end{bmatrix} \qquad \mathscr{v}(k-1) = \begin{bmatrix} v(k-N) \\ \vdots \\ v(k-2) \\ v(k-1) \end{bmatrix} \tag{5.3-43}$$

Unfortunately, Eq. (5.3-42) is considerably more complex than Eq. (5.3-26) in that the noise term $\mathscr{V}(k-1)\mathbf{a}$ is correlated from sample to sample. Thus the expression

$$\gamma(k-1) = w(k-1) + \mathscr{v}^{\mathrm{T}}(k-1)\mathbf{a} \tag{5.3-44}$$

is not white since

$$\operatorname{cov}\{\gamma(k-1), \gamma(j-1)\} \begin{cases} \neq 0, & |k-j| \leqslant N \\ = 0, & |k-j| > N \end{cases}$$

Therefore we must be cautious in trying to minimize

$$J = \mathscr{E}\{\gamma^2(k-1)\} \tag{5.3-45}$$

$$\gamma(k-1) = z(k) + \mathscr{z}^{\mathrm{T}}(k-1)\mathbf{a} \tag{5.3-46}$$

because of the correlation in $\gamma(k-1)$. We may attempt to do this using the stochastic approximation algorithms by restricting k such that $k = D, N+1, 2N+2$, etc., such that

$$\operatorname{cov}\{\gamma(k-1), \gamma(j-1)\} = 0, \qquad j \geqslant (N+1)k$$

Minimization of Eq. (5.3-45) leads, of course, to the same result as minimization of Eq. (5.3-27). The measurement noise is, in effect, ignored. We obtain

$$\hat{\mathbf{a}}^* = -[\mathscr{E}\{\mathscr{z}(k-1)\,\mathscr{z}^{\mathrm{T}}(k-1)\}]^{-1}\,\mathscr{E}\{\mathscr{z}(k-1)\,z(k)\} \tag{5.3-47}$$

But substitution of Eq. (5.3-42) yields

$$\hat{\mathbf{a}}^* = \mathbf{a} - [\mathscr{E}\{\mathscr{z}(k-1)\,\mathscr{z}^{\mathrm{T}}(k-1)\}]^{-1}\,V_v\mathbf{a} \tag{5.3-48}$$

so we see that optimization by means of Eq. (5.3-45) will of necessity lead to a biased result. The bias term, if subtracted from the stochastic approximation algorithms resulting from use of Eq. (5.3-45), will lead to an unbiased result. In order to have a meaningful stochastic approximation algorithm, it is still necessary that the update of the $\mathbf{a}$

estimate be spaced $N + 1$ samples apart. The stochastic approximation algorithm is therefore

$$\hat{a}(k + N + 1) = \hat{a}(k) - K(k/N + 1) \times \{z(k + N)[z(k + N + 1) + z^{\mathrm{T}}(k + N)\,\hat{a}(k)] - V_v\hat{\mathbf{a}}(k)\} \tag{5.3-49}$$

where $k = 0, N + 1, 2N + 2, \ldots$. This algorithm is unbiased as may readily be verified.

$K(k/N + 1)$ is a gain selected to satisfy the usual stochastic approximation requirements. We note that, in the formulation, it is necessary to know the measurement noise variance but not (directly) the plant noise variance. By analogy to the results of the previous subsection, it would appear that an appropriate and near optimum K is

$$K\left(\frac{k}{N + 1}\right) = \frac{\mathscr{P}(k + N + 1)}{1 + z^{\mathrm{T}}(k + N)\,\mathscr{P}(k)\,z^{\mathrm{T}}(k + N)} \tag{5.3-50}$$

where

$$\mathscr{P}(k + N + 1) = \mathscr{P}(k) - \mathscr{P}(k)\,z(k + N) \times [z^{\mathrm{T}}(k + N)\,\mathscr{P}(k)\,z^{\mathrm{T}}(k + N) + 1]^{-1}\,z^{\mathrm{T}}(k + N)\,\mathscr{P}(k) \tag{5.3-51}$$

Initial conditions for $\hat{\mathbf{a}}$ and $\mathscr{P}$ are not critical and, if desired, may be selected as in the previous subsection.

Example 5.3-1. The simplest example which can be considered here is that of identifying $\boldsymbol{\Phi}$ for the scalar system

$$x(k + 1) = \Phi x(k) + w(k)$$

$$y(k) = x(k)$$

$$z(k) = x(k) + v(k)$$

where $w(k)$ and $v(k)$ are zero mean and white random noise terms. The system is of first order, so $N = 1$. The specific stochastic approximation algorithms corresponding to Eqs. (5.3-49)–(5.3-51) become

$$\hat{\Phi}(k + 2) = \hat{\Phi}(k) + K(k/2)\{z(k + 1)[z(k + 2) - z(k + 1)\,\hat{\Phi}(k)] + V_v\hat{\Phi}(k)\}$$

where

$$K(k/2) = \frac{\mathscr{P}(k+2)}{1 + z^2(k+1)\,\mathscr{P}(k)}, \qquad \mathscr{P}(k+2) = \frac{\mathscr{P}(k)}{1 + z^2(k+1)\,\mathscr{P}(k)}$$

for $k = 0, 2, 4, 6, \ldots$. Figures 5.3-1 and 5.3-2 illustrate convergence

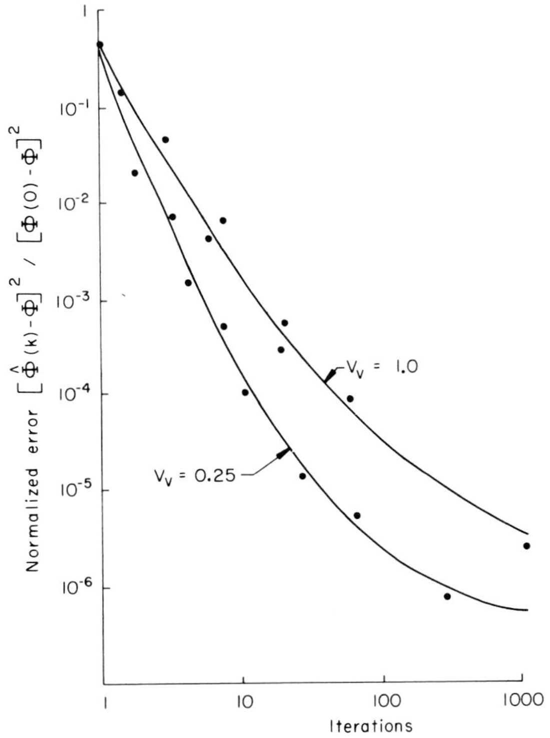

FIG. 5.3-1. Sequential identification of Φ_1, Example 5.3-1.

of the algorithms for the case in which plant and measurement noises are Gaussian. In each case, the true value of Φ is 0.8. The assumed initial value of Φ is 0; the measurement noise variance is 0.25 and 1.0. The initial bahavior of the curves is quite dependent upon the value of $\mathscr{P}(0)$ used. After a few iterations (3–5), the results are essentially independent of the value of $\mathscr{P}(0)$ used to start the computation. It is interesting to note that the cost function is decreasing slightly, even after 1000 iterations. The convergence of Φ to (close to) the true value is reasonably rapid, however.

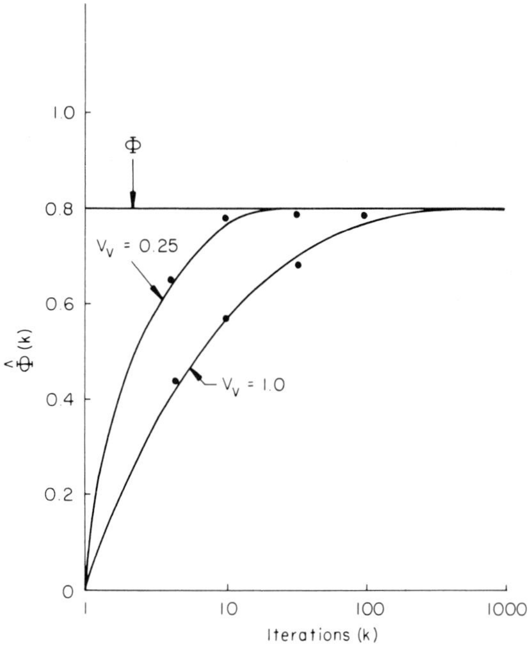

FIG. 5.3-2. Sequential identification of Φ, Example 5.3-1.

Forced systems—numerator dynamics and measurement noise. We now conclude our development by presenting algorithms for identification by stochastic approximation of the parameters **a** and **b** forming the model for the system

$$\mathbf{x}(k+1) = \boldsymbol{\Upsilon}\mathbf{x}(k) + \boldsymbol{\Gamma}w(k) + \mathbf{d}u(k) \tag{5.3-52}$$

$$z(k) = \mathbf{h}^{\mathrm{T}}\mathbf{x}(k) + v(k) \tag{5.3-53}$$

where **d** is an arbitrary but known N-dimensional vector. It is assumed that $u(k)$ is a sequence of independent unobserved zero mean white random variables. $w(k)$ is a sequence of zero mean white random variables. $v(t)$ is the zero mean white measurement noise. $u(k)$, $w(k)$, and $v(k)$ are assumed to be mutually independent.

The response of Eq. (5.3-52) is

$$\mathbf{x}(k) = \sum_{i=0}^{\infty} \boldsymbol{\psi}[\boldsymbol{\Gamma}w(k-i-1) + \mathbf{d}u(k-i-1)] \tag{5.3-54}$$

We define the $2N$-dimensional vectors

$$\boldsymbol{\theta} = \begin{bmatrix} \mathbf{h}^{\mathrm{T}}\boldsymbol{\Upsilon}^{2N-1}\boldsymbol{\Gamma} \\ \vdots \\ \mathbf{h}^{\mathrm{T}}\boldsymbol{\Upsilon}^{2}\boldsymbol{\Gamma} \\ \mathbf{h}^{\mathrm{T}}\boldsymbol{\Upsilon}\boldsymbol{\Gamma} \\ \mathbf{h}^{\mathrm{T}}\boldsymbol{\Gamma} \end{bmatrix}, \qquad \boldsymbol{\vartheta} = \begin{bmatrix} \mathbf{h}^{\mathrm{T}}\boldsymbol{\Upsilon}^{2N-1}\mathbf{d} \\ \vdots \\ \mathbf{h}^{\mathrm{T}}\boldsymbol{\Upsilon}^{2}\mathbf{d} \\ \mathbf{h}^{\mathrm{T}}\boldsymbol{\Upsilon}\mathbf{d} \\ \mathbf{h}^{\mathrm{T}}\mathbf{d} \end{bmatrix}, \tag{5.3-55}$$

$$\mathscr{u}(k) = \begin{bmatrix} u(k-2N+1) \\ \vdots \\ u(k-1) \\ u(k) \end{bmatrix}, \qquad \mathscr{w}(k) = \begin{bmatrix} w(k-2N+1) \\ \vdots \\ w(k-1) \\ w(k) \end{bmatrix}$$

It is clear that $\boldsymbol{\theta}$ represents the impulse response of the system, and thus we are reconsidering the problem considered in Sect. 2–2. We may now write the observation as

$$z(k) = \mathscr{w}^{\mathrm{T}}(k)\boldsymbol{\theta} + \mathscr{w}^{\mathrm{T}}(k-1)\boldsymbol{\vartheta} + \gamma(k) \tag{5.3-56}$$

where

$$\gamma(k) = v(k) + \mathbf{h}^{\mathrm{T}} \sum_{i=2N}^{\infty} \boldsymbol{\Upsilon}^{i}[\boldsymbol{\Gamma}w(k-i-1) + \mathbf{d}u(k-i+1)] \tag{5.3-57}$$

We define the estimate of the parameter vector as the vector which minimizes the mean square error criterion

$$J = \mathscr{E}\{[z(k) - \mathscr{w}^{\mathrm{T}}(k)\boldsymbol{\theta}]^2\} \tag{5.3-58}$$

The resulting estimator of $\boldsymbol{\theta}$ is unbiased as may be verified. The stochastic approximation algorithms for minimization of the foregoing are [assuming that $w(k)$ is directly observed]

$$\hat{\boldsymbol{\theta}}(k+2N+1) = \hat{\boldsymbol{\theta}}(k) + K(k/2N+1)\,\mathscr{w}(k+2N) \times [z(k+2N+1) - \mathscr{w}^{\mathrm{T}}(k+2N)\,\hat{\boldsymbol{\theta}}(k)] \tag{5.3-59}$$

for $k = 0, 2N+1, 4N+2$, etc. Samples are spaced $2N+1$ apart ($\mathbf{x}$ is an N vector) because the correlation in $\gamma(k)$ extends over $2N$

samples as indicated in Eq. (5.3-57). If only additive noise corrupted versions of $w(k)$ are available,

$$m(k) = w(k) + \nu(k) \tag{5.3-60}$$

where $\nu(k)$ is zero mean and white, it is possible to repeat the arguments leading to Eq. (5.3-59) and incorporate the bias term of the previous subsection to obtain

$$\begin{aligned}\hat{\boldsymbol{\theta}}(k+2N+1) = \hat{\boldsymbol{\theta}}(k) + K(k/2N+1)\{\boldsymbol{m}(k+2N)\\ \times[z(k+2N+1) - \boldsymbol{m}^{\mathrm{T}}(k+2N)\,\hat{\boldsymbol{\theta}}(k)] + V_v\hat{\boldsymbol{\theta}}(k)\}\end{aligned} \tag{5.3-61}$$

where

$$\boldsymbol{m}^{\mathrm{T}}(k) = [m(k-2N+1) \cdots m(k-1)\; m(k)]^{\mathrm{T}} \tag{5.3-62}$$

It should be noted that it is necessary to have input observation in order to implement these identification algorithms for the determination of the impulse response $\boldsymbol{\theta}(k)$ or the control distribution term $\boldsymbol{\Gamma}$. The result of the identification is the optimum estimate of the parameter vector $\hat{\boldsymbol{\theta}}$. From the first equation in Eq. (5.3-55), we have

$$\hat{\boldsymbol{\Gamma}} = \begin{bmatrix} \hat{\theta}_N \\ \hat{\theta}_{N-1} \\ \vdots \\ \hat{\theta}_1 \end{bmatrix}, \qquad \hat{\mathbf{a}} = \begin{bmatrix} \hat{\theta}_N & \hat{\theta}_{N+1} & \cdots & \hat{\theta}_{2N-1} \\ \hat{\theta}_{N-1} & \hat{\theta}_N & & \vdots \\ \vdots & \vdots & & \hat{\theta}_{N+1} \\ \hat{\theta}_1 & \hat{\theta}_2 & & \hat{\theta}_N \end{bmatrix}^{-1} \begin{bmatrix} \hat{\theta}_{2N} \\ \hat{\theta}_{2N-1} \\ \vdots \\ \hat{\theta}_{N+1} \end{bmatrix}$$

It is possible to develop a variety of similar stochastic approximation algorithms to solve the problem posed in this subsection. In particular, the reader is referred to the excellent work of Saridis and Stein (1968) and Holmes (1969) for a discussion of other slightly different approaches to the stochastic approximation identification of linear systems as well as results of some interesting computational studies.

5.4. Summary

In this chapter we have presented a heuristic approach to the subject of stochastic approximation. Fixed interval stochastic approximation results in algorithms very similar to those which we obtained in the last chapter concerning gradient methods. The real time or sequential stochastic approximation algorithms are very similar in form to those which result from linear filter theory for linear problems or from the invariant imbedding sequential solution to two-point boundary value problems, as we will see in Chap. 7.

6

QUASILINEARIZATION

6.1. Introduction

In the two preceding chapters, direct computational methods for system identification have been developed. Here we will examine an indirect computational method known as quasilinearization for solving system identification problems; the reader will recall that indirect methods are characterized as methods which attack the two-point boundary value problems which arise from optimization theory. The quasilinearization technique, which is often referred to as a generalized Newton–Raphson technique, owes its primary development to Bellman and Kalaba (1965). Detchmendy and Sridhar (1965) and Kumar and Sridhar (1964) made some of the early applications of the theory to system identification; the work of Sage and Eisenberg (1965) considers some extensions of the previous work especially with regard to certain system modeling problems. The discrete form of quasilinearization has received somewhat less treatment, although Henrici (1962) has developed convergence proofs for certain classes of problems. Sage and Burt (1965) and Sage and Smith (1966) have examined the application of discrete quasilinearization to system identification problems.

In this chapter, we will develop the methods of quasilinearization for both continuous and discrete models. The continuous case will be considered first in Sect. 6.2, while the discrete case will follow in Sect. 6.3. Particular attention will be given to the development of algorithms for solving the two-point boundary value problem formulated in Chap. 3. It will also be shown how quasilinearization may be used as a direct computational algorithm for solving certain classes of identification problems.

The quasilinearization algorithm is an iterative approach which often requires a very good initialization trajectory in order to converge. One method which can be used to initialize the quasilinearization is known as *differential approximation* and is discussed in Sect. 6.4. Unfortunately, the method is not too useful because of some rather severe restrictions; nevertheless, it can on occasion provide an effective technique for beginning the quasilinearization algorithm.

6.2. Continuous Systems

Let us consider the solution of the following multipoint boundary value problem (MPBVP). We wish to obtain the trajectory $\boldsymbol{\gamma}(t)$, where $t \in [t_0, t_f]$ for the N-dimensional nonlinear time-varying system

$$\dot{\boldsymbol{\gamma}}(t) = \boldsymbol{\Gamma}[\boldsymbol{\gamma}(t), t] \tag{6.2-1}$$

which will satisfy (in some sense) the set of linear boundary conditions

$$\mathbf{C}(t_j)\, \boldsymbol{\gamma}(t_j) = \mathbf{b}(t_j), \qquad j = 1, 2,..., m \tag{6.2-2}$$

where $t_j \in [t_0, t_f]$. For simplicity, we will assume that the times t_j are ordered so that $t_j < t_k$, if $j < k$. For the moment, we will not be concerned about the compatibility of the boundary condition of Eq. (6.2-2) and the differential equation (6.2-1). It should be obvious that we may not select the boundary conditions arbitrarily; we will assume that the boundary conditions have been properly selected so that a solution to Eq. (6.2-1) does exist which will satisfy the boundary conditions. We will return to a discussion of the compatibility of Eqs. (6.2-1) and (6.2-2).

If N independent boundary values were given at any single instant of time, then the problem would be trivial, since Eq. (6.2-1) could be integrated forward and/or backward from that point to obtain the desired solution. The problem which is of interest here is the case where the independent boundary values are distributed in time so that one does not know all of them at any one instant. For example, a case which is of particular significance is when the N values are distributed so that $N/2$ are given at t_0 and the remaining $N/2$ are given at t_f. This is the two-point boundary value problem (TPBVP) upon which we will elaborate later.

When the boundary values are distributed in time, it is no longer possible to solve Eq. (6.2-1) in any but the most trivial case, which occurs where the differential equation is linear, and we must normally resort to iterative techniques in order to obtain a solution which satisfies the boundary conditions. Let us assume that we have a trajectory $\boldsymbol{\gamma}^i(t)$ which does not satisfy either or both of Eqs. (6.2-1) and (6.2-2) but does approximate a solution of both. We will discuss how initial trajectories may be selected in the sequel. If we expand $\boldsymbol{\Gamma}[\boldsymbol{\gamma}(t), t]$ in a Taylor series about $\boldsymbol{\gamma}^i(t)$, then Eq. (6.2-1) becomes

$$\dot{\boldsymbol{\gamma}}^{i+1}(t) = \boldsymbol{\Gamma}[\boldsymbol{\gamma}^i(t), t] + \frac{\partial \boldsymbol{\Gamma}[\boldsymbol{\gamma}^i(t), t]}{\partial \boldsymbol{\gamma}^i(t)} [\boldsymbol{\gamma}^{i+1}(t) - \boldsymbol{\gamma}^i(t)] + \text{higher order terms} \tag{6.2-3}$$

Here the superscript $i + 1$ has been added to $\boldsymbol{\gamma}(t)$ to indicate the iterative nature of the solution which we will obtain.

We assume that the initial trajectory is close to $\boldsymbol{\gamma}^{i+1}(t)$ and drop the higher-order terms in Eq. (6.2-3), so that we obtain the following linear time-varying, inhomogeneous equation for $\boldsymbol{\gamma}^{i+1}(t)$:

$$\dot{\boldsymbol{\gamma}}^{i+1}(t) = \frac{\partial \boldsymbol{\Gamma}[\boldsymbol{\gamma}^i(t), t]}{\partial \boldsymbol{\gamma}^i(t)} \boldsymbol{\gamma}^{i+1}(t) + \left[\boldsymbol{\Gamma}[\boldsymbol{\gamma}^i(t), t] - \frac{\partial \boldsymbol{\Gamma}[\boldsymbol{\gamma}^i(t), t]}{\partial \boldsymbol{\gamma}^i(t)} \boldsymbol{\gamma}^i(t)\right] \tag{6.2-4}$$

The solution of this equation takes the form

$$\boldsymbol{\gamma}^{i+1}(t) = \boldsymbol{\Omega}^{i+1}(t)\, \boldsymbol{\gamma}^{i+1}(t_0) + \mathbf{p}^{i+1}(t) \tag{6.2-5}$$

where $\boldsymbol{\Omega}^{i+1}(t)$ satisfies the differential equation

$$\dot{\boldsymbol{\Omega}}^{i+1}(t) = \frac{\partial \boldsymbol{\Gamma}[\boldsymbol{\gamma}^i(t), t]}{\partial \boldsymbol{\gamma}^i(t)} \boldsymbol{\Omega}^{i+1}(t) \tag{6.2-6}$$

$$\boldsymbol{\Omega}^{i+1}(t_0) = \mathbf{I} \tag{6.2-7}$$

while the particular solution satisfies the equation

$$\dot{\mathbf{p}}^{i+1}(t) = \boldsymbol{\Gamma}[\boldsymbol{\gamma}^i(t), t] - \frac{\partial \boldsymbol{\Gamma}[\boldsymbol{\gamma}^i(t), t]}{\partial \boldsymbol{\gamma}^i(t)} [\boldsymbol{\gamma}^i(t) - \mathbf{p}^{i+1}(t)] \tag{6.2-8}$$

with

$$\mathbf{p}^{i+1}(t_0) = \mathbf{0} \tag{6.2-9}$$

We see that the solution for $\boldsymbol{\gamma}^{i+1}(t)$ has been reduced to a set of initial condition problems which are easily solved, with the single parameter $\gamma^{i+1}(t_0)$. We must select $\gamma^{i+1}(t_0)$ in order to satisfy the boundary conditions of Eq. (6.2-2). If we substitute the solution of Eq. (6.2-5) into Eq. (6.2-2), we obtain

$$\mathbf{C}(t_j)[\boldsymbol{\Omega}^{i+1}(t_j)\,\boldsymbol{\gamma}^{i+1}(t_0) + \mathbf{p}^{i+1}(t_j)] = \mathbf{b}(t_j), \qquad j = 1, 2,..., m \tag{6.2-10}$$

or

$$\mathbf{C}(t_j)\,\boldsymbol{\Omega}^{i+1}(t_j)\,\boldsymbol{\gamma}^{i+1}(t_0) = \mathbf{b}(t_j) - \mathbf{C}(t_j)\,\mathbf{p}^{i+1}(t_j), \qquad j = 1, 2,..., m \tag{6.2-11}$$

By combining all of these m equations, we can obtain one linear algebraic equation for $\gamma^{i+1}(t_0)$ of the form

$$\mathbf{A}\boldsymbol{\gamma}^{i+1}(t_0) = \mathbf{b} \tag{6.2-12}$$

where

$$\mathbf{A} = \begin{bmatrix} \mathbf{C}(t_1)\,\boldsymbol{\Omega}^{i+1}(t_1) \\ \hline \mathbf{C}(t_2)\,\boldsymbol{\Omega}^{i+1}(t_2) \\ \hline \cdots\cdots \\ \hline \mathbf{C}(t_m)\,\boldsymbol{\Omega}^{i+1}(t_m) \end{bmatrix} \tag{6.2-13}$$

$$\mathbf{b} = \begin{bmatrix} \mathbf{b}(t_1) - \mathbf{C}(t_1)\,\mathbf{p}^{i+1}(t_1) \\ \hline \mathbf{b}(t_2) - \mathbf{C}(t_2)\,\mathbf{p}^{i+1}(t_2) \\ \hline \cdots\cdots \\ \hline \mathbf{b}(t_m) - \mathbf{C}(t_m)\,\mathbf{p}^{i+1}(t_m) \end{bmatrix} \tag{6.2-14}$$

If the boundary conditions of Eq. (6.2-2) and the dynamic system of Eq. (6.2-1) are compatible, then a unique solution of Eq. (6.2-12) for $\gamma^{i+1}(t_0)$ will exist, in the sense that the rank of the matrix $\mathbf{A}$ and the rank of the augmented matrix $[\mathbf{A} \mid \mathbf{b}]$ are equal. Hence we may solve for $\gamma^{i+1}(t_0)$:

$$\boldsymbol{\gamma}^{N+1}(t_0) = \mathbf{A}^{-1}\mathbf{b} \tag{6.2-15}$$

which may then be used in Eq. (6.2-5) to obtain the new trajectory

$\boldsymbol{\gamma}^{i+1}(t)$, $t \in [t_0, t_f]$. This trajectory will satisfy the boundary condition of Eq. (6.2-2) but will not in general satisfy Eq. (6.2-1), since we used a linearization assumption in deriving Eq. (6.2-4). We may, however, use $\boldsymbol{\gamma}^{i+1}(t)$ to replace the initial guess for the trajectory $\boldsymbol{\gamma}^i(t)$ and then repeat the above procedure in an iterative fashion.

By using this iterative procedure, we generate a sequence of trajectories $\{\boldsymbol{\gamma}^i(t)\}$ which should converge to the correct solution of the MPBVP. Convergence can be checked by examining the rate of change of the initial condition $\boldsymbol{\gamma}^i(t_0)$. It can be shown (Bellman and Kalaba, 1965) that if this procedure converges, then the convergence is quadratic and hence quite rapid. On the other hand, it is also easy to show by example that convergence often requires a very good initial estimate of the solution. Because of this fact, gradient or differential approximation procedures are frequently used to form the initial estimate for the quasilinearization algorithm.

Unfortunately the procedure developed above has a serious computer storage problem for any nontrivial practical problem. Note that it is necessary to store both $\boldsymbol{\Omega}^{i+1}(t)$ and $\mathbf{p}^{i+1}(t)$ in order to use Eq. (6.2-5) to generate $\boldsymbol{\gamma}^{i+1}(t)$ from $\boldsymbol{\gamma}^{i+1}(t_0)$. Let us suppose that the time interval of interest is 10 sec and that an integration step size of $T = 1/100$ sec is used. Now if the dimension of $\boldsymbol{\gamma}$ is ten, then we must use $(10 \times 10) \times (10) \times 100 = 100{,}000$ values just to store $\boldsymbol{\Omega}^{i+1}(t)$; this is clearly prohibitive.

We may however, fortunately, use an alternate procedure which considerably reduces the storage requirements. Rather than storing the values of $\boldsymbol{\Omega}^{i+1}(t)$ and $\mathbf{p}^{i+1}(t)$ as they are computed, we simply use the appropriate values to generate $\mathbf{A}$ and $\mathbf{b}$ by the use of Eqs. (6.2-13) and (6.2-14); all of the other values are discarded. After $\boldsymbol{\gamma}^{i+1}(t_0)$ is computed by solving Eq. (6.2-12), $\boldsymbol{\Omega}^{i+1}(t)$ and $\mathbf{p}^{i+1}(t)$ are recomputed, and $\boldsymbol{\gamma}^{i+1}$ is computed and stored, while the values of $\boldsymbol{\Omega}^{i+1}(t)$ and $\mathbf{p}^{i+1}(t)$ are again discarded. Now we are ready to start the iterative process again using the improved estimate of the trajectory. Using this procedure, we need only store $\boldsymbol{\gamma}^{i+1}(t)$; this would require only 10,000 values rather than 100,000 plus values in the direct application of the algorithm. Of course we must pay for this reduction in storage by the need to solve Eqs. (6.2-6) and (6.2-8) twice for each iteration.

In some problems, the need to store $\boldsymbol{\gamma}^{i+1}(t)$ may still pose a storage problem. If, for example, the time interval of interest in the case cited above were 100 sec (still a modest time interval), the storage required for $\boldsymbol{\gamma}^{i+1}(t)$ would become 100,000 values. We could store only some of the values (say every tenth one) and then use an interpolation procedure to generate the missing values. However if this approach is successful, we are probably using a step size which is too small, so the difficulty could be avoided by increasing the step size. It is still possible to use quasilinearization in this problem, although it is necessary to modify the algorithm slightly.

Instead of using Eq. (6.2-5) to compute $\gamma^{i+1}(t)$, we may simply integrate Eq. (6.2-1) using the initial condition $\gamma^{i+1}(t_0)$ determined as before from Eq. (6.2-12). Because Eq. (6.2-1) can be integrated at the same time as Eqs. (6.2-6) and (6.2-8), it is not necessary to store $\boldsymbol{\Omega}^{i+1}(t)$, $\mathbf{p}^{i+1}(t)$, or $\gamma^{i+1}(t)$; hence, storage is minimal. On the other hand, we need only integrate Eqs. (6.2-6) and (6.2-8) once for each integration; hence the computational time is increased only slightly above the minimal time used in the direct technique.

From these two statements, it may appear that this procedure is superior to the two other approaches discussed above. The appearance is deceptive, however, since one often experiences greater convergence problems when Eq. (6.2-1) is used to obtain the trajectory. This difficulty can be explained quite simply. When Eq. (6.2-5) is used, the trial trajectory satisfies the boundary conditions of Eq. (6.2-2) although it does not satisfy Eq. (6.2-1). When the trial trajectory is obtained by solving Eq. (6.2-1), it does not satisfy the boundary conditions of Eq. (6.2-2). If the trial trajectory satisfies the boundary conditions, then it must match the correct solution at these points and, in addition, be approximately correct in the neighborhood of these points if the solution is continuous. On the other hand, if the solution simply satisfies Eq. (6.2-1), then there is no reason to suspect that it is even close to the correct solution at any point. Because of this property, it is usually desirable to use Eq. (6.2-5) to compute the trial trajectory if there is sufficient storage to permit it.

As noted earlier, one of the fundamental difficulties with the use of the quasilinearization algorithm is the extremely narrow range of convergence. For this reason, it is often necessary to use another algorithm such as the gradient or differential approximation to generate a good initial guess.

Let us consider the use of the quasilinearization algorithm to solve the two-point boundary value problems associated with system identification. A number of TPBVP were developed in Chap. 3; we will consider the most basic form here in order to minimize the notational difficulties. The more general problems are handled in a similar way. The problem is given by Eqs. (3.2-37)–(3.2-39) which are repeated here for ease of reference:

$$\dot{\hat{\mathbf{x}}}(t) = \mathbf{f}[\hat{\mathbf{x}}(t), t] - \mathbf{G}[\hat{\mathbf{x}}(t), t]\, \boldsymbol{\Psi}_{\mathrm{w}}(t)\, \mathbf{G}^{\mathrm{T}}[\hat{\mathbf{x}}(t), t]\, \boldsymbol{\lambda}(t) \tag{3.2-37}$$

$$\begin{aligned}\dot{\boldsymbol{\lambda}}(t) = {} & \frac{\partial \mathbf{h}^{\mathrm{T}}[\hat{\mathbf{x}}(t), t]}{\partial \hat{\mathbf{x}}(t)}\, \boldsymbol{\Psi}_{\mathrm{v}}^{-1}(t)\{\mathbf{z}(t) - \mathbf{h}[\hat{\mathbf{x}}(t), t]\} - \frac{\partial \mathbf{f}^{\mathrm{T}}[\hat{\mathbf{x}}(t), t]}{\partial \hat{\mathbf{x}}(t)}\, \boldsymbol{\lambda}(t) \\ & + \frac{\partial\{\boldsymbol{\lambda}^{\mathrm{T}}(t)\, \mathbf{G}[\hat{\mathbf{x}}(t), t]\, \boldsymbol{\Psi}_{w}(t)\, \mathbf{G}^{\mathrm{T}}[\hat{\mathbf{x}}(t), t]\}}{\partial \hat{\mathbf{x}}(t)}\, \boldsymbol{\lambda}(t)\end{aligned} \tag{3.2-38}$$

$$\boldsymbol{\lambda}(t_0) = \mathbf{V}_{\mathrm{x}}^{-1}(t_0)[\hat{\mathbf{x}}(t_0) - \boldsymbol{\mu}_{\mathrm{x}}(t_0)], \qquad \boldsymbol{\lambda}(t_{\mathrm{f}}) = \mathbf{0} \tag{3.2-39}$$

Let us write Eqs. (3.2-37) and (3.2-38) in the form

$$\dot{\hat{\mathbf{x}}}(t) = \boldsymbol{\Gamma}_1[\hat{\mathbf{x}}(t), \boldsymbol{\lambda}(t), t], \qquad \dot{\boldsymbol{\lambda}}(t) = \boldsymbol{\Gamma}_2[\hat{\mathbf{x}}(t), \boldsymbol{\lambda}(t), t] \tag{6.2-16}$$

where the definitions of $\boldsymbol{\Gamma}_1$ and $\boldsymbol{\Gamma}_2$ are obvious. The reader is reminded that the "state" $\hat{\mathbf{x}}(t)$ may actually represent system parameters as well as the usual system state. Note that the observation just takes the role of a known time function in $\boldsymbol{\Gamma}_1$ and $\boldsymbol{\Gamma}_2$.

We may put this problem into the framework of Eq. (6.2-1) if we let

$$\boldsymbol{\gamma}(t) = \begin{bmatrix} \hat{\mathbf{x}}(t) \\ \hline \boldsymbol{\lambda}(t) \end{bmatrix}, \qquad \boldsymbol{\Gamma}[\boldsymbol{\gamma}(t), t] = \begin{bmatrix} \boldsymbol{\Gamma}_1[\boldsymbol{\gamma}(t), t] \\ \hline \boldsymbol{\Gamma}_2[\boldsymbol{\gamma}(t), t] \end{bmatrix} \tag{6.2-17}$$

The boundary conditions of Eq. (6.2-2) become

$$\mathbf{C}(t_0)\, \boldsymbol{\gamma}(t_0) = \mathbf{b}(t_0) \tag{6.2-18}$$

$$\mathbf{C}(t_{\mathrm{f}})\, \boldsymbol{\gamma}(t_{\mathrm{f}}) = \mathbf{b}(t_{\mathrm{f}}) \tag{6.2-19}$$

where

$$\mathbf{C}(t_0) = \begin{bmatrix} -\mathbf{V}_{\mathrm{x}}^{-1}(t_0) \\ \hline \mathbf{I} \end{bmatrix}, \qquad \mathbf{b}(t_0) = -\mathbf{V}_{\mathrm{x}}^{-1}(t_0)\, \boldsymbol{\mu}_{\mathrm{x}}(t_0) \tag{6.2-20}$$

and

$$\mathbf{C}(t_{\mathrm{f}}) = \begin{bmatrix} \mathbf{0} \\ \hline \mathbf{I} \end{bmatrix}, \qquad \mathbf{b}(t_{\mathrm{f}}) = \mathbf{0} \tag{6.2-21}$$

The TPBVP has now been cast into the form of Eqs. (6.2-1) and (6.2-2) and the quasilinearization algorithm may be used to obtain an iterative solution. For this problem, Eqs. (6.2-6) and (6.2-8) become

$$\dot{\mathbf{\Omega}}^{i+1}(t) = \left[\begin{array}{c|c} \dfrac{\partial \mathbf{\Gamma}_1[\boldsymbol{\gamma}^i(t), t]}{\partial \hat{\mathbf{x}}^i(t)} & \dfrac{\partial \mathbf{\Gamma}_2[\boldsymbol{\gamma}^i(t), t]}{\partial \hat{\mathbf{x}}^i(t)} \\ \hline \dfrac{\partial \mathbf{\Gamma}_1[\boldsymbol{\gamma}^i(t), t]}{\partial \boldsymbol{\lambda}^i(t)} & \dfrac{\partial \mathbf{\Gamma}_2[\boldsymbol{\gamma}^i(t), t]}{\partial \boldsymbol{\lambda}^i(t)} \end{array}\right] \mathbf{\Omega}^{i+1}(t) \tag{6.2-22}$$

$$\dot{\mathbf{p}}^{i+1}(t) = \left[\begin{array}{c} \mathbf{\Gamma}_1[\boldsymbol{\gamma}^i(t), t] \\ \hline \mathbf{\Gamma}_2[\boldsymbol{\gamma}^i(t), t] \end{array}\right] - \left[\begin{array}{c|c} \dfrac{\partial \mathbf{\Gamma}_1[\boldsymbol{\gamma}^i(t), t]}{\partial \hat{\mathbf{x}}^i(t)} & \dfrac{\partial \mathbf{\Gamma}_2[\boldsymbol{\gamma}^i(t), t]}{\partial \hat{\mathbf{x}}^i(t)} \\ \hline \dfrac{\partial \mathbf{\Gamma}_1[\boldsymbol{\gamma}^i(t), t]}{\partial \boldsymbol{\lambda}^i(t)} & \dfrac{\partial \mathbf{\Gamma}_2[\boldsymbol{\gamma}^i(t), t]}{\partial \boldsymbol{\lambda}^i(t)} \end{array}\right] [\boldsymbol{\gamma}^i(t) - \mathbf{p}^{i+1}(t)] \tag{6.2-23}$$

Note that $\mathbf{\Omega}^{i+1}(t)$ is a $2N \times 2N$ matrix and $\mathbf{p}^{i+1}(t)$ is a $2N$-dimensional vector. Let us partition the solutions of these equations in the following manner:

$$\mathbf{\Omega}^{i+1}(t) = \left[\begin{array}{c|c} \mathbf{\Omega}_{11}^{i+1}(t) & \mathbf{\Omega}_{12}^{i+1}(t) \\ \hline \mathbf{\Omega}_{21}^{i+1}(t) & \mathbf{\Omega}_{22}^{i+1}(t) \end{array}\right] \tag{6.2-24}$$

$$\mathbf{p}^{i+1}(t) = \left[\begin{array}{c} \mathbf{p}_1^{i+1}(t) \\ \hline \mathbf{p}_2^{i+1}(t) \end{array}\right] \tag{6.2-25}$$

where $\mathbf{\Omega}_{ij}^{i+1}(t)$ are all $N \times N$ matrices and $\mathbf{p}_1^{i+1}(t)$ and $\mathbf{p}_2^{i+1}(t)$ are N dimensional. The matrices $\mathbf{A}$ and $\mathbf{b}$ in Eq. (6.2-12) are given by

$$\mathbf{A} = \left[\begin{array}{c|c} -\mathbf{V}_x^{-1}(t_0) & \mathbf{I} \\ \hline \mathbf{\Omega}_{21}(t_f) & \mathbf{\Omega}_{22}(t_f) \end{array}\right] \tag{6.2-26}$$

$$\mathbf{b} = \left[\begin{array}{c} -\mathbf{V}_x^{-1}(t_0)\, \boldsymbol{\mu}_x(t_0) \\ \hline \mathbf{p}_2(t_f) \end{array}\right] \tag{6.2-27}$$

The initial condition $\boldsymbol{\gamma}(t_0)$ may now be obtained by the use of Eq. (6.2-12) and then either Eq. (6.2-1) or (6.2-5) used to obtain the next trial trajectory depending on storage limitations as discussed above. In most practical identification problems, the $2N \times 2N$ matrix $\partial\boldsymbol{\Gamma}/\partial\boldsymbol{\gamma}$ has many zero elements, and the actual use of Eqs. (6.2-22) and (6.2-23) may be considerably simplified by taking advantage of this fact. There are many variations of the above identification problem which may be treated depending on special circumstances. Rather than discuss the variations in terms of a general framework, let us consider a specific case in which significant reduction in computational effort can be achieved.

The algorithm developed above has some serious computational problems if the original system model is of high order. If the system model is Nth order, then it is necessary to solve $4N^2$ differential equations to obtain $\boldsymbol{\Omega}^{i+1}(t)$ from Eq. (6.2-22). It should be remembered that each unknown parameter to be identified increases the order of the system model by at least one order. Hence for a fifth-order system with five unknown parameters, one could develop a tenth-order system model so that Eq. (6.2-22) would become four hundred first-order coupled differential equations. The process of setting up the Jacobian type matrix $\partial\boldsymbol{\Gamma}/\partial\boldsymbol{\gamma}$ in this case is not insignificant.

Another difficulty with the above algorithm is the need to store a continuous observation $\mathbf{z}(t)$, for all $t_0 \leqslant t \leqslant t_f$. One often takes discrete samples of the system output rather than a continuous sample. It is possible to set up an identification problem directly as a multipoint boundary value problem based on discrete observations and avoid the use of the maximum principle and the inherent increase in dimensionality. Unfortunately, it is necessary that the plant noise $\mathbf{w}$ be negligibly small for this technique to work; however this is often not a serious limitation, and hence the technique can be very effective.

We assume that the system is described as usual by the first-order N-vector differential equation

$$\dot{\mathbf{x}}(t) = \mathbf{f}[\mathbf{x}(t), t] \tag{6.2-28}$$

Any unknown system parameters are modeled as part of the system as before. Initially we will assume that $m \gg N$ noisy linear observations of the form

$$\mathbf{z}(t_j) = \mathbf{C}(t_j)\,\mathbf{x}(t_j) + \mathbf{v}(t_j), \qquad j = 1, 2, 3, \ldots, m \tag{6.2-29}$$

where $t_j \in [t_0, t_f]$ and $\mathbf{v}(t_j)$ is zero-mean, independent noise with $\operatorname{var}\{\mathbf{v}(t_j)\} = \mathbf{V}_v(t_j)$. Based on these observations, we wish to find an estimate of the system trajectory $\hat{\mathbf{x}}(t)$, $t_0 \leqslant t \leqslant t_f$, such that we have a least squared error fit with the data, that is, the performance index

$$J = \sum_{j=1}^{m} \| \mathbf{z}(t_j) - \mathbf{C}(t_j)\, \hat{\mathbf{x}}(t_j) \|^2_{\mathbf{Q}(t)} \tag{6.2-30}$$

is minimized. The weighting matrix $\mathbf{Q}$ may be any nonnegative definite matrix, although we will often let $\mathbf{Q}(t_i) = \mathbf{V}_{\mathbf{v}}(t_i)$ to obtain a maximum a posteriori estimate. Note that, because there is no random input in Eq. (6.2-28), the entire trajectory $\hat{\mathbf{x}}(t)$ may be generated by a knowledge of the initial condition $\hat{\mathbf{x}}(t_0)$. It is also possible to let $\mathbf{Q}$ be a function of the iteration $\mathbf{Q}^i(t)$, although we will not consider this here.

Once again, we assume that an initial estimate $\hat{\mathbf{x}}^i(t)$ is known, and we wish to obtain a new estimate $\hat{\mathbf{x}}^{i+1}(t)$ which more nearly minimizes the J of Eq. (6.2-30). We use the quasilinearization techniques developed above with $\hat{\mathbf{x}}(t) = \boldsymbol{\gamma}(t)$ and $\mathbf{f} = \boldsymbol{\Gamma}$ to write $\hat{\mathbf{x}}^{i+1}(t)$ as

$$\hat{\mathbf{x}}^{i+1}(t) = \boldsymbol{\Omega}^{i+1}(t)\, \hat{\mathbf{x}}^{i+1}(t_0) + \mathbf{p}^{i+1}(t) \tag{6.2-31}$$

where $\boldsymbol{\Omega}^{i+1}(t)$ is the solution of the equation

$$\dot{\boldsymbol{\Omega}}^{i+1}(t) = \frac{\partial \mathbf{f}[\hat{\mathbf{x}}^i(t), t]}{\partial \hat{\mathbf{x}}^i(t)}\, \boldsymbol{\Omega}^{i+1}(t) \tag{6.2-32}$$

with $\boldsymbol{\Omega}^{i+1}(t) = \mathbf{I}$, and $\mathbf{p}^{i+1}(t)$ is the solution of

$$\dot{\mathbf{p}}^{i+1}(t) = \mathbf{f}[\hat{\mathbf{x}}^i(t), t] - \frac{\partial \mathbf{f}[\hat{\mathbf{x}}^i(t), t]}{\partial \hat{\mathbf{x}}^i(t)}\, [\hat{\mathbf{x}}^i(t) - \mathbf{p}^{i+1}(t)] \tag{6.2-33}$$

with $\mathbf{p}^i(t_0) = \mathbf{0}$. Note that $\boldsymbol{\Omega}(t)$ in this case is only an $N \times N$ matrix rather than $2N \times 2N$ as before. The problem is now to select $\hat{\mathbf{x}}^{i+1}(t_0)$ in Eq. (6.2-31) to minimize J of Eq. (6.2-30).

If we substitute $\hat{\mathbf{x}}^{i+1}(t)$ from Eq. (6.2-31) into J, we obtain

$$J = \sum_{j=1}^{m} \| \mathbf{z}(t_j) - \mathbf{C}(t_j)[\boldsymbol{\Omega}^{i+1}(t_j)\, \hat{\mathbf{x}}^{i+1}(t_0) + \mathbf{p}^{i+1}(t_j)] \|^2_{\mathbf{Q}(t)}$$

By taking the partial derivative of J with respect to $\hat{\mathbf{x}}^{i+1}(t_0)$ and setting the result equal to zero, we find that

$$\left\{ \sum_{j=1}^{m} [\boldsymbol{\Omega}^{i+1}(t_j)]^T\, \mathbf{C}^T(t_j)\, \mathbf{Q}(t_j)\, \mathbf{C}(t_j)\, \boldsymbol{\Omega}^{i+1}(t_j) \right\} \hat{\mathbf{x}}^{i+1}(t_0)$$

$$= \sum_{j=1}^{m} [\boldsymbol{\Omega}^{i+1}(t_j)]^T\, \mathbf{C}^T(t_j)\, \mathbf{Q}(t_j)[\mathbf{z}(t_j) - \mathbf{C}(t_j)\, \mathbf{p}^{i+1}(t_j)] \tag{6.2-34}$$

Now let us define

$$\mathbf{M}^{i+1} = \sum_{j=1}^{m} [\boldsymbol{\Omega}^{i+1}(t_j)]^{\mathrm{T}} \mathbf{C}^{\mathrm{T}}(t_j)\, \mathbf{Q}(t_j)\, \mathbf{C}(t_j)\, \boldsymbol{\Omega}^{i+1}(t_j) \tag{6.2-35}$$

and

$$\mathbf{N}^{i+1} = \sum_{j=1}^{m} [\boldsymbol{\Omega}^{i+1}(t_j)]^{\mathrm{T}} \mathbf{C}^{\mathrm{T}}(t_j)\, \mathbf{Q}(t_j)[\mathbf{z}(t_j) - \mathbf{C}(t_j)\, \mathbf{p}^{i+1}(t_j)] \tag{6.2-36}$$

so that Eq. (6.2-34) becomes

$$\mathbf{M}^{i+1}\hat{\mathbf{x}}^{i+1}(t_0) = \mathbf{N}^{i+1}$$

which we may solve to obtain $\hat{\mathbf{x}}(t_0)$ as

$$\hat{\mathbf{x}}^{i+1}(t_0) = [\mathbf{M}^{i+1}]^{-1}\, \mathbf{N}^{i+1} \tag{6.2-37}$$

With $\hat{\mathbf{x}}^{i+1}(t_0)$ known, one may now determine the entire trial trajectory $\hat{\mathbf{x}}^{i+1}(t)$ and repeat the process until the solution changes by only a small amount from iteration to iteration.

One of the computational difficulties encountered with the above method is the singularity (or near singularity) of the matrix $\mathbf{M}^{i+1}$. This most often occurs in the early iterations when the trajectory is still a poor estimate of the solution. This problem can be overcome in some cases by using only a small number of the available observations in the early iterations.

If the observations are nonlinear, then we simply use a quasilinearization procedure to linearize them and apply the algorithm developed above. Let us consider the nonlinear observation

$$\mathbf{z}(t_j) = \mathbf{h}[\mathbf{x}(t_j), t_j] + \mathbf{v}(t_j) \tag{6.2-38}$$

Once again we assume that an initial trajectory $\hat{\mathbf{x}}^i(t)$ is known, and we make Taylor series expansion $\mathbf{h}[\mathbf{x}(t_j), t_j]$ about $\hat{\mathbf{x}}^i(t)$ and retain only linear terms to obtain

$$\mathbf{z}(t_j) = \mathbf{h}[\hat{\mathbf{x}}^i(t_j), t_j] + \frac{\partial \mathbf{h}[\hat{\mathbf{x}}^i(t_j), t]}{\partial \hat{\mathbf{x}}^i(t_j)} [\hat{\mathbf{x}}^{i+1}(t_j) - \hat{\mathbf{x}}^i(t_j)] + \mathbf{v}(t_j)$$

Now if we make the following definitions:

$$\mathbf{z}'(t_j) = \mathbf{z}(t_j) - \mathbf{h}[\hat{\mathbf{x}}^i(t_j), t_j] + \mathbf{C}^i(t_j)\, \hat{\mathbf{x}}^i(t_j) \tag{6.2-39}$$

where

$$\mathbf{C}^i(t_j) = \frac{\partial \mathbf{h}[\hat{\mathbf{x}}^i(t_j), t_j]}{\partial \hat{\mathbf{x}}^i(t_j)} \tag{6.2-40}$$

then the nonlinear observation takes the form of a linear observation

$$\mathbf{z}'(t_j) = \mathbf{C}^i(t_j)\,\mathbf{x}^{i+1}(t_j) + \mathbf{v}(t_j) \tag{6.2-41}$$

By substituting $\mathbf{z}'(t_j)$ for $\mathbf{z}(t_j)$ and $\mathbf{C}^i(t_j)$ for $\mathbf{C}(t_j)$ in the above algorithm, it is possible to handle nonlinear observations.

We are, in effect, replacing the nonquadratic cost function

$$J = \sum_{j=1}^{m} \| \mathbf{z}(t_f) - \mathbf{h}[\mathbf{x}(t_j), t_j] \|^2_{\mathbf{Q}(t_j)} \tag{6.2-42}$$

by the quadratic approximation

$$J = \sum_{j=1}^{m} \left\| \mathbf{z}(t_j) - \mathbf{h}[\hat{\mathbf{x}}^i(t_j), t_j] - \frac{\partial \mathbf{h}[\hat{\mathbf{x}}^i(t_j), t_j]}{\partial \hat{\mathbf{x}}^i(t_j)} [\hat{\mathbf{x}}^{i+1}(t_j) - \hat{\mathbf{x}}^i(t_j)] \right\|^2_{\mathbf{Q}(t_j)}$$

$$= \sum_{j=1}^{m} \| \mathbf{z}'(t_j) - \mathbf{C}^i(t_j)\mathbf{x}^{i+1}(t_j) \|^2_{\mathbf{Q}(t_j)} \tag{6.2-43}$$

and then applying the just developed quasilinearization results.

In this section, we have developed the technique of quasilinearization for solving identification problems for continuous systems. In the next sections, these same techniques are developed for discrete models. In terms of practical applications, the problems of interest often fall between these two classes. Observations are usually taken in discrete time and a digital computer is normally used to apply the algorithms so that a discrete format is desirable. On the other hand, the basic system model is quite often continuous, so that the continuous format also is applicable.

Example 6.2-1. Consider the system shown in Fig. 6.2-1. A model of the form

$$\dot{\mathbf{x}} = \mathbf{A}(t)\mathbf{x} + \mathbf{u}(t)$$

where

$$\mathbf{x}(t) = \begin{bmatrix} x_1(t) \\ x_2(t) \end{bmatrix}, \quad \mathbf{u}(t) = \begin{bmatrix} 0 \\ \sin(0.8\pi t) \end{bmatrix}, \quad \mathbf{A}(t) = \begin{bmatrix} 0 & 1 \\ -4 & -(a + b \sin t) \end{bmatrix}$$

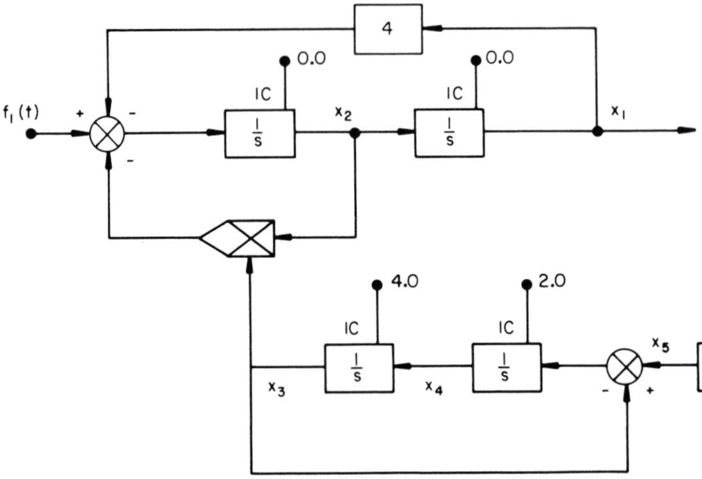

FIG. 6.2-1. Block diagram description of system studied in Example 6.2-1. $f_1(t) = \sin(0.8\pi t)$, $f_2(t) = x_5(t) = 4\mu_{-1}(t)$.

is chosen from a physical knowledge of the plant. We desire to determine the values of a and b so that the model for the dynamic system matches the true system response. This identification problem will be solved using quasilinearization. The adjoined system equation becomes

$$\begin{aligned}
\dot{\hat{x}}_1 &= \hat{x}_2(t) \\
\dot{\hat{x}}_2 &= -4\hat{x}_1(t) - \hat{x}_3(t)\,\hat{x}_2(t) + \sin(0.8\pi t) \\
\dot{\hat{x}}_3 &= \hat{x}_4(t) \\
\dot{\hat{x}}_4 &= \hat{x}_5(t) - \hat{x}_3(t) \\
\dot{\hat{x}}_5 &= 0
\end{aligned}$$

Linearizing this fifth-order system about the trajectory $(\mathbf{x}^N)$ results in the following linear, time-varying vector differential equation for the $(N+1)$th approximation:

$$\dot{\mathbf{Y}}(t) = \mathbf{B}(t)\,\mathbf{Y}(t) + \mathbf{r}(t)$$

where

$$\mathbf{Y}(t) = \begin{bmatrix} \hat{x}_1^{N+1}(t) \\ \hat{x}_2^{N+1}(t) \\ \hat{x}_3^{N+1}(t) \\ \hat{x}_4^{N+1}(t) \\ \hat{x}_5^{N+1}(t) \end{bmatrix}, \qquad \mathbf{r}(t) = \begin{bmatrix} 0 \\ \hat{x}_5{}^N(t)\,\hat{x}_2{}^N(t) + \sin(0.8\pi t) \\ 0 \\ 0 \\ 0 \end{bmatrix}$$

and

$$\mathbf{B}(t) = \begin{bmatrix} 0 & 1 & 0 & 0 & 0 \\ -4 & -x_5{}^N(t) & -x_2{}^N(t) & 0 & 0 \\ 0 & 0 & 0 & 1 & 0 \\ 0 & 0 & -1 & 0 & 1 \end{bmatrix}$$

A computer solution of the quasilinearization technique was carried out over the time interval 0–10 sec. Ten equally spaced measurements of the true system response were used in the least squares curve fitting, $J = \sum_{t=1}^{10} [x_1(t) - \hat{x}_1(t)]^2$. The solution quickly converged to

$$\hat{x}_5(0) = \hat{a} = 3.999989$$

$$\hat{x}_4(0) = \hat{b} = 2.000535$$

The true plant parameters were

$$a = 4.000000$$

$$b = 2.000000$$

A comparison of these results indicates the inherent accuracy of this method. We note that, in this example, a great deal of a priori knowledge of the plant dynamics is necessary. In addition, the time-varying coefficients must obey simple differential equations of known form.

A more general means of identifying nonstationary systems will now be investigated. A system described by

$$\dot{\mathbf{x}} = \mathbf{f}[\mathbf{x}(t), \mathbf{u}(t), \mathbf{p}(t), t]$$

is approximated over small intervals by

$$\dot{\mathbf{y}} = \mathbf{g}[\mathbf{y}(t), \mathbf{u}(t), \mathbf{b}, t]$$

The constant parameter vector $\mathbf{b}$ is then identified over these small intervals. We may reduce the computational load by utilizing the past calculations to obtain the present value of $\mathbf{b}$.

The system depicted in Fig. 6.2-1 was identified using this "tracking" method. The system is modeled by

$$\dot{\hat{x}}_1 = \hat{x}_2$$

$$\dot{\hat{x}}_2 = -4\hat{x}_1 - a\hat{x}_2 + \sin(0.8\pi t)$$

The parameter a was assumed constant over 0.1 sec intervals. Adjoining the differential equation

$$\dot{a} = 0$$

to the above model and linearizing the third-order adjoined system about the trajectory (x^N, a^N) results in the differential equation for the $(N+1)$th approximation:

$$\dot{\mathbf{z}}(t) = \mathbf{B}(t)\,\mathbf{z}(t) + \mathbf{r}(t)$$

where

$$\mathbf{z}(t) = \begin{bmatrix} \hat{x}_1^{N+1} \\ \hat{x}_2^{N+1} \\ \hat{x}_3^{N+1} \end{bmatrix}, \qquad \mathbf{r}(t) = \begin{bmatrix} 0 \\ \hat{x}_3{}^N \hat{x}_2{}^N + \sin(0.8\pi t) \\ 0 \end{bmatrix}$$

where $\mathbf{B}(t)$ is given by

$$\mathbf{B}(t) = \begin{bmatrix} 0 & 1 & 0 \\ -4 & -\hat{x}_3{}^N & -\hat{x}_2{}^N \\ 0 & 0 & 0 \end{bmatrix}$$

The quasilinearization algorithm is processed for this system over the interval of time $t \in (\tau, \tau + 0.1)$, where $\tau = 0, 0.1, 0.2, \ldots 10$. The identified parameter $a(t)$ is compared to its true value in Fig. 6.2-2. Figure 6.2–3 is a comparison of the model and system responses.

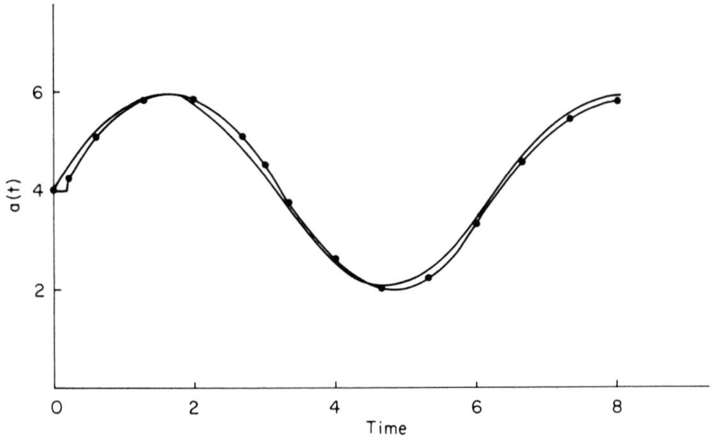

FIG. 6.2-2. The parameter $a(t)$ via tracking. ——— true solution, -•—•- quasilinearization.

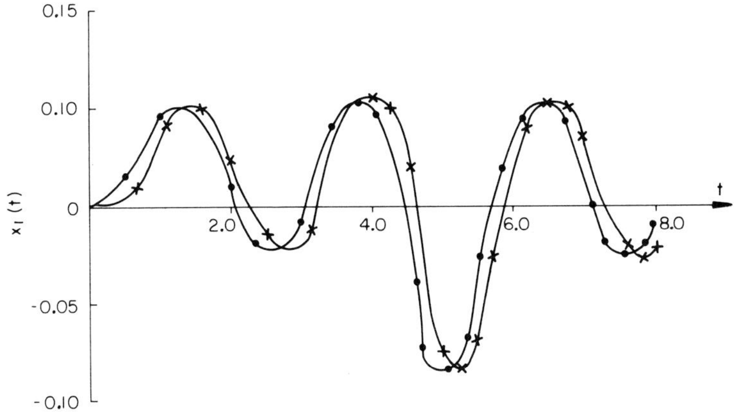

FIG. 6.2-3. Response of the nonstationary system. –•—•– true solution, —*— tracking solution.

For a large class of systems, the plant parameters, normally of a stationary nature, are subjected to external random disturbances such as wind gusts, load variations, etc. It is feasible to employ quasilinearization to sense these disturbances. A second-order, linear system having one parameter sensitive to environmental conditions is studied. At time t_0, a disturbance is introduced into the system. The measurable operating record is shown in Fig. 6.2–4. The tracking method, previously described, used this available record to determine the parameter variations. The excellent results ase shown in Fig. 6.2–5. It is of interest to determine the most rapid variations detectable by the tracking method. Our system is subjected to a rapidly changing environment and the parameter variation chosen to represent this environment consists of series of pulses having varying widths as shown in Fig. 6.2–6. The results, also shown in this figure, indicate that changes having variations on the order of two times the basic identification interval will produce acceptable results.

6.3. Discrete Systems

The development of the quasilinearization algorithm for discrete systems follows quite closely the development for continuous systems presented in the previous section. We wish to obtain the trajectory

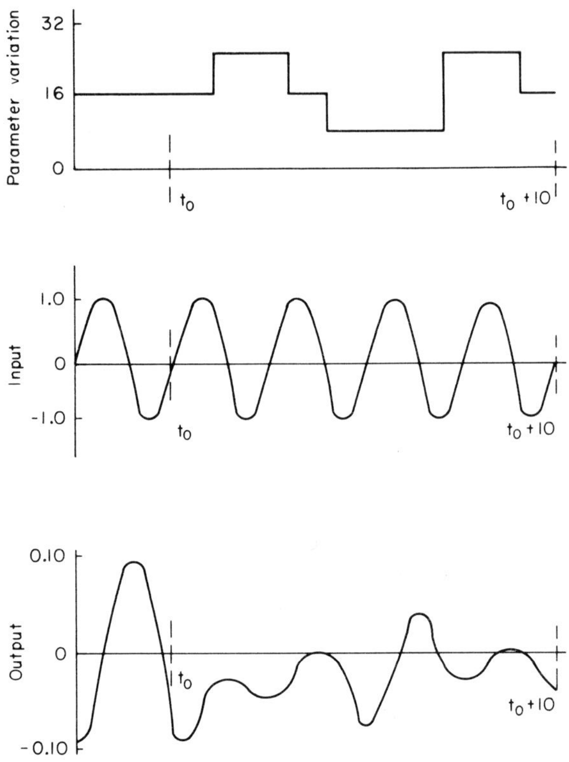

FIG. 6.2-4. Operating record of second-order system subject to environmental changes in ***a***.

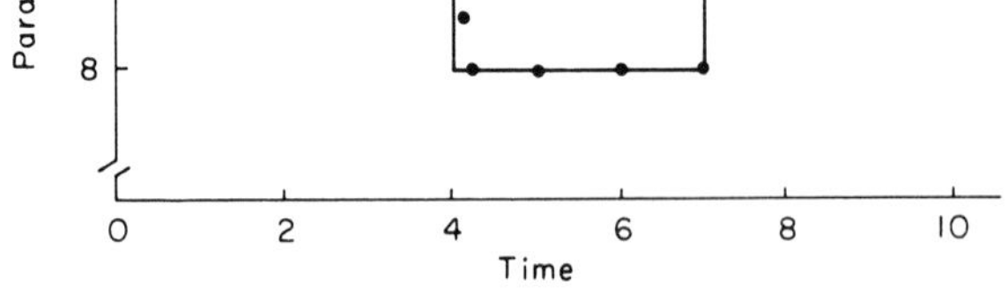

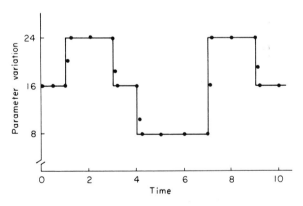

FIG. 6.2-5. Parameter variations. ——— true solution, • tracking solution.

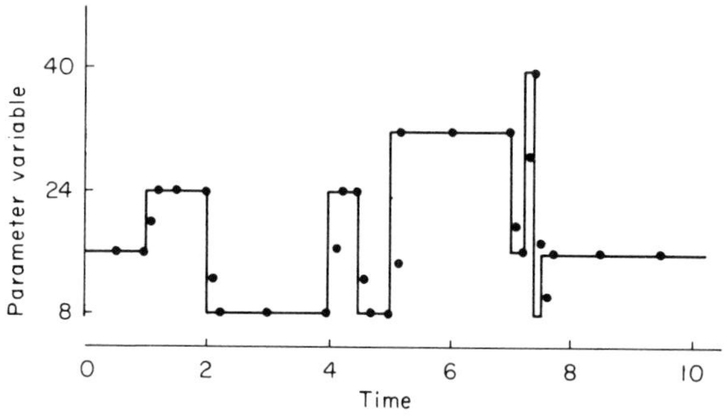

FIG. 6.2-6. Study of rapid variations. ——— true solution, • tracking solution.

$\gamma(k)$, where[1] $k \in [k_0, k_f]$ which satisfies the nonlinear, time-varying, N-vector, difference equation

$$\gamma(k+1) = \phi[\gamma(k), k] \tag{6.3-1}$$

and the set of linear boundary conditions

$$\mathbf{C}(k_j)\,\gamma(k_j) = \mathbf{b}_j, \qquad j = 1, 2, \ldots, m \tag{6.3-2}$$

where $k_j \in [k_0, k_f]$. We will assume that the k_j are ordered so that $k_j < k_l$, if $j < l$, and that the boundary conditions of Eq. (6.3-2) are consistent with a unique solution of Eq. (6.3-1).

We begin with an initial trial trajectory $\gamma^i(t)$ which, although not a solution of the above MPBVP, is an approximate solution. Once again we use a Taylor series expansion and retain only linear terms, so that Eq. (6.3-1) becomes

$$\gamma^{i+1}(k+1) = \phi[\gamma^i(k), k] + \frac{\partial\phi[\gamma^i(k), k]}{\partial\gamma^i(k)}\,[\gamma^{i+1}(k) - \gamma^i(k)] \tag{6.3-3}$$

If we rearrange this equation, we see that it takes the form of a linear, time-varying, inhomogeneous difference equation for the new trial trajectory $\gamma^{i+1}(k)$:

$$\gamma^{i+1}(k) = \frac{\partial\phi[\gamma^i(k), k]}{\partial\gamma^i(k)}\,\gamma^{i+1}(k) + \left[\phi[\gamma^i(k), k] - \frac{\partial\phi[\gamma^i(k), k]}{\partial\gamma^i(k)}\,\gamma^i(k)\right] \tag{6.3-4}$$

[1] For notational simplicity, we will drop the sample interval T in the discrete expression of this section. The stage argument will then take on only integer values.

The solution of this equation is easily shown to be

$$\boldsymbol{\gamma}^{i+1}(k) = \boldsymbol{\Omega}^{i+1}(k)\,\boldsymbol{\gamma}^{i+1}(k_0) + \mathbf{p}^{i+1}(k) \tag{6.3-5}$$

where $\boldsymbol{\Omega}^{i+1}(k)$ is the solution of the difference equation

$$\boldsymbol{\Omega}^{i+1}(k+1) = \frac{\partial \boldsymbol{\phi}[\boldsymbol{\gamma}^i(k), k]}{\partial \boldsymbol{\gamma}^i(k)}\,\boldsymbol{\Omega}^{i+1}(k) \tag{6.3-6}$$

with $\boldsymbol{\Omega}^{i+1}(k_0) = \mathbf{I}$, and $\mathbf{p}^{i+1}(k)$ is the solution of the equation

$$\mathbf{p}^{i+1}(k) = \boldsymbol{\phi}[\boldsymbol{\gamma}^i(k), k] - \frac{\partial \boldsymbol{\phi}[\boldsymbol{\gamma}^i(k), k]}{\partial \boldsymbol{\gamma}^i(k)}\,[\boldsymbol{\gamma}^i(k) - \mathbf{p}^{i+1}(k)] \tag{6.3-7}$$

with the initial condition $\mathbf{p}^{i+1}(k_0) = \mathbf{0}$.

Here again we see that the solution for $\boldsymbol{\gamma}^{i+1}(k)$ has been changed into a pair of initial condition problems, which may be easily solved, and has a single parameter $\boldsymbol{\gamma}^{i+1}(k_0)$. Now we wish to select $\boldsymbol{\gamma}^{i+1}(k_0)$ so that Eq. (6.3-2) is satisfied; this may be accomplished by substituting Eq. (6.3-5) into Eq. (6.3-2); so we have

$$\mathbf{C}(k_j)[\boldsymbol{\Omega}^{i+1}(k_j)\,\boldsymbol{\gamma}^{i+1}(k_0) + \mathbf{p}^{i+1}(k_j)] = \mathbf{b}_j\,, \qquad j = 1, 2,\ldots, m \tag{6.3-8}$$

We may rewrite this result so that it takes the form of a set of m linear algebraic equations

$$[\mathbf{C}(k_j)\,\boldsymbol{\Omega}^{i+1}(k_j)]\,\boldsymbol{\gamma}^{i+1}(k_0) = \mathbf{b}(k_j) - \mathbf{C}(k_j)\,\mathbf{p}^{i+1}(k_j), \qquad j = 1, 2,\ldots, m$$

which may be written as one linear algebraic equation as

$$\mathbf{A}\boldsymbol{\gamma}^{i+1}(k_0) = \mathbf{b} \tag{6.3-9}$$

which has the solution

$$\boldsymbol{\gamma}^{i+1}(k_0) = \mathbf{A}^{-1}\mathbf{b} \tag{6.3-10}$$

where

$$\mathbf{b} = \begin{bmatrix} \mathbf{b}(k_1) - \mathbf{C}(k_1)\,\mathbf{p}^{i+1}(k_1) \\ \hline \mathbf{b}(k_2) - \mathbf{C}(k_2)\,\mathbf{p}^{i+1}(k_2) \\ \hline \cdots\cdots \\ \hline \mathbf{b}(k_m) - \mathbf{C}(k_m)\,\mathbf{p}^{i+1}(k_m) \end{bmatrix}, \qquad \mathbf{A} = \begin{bmatrix} \mathbf{C}(k_1)\,\boldsymbol{\Omega}^{i+1}(k_1) \\ \hline \mathbf{C}(k_2)\,\boldsymbol{\Omega}^{i+1}(k_2) \\ \hline \cdots\cdots \\ \hline \mathbf{C}(k_m)\,\boldsymbol{\Omega}^{i+1}(k_m) \end{bmatrix} \tag{6.3-11}$$

Note that the form for **A** and **b** is identical to the result of the preceding section, except that the continuous time variable has been replaced by the discrete time variable.

By solving Eq. (6.3-9), the initial condition $\gamma^{i+1}(k_0)$ is obtained from which the entire new trial trajectory may be determined. Hence we have an iterative procedure for solving the discrete MPBVP of Eqs. (6.3-1) and (6.3-2). The reader is directed to the preceding section for a discussion of convergence as well as techniques of programming the algorithm for use on a digital computer.

The use of the discrete quasilinearization algorithm to solve the discrete two-point boundary-value problems of Chap. 3 associated with system identification follows so closely the development for continuous system presented in the preceding section that the development is not repeated here. The direct application of the quasilinearization algorithm to system identification with no plant noise for the discrete case is also identical to the continuous case. Equations (6.2-35)–(6.2-37) may be used in the discrete case if the continuous sample points t_j are replaced by the discrete time k_j.

The discrete quasilinearization algorithm unfortunately suffers from the same convergence problems as the continuous algorithm. When the algorithm converges, the convergence is quadratic and quite fast. However, it is usually necessary to have a good estimate of the trajectory to initialize the procedure in order to obtain convergence. One approach is to use the gradient techniques of Chap. 4 to obtain an initial trajectory, since the gradient algorithm is usually capable of converging close to the correct solution from a poor initial estimate in a modest computing time. Another approach which may be used to obtain a good initial estimate is the differential or difference approximation procedure. This is the subject of the next section.

Example 6.3-1. In this example we shall use the quasilinearization technique to estimate the pulse transfer function for a linear system from normal operating records (Schultz, 1968). We assume the identification problem of Fig. 6.3-1, in which the model is driven by a noise free measurement of the plant input $w(t)$ and is characterized by a sampled data (pulse) transfer function $N(z)/D(z)$, where

$$N(z) = a_0 + a_1 z^{-1} + \cdots + a_m z^{-m}$$
$$D(z) = 1 + b_1 z^{-1} + \cdots + b_n z^{-n}$$

The a's and b's are constant parameters which we desire to identify.

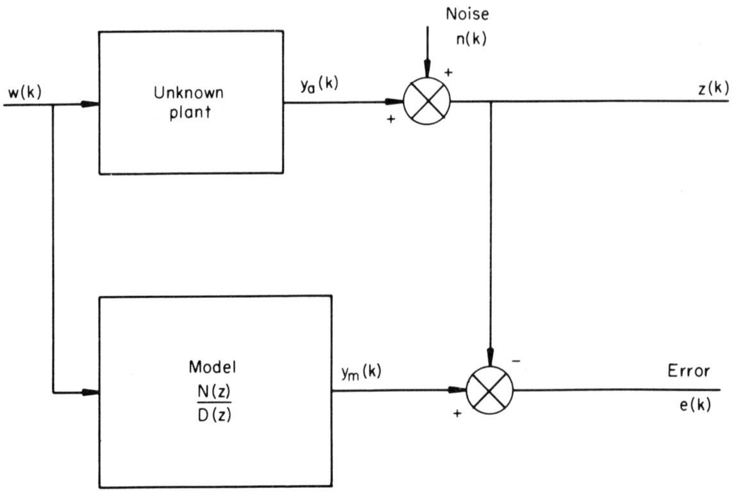

FIG. 6.3-1. Block diagram of identification problem, Example 6.3-1.

The order of the system is assumed known. Thus we know m and n and assume $m \leqslant n$. It is convenient to assume zero initial conditions.

We desire to find the a_i and b_i such that the cost function for K stages

$$J = \sum_{k=1}^{K} e^2(k), \qquad e(k) = y_m(k) - z(k)$$

is a minimum. We develop an iterative method of accomplishing this using quasilinearization. We wish to accomplish this minimization subject to the equality constraint

$$\frac{y_m(z)}{w(z)} = \frac{N(z)}{D(z)}$$

which is the model constraint from Fig. (6.3-1). This is rewritten as

$$y_m(z)\, D(z) = w(z)\, N(z)$$

We quasilinearize about the Nth iteration to obtain for the foregoing nonlinear model equation

$$y_m{}^N(z)\, D^N(z) + y_m{}^N(z)[D^{N+1}(z) - D^N(z)] + D^N(z)[y_m^{N+1}(z) - y_m{}^N(z)]$$
$$= N^{N+1}(z)\, w(z)$$

In this relation, $y_m^{N+1}(z)$ is a linear approximation to the $y_m(z)$ that results from the $(N+1)$th iteration. We solve the foregoing for $y_m^{N+1}(z)$ and subtract $z(z)$ to give the quasilinearized expression for the error on the $(N+1)$th iteration as

$$\begin{aligned} e^{N+1}(z) &= y_m^{N+1}(z) - z(z) \\ &= \frac{N^{N+1}(z)}{D^N(z)}\, w(z) + \left[1 - \frac{D^{N+1}(z)}{D^i(z)}\right] y_m^{\,2}(z) - z(z) \end{aligned}$$

It is now convenient to define

$$W^N(z) = \frac{w(z)}{D^N(z)}, \qquad Y_m^{\,N}(z) = \frac{y_m(z)}{D^N(z)}$$

such that the error equation becomes

$$e^{N+1}(z) = N^{N+1}(z)\, W^N(z) + y_m^{\,N}(z) - D^{N+1}(z)\, Y_m^{\,N}(z) - z(z)$$

Also, we define

$$\begin{aligned} \mathbf{P} &= [a_0 a_1 \cdots a_m \;\; -b_1 \;\; -b_2 \cdots -b_n]^{\mathrm{T}} \\ \mathbf{Q}_k &= [W^N(k)\; W^N(k-1) \cdots W^N(k-m) \\ &\quad \times Y_m^{\,N}(k-1)\; Y_m^{\,N}(k-2) \cdots Y_m^{\,N}(k-n)]^{\mathrm{T}} \end{aligned}$$

such that the inverse z transform of the error equation for $e^{N+1}(z)$ yields

$$e^{N+1}(k) = \mathbf{Q}_k^{\mathrm{T}} \mathbf{P}^{N+1} + y_m^{\,N}(z) - Y_m^{\,N}(k) - z(k)$$

We now have a simple form for the time domain error which we may substitute into the cost function and minimize with respect to $\mathbf{P}^{N+1}$ to obtain

$$\mathbf{P}^{N+1} = \Xi^{-1} \sum_{k=1}^{K} [Y_m^{\,N}(k) + z(k) - y_m^{\,N}(k)]\, \mathbf{Q}_k$$

$$\Xi = \sum_{k=1}^{K} \mathbf{Q}_k \mathbf{Q}_k^{\mathrm{T}}$$

The procedure for using the identification algorithms is relatively simple. The defining equations for $W^N(z)$ and $Y_m^{\,N}(z)$ are used to obtain the $\mathbf{P}$ and $\mathbf{Q}_k$ vectors which are then processed iteratively using data from the output of the actual system and the iterated model in accordance with the foregoing two equations. Schultz (1968) presents details of computer simulations for this method which is an adaptation of the quasilinearization algorithms for a problem proposed by Steiglitz and McBride (1965) and solved by them using the gradient procedure.

Example 6.3-2. We will now illustrate the use of the discrete quasilinearization algorithms in order to identify system parameters. It is desired that a digital system output, $\mathbf{y}_d(t)$ approach an analog system output $\mathbf{y}_a(t)$, where $\mathbf{y}_a(t)$ and $\mathbf{y}_d(t)$ are m vectors describing the system output state for a given input. It is assumed that the analog state vector output $\mathbf{y}_a(t)$ is completely known, as is the input to the system. The form of the digital system has been determined and is known except for a certain number of constant parameters $\mathbf{P}$, which are to be determined. $\mathbf{P}$ will be interpreted as a p-vector and will be adjusted to minimize the cost function

$$J = \frac{1}{2} \sum_{k=0}^{N-1} \| \mathbf{y}_a(kT) - \mathbf{y}_d(kT) \|_{\mathbf{R}}^2 \tag{1}$$

subject to the constraint that

$$\mathbf{y}_d(\overline{n+1}\,T) = \mathbf{f}[\mathbf{y}_d(nT), \mathbf{P}] \tag{2}$$

$$\mathbf{P}(\overline{n+1}\,T) = \mathbf{P}(nT) \tag{3}$$

Application of standard variational calculus procedures (Sage, 1968) demonstrates that the optimum parameter vector $\mathbf{P}$ is determined by solution of difference equations (2) and (3) together with the adjoint vector difference equations

$$\boldsymbol{\lambda}_{\mathbf{y}}(nT) = \frac{\partial \mathbf{f}[\mathbf{y}_d(nT), \mathbf{P}]}{\partial \mathbf{y}_d(nT)} \boldsymbol{\lambda}_{\mathbf{y}}(\overline{n+1}\,T) + \mathbf{R}[\mathbf{y}_a(nT) - \mathbf{y}_d(nT)] \tag{4}$$

$$\boldsymbol{\lambda}_{\mathbf{p}}(nT) = \frac{\partial \mathbf{f}[\mathbf{y}_d(nT), \mathbf{P}]}{\partial \mathbf{y}_d(nT)} \boldsymbol{\lambda}_{\mathbf{y}}(\overline{n+1}\,T) + \boldsymbol{\lambda}_{\mathbf{p}}(\overline{n+1}\,T) \tag{5}$$

and associated two-point boundary conditions

$$\boldsymbol{\lambda}_{\mathbf{y}}(NT) = \boldsymbol{\lambda}_{\mathbf{p}}(NT) = \boldsymbol{\lambda}_{\mathbf{p}}(0) = \boldsymbol{\lambda}_{\mathbf{y}}(0) = \mathbf{0} \tag{6}$$

Equations (2)–(6) represent a two-point nonlinear boundary value problem which may be conveniently solved by the method of quasilinearization. A new $2(m + p)$-vector

$$\mathbf{x}(\overline{n + 1\ T}) = \mathbf{q}[\mathbf{x}(nT)] \tag{7}$$

with the boundary conditions

$$\langle \mathbf{C}_i(jT), \mathbf{x}(jT) \rangle = b_i(jT), \qquad j = 0, N, \qquad i = 1, 2, \ldots, (m + p) \tag{8}$$

where $\mathbf{C}$ and $\mathbf{x}$ are $2(m + p)$-dimensional vectors and $\langle\ ,\ \rangle$ denotes the inner product. If $\mathbf{x}^0(nT)$ is the initial guess to the solution of Eq. (7), the $(N + 1)$th approximation is obtained from the Nth by

$$\mathbf{x}^{N+1}(\overline{n + 1\ T}) = \mathbf{q}[\mathbf{x}^N(nT)] + J\{\mathbf{q}[\mathbf{x}^N(nT)]\}\{\mathbf{x}^{N+1}(nT) - \mathbf{x}^N(nT)\} \tag{9}$$

where J is the Jacobian matrix whose ijth element is the partial derivative $\partial q_i / \partial x_j$.

The application of this method to the discrete representation of continuous systems allows the determination of the "best" discrete representation of the continuous system. Since this best approximation is a function of the analog system input and any time varying parameters in the analog model, an adaptive loop which would adjust the parameter vector $\mathbf{P}$ so as to continuously optimize in an on-line fashion the discrete model output to the analog model output would improve the discrete model further.

Consider the nonlinear system represented in block diagram form in Fig. 6.3-2. The three point integration rule could be used to simulate

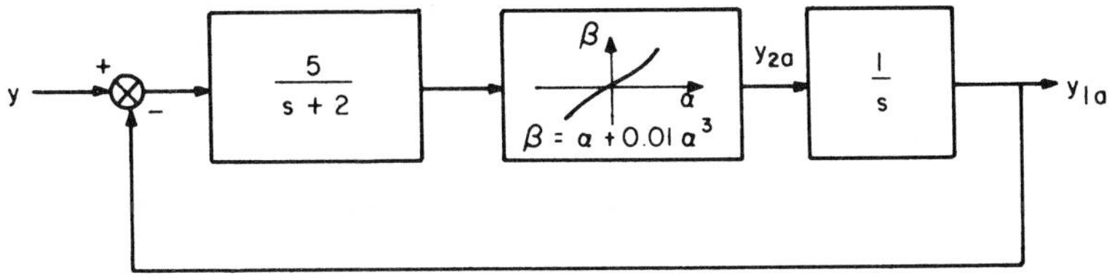

FIG. 6.3-2. Feedback system with limiter.

the continuous integration terms in the figure. Gains before and after the nonlinearity would then be adjusted by the discrete quasilinearization procedure. An alternate approach, which yields a less exact approximation but which results in a lower order discrete system, will be used here. The integration $1/s$ is replaced by the rectangular discrete integration rule and the time constant term with its z transform equivalent

$$\frac{5}{s+2} \Leftrightarrow \frac{5Tz^{-1}}{1-e^{-2T}z^{-1}}$$

Gain parameters p_1 and p_2 are placed before and after the nonlinearity. Thus the discrete model is as shown in Fig. 6.3-3. The input is

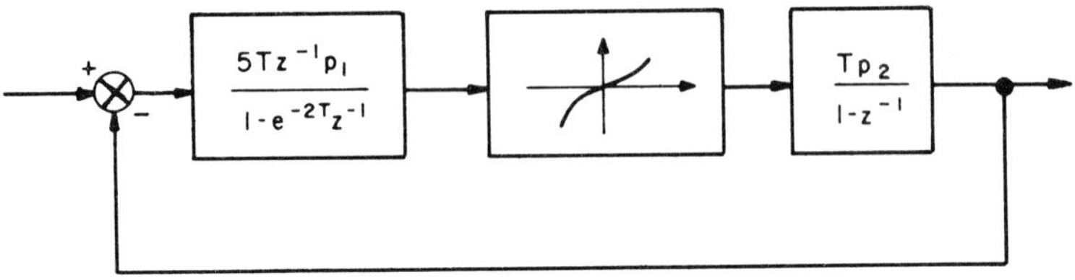

FIG. 6.3-3. Discrete model of Fig. 6.3-2.

assume to be a step of amplitude A. Thus the difference equation describing the discrete model is ($R = I$),

$$y_1(\overline{n+1\ T}) = y_1(nT) + Tp_2(\overline{n+1\ T})[y_2(\overline{n+1\ T}) + 0.01y_2^3(\overline{n+1\ T})]$$

$$y_2(\overline{n+1\ T}) = -5Tp_1(nT)\,y_1(nT) + e^{-2T}y_2(nT) + 5Tp_1(nT)A$$

$$p_1(\overline{n+1\ T}) = p_1(nT)$$

$$p_2(\overline{n+1\ T}) = p_2(nT)$$

The adjoint difference equation is written and the discrete model equations and adjoint equations are quasilinearized. Figure 6.3-4 illustrates computational results concerning the optimum parameters p_1 and p_2 for various sampling periods. Figure 6.3-5 illustrates the optimum parameter vector as a function of input step amplitude.

6.4. DIFFERENCE AND DIFFERENTIAL APPROXIMATION

Although the technique of difference or differential approximation has some serious limitations, it can on occasion be used to develop a first approximation to a solution. As such, its most valuable application is in initialization of gradient or quasilinearization approaches. The

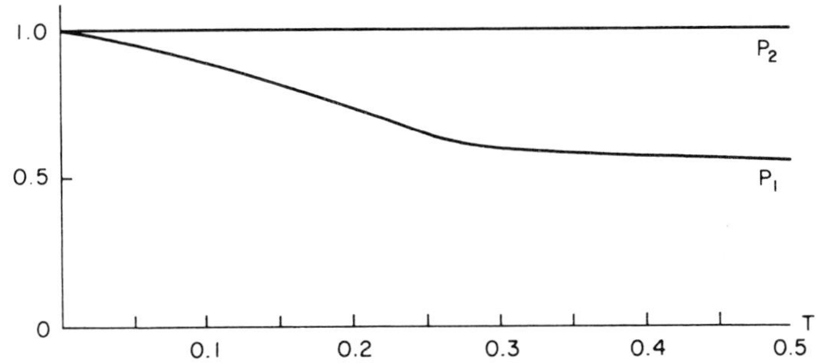

FIG. 6.3-4. Computer results for Example 6.3-2.

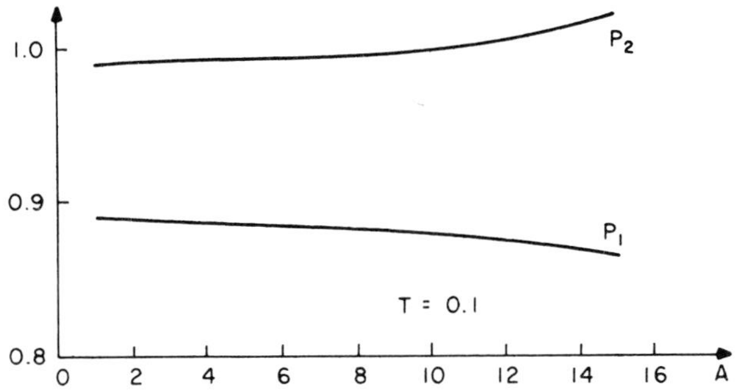

FIG. 6.3-5. Computer results for Example 6.3-2.

algorithm is noniterative, using only one iteration to obtain a solution. We will consider first the discrete version of the algorithm known as difference approximation and then consider the continuous case known as differential approximation. The algorithm is quite simple to develop and has been introduced and reintroduced by numerous authors, although Bellman is commonly acknowledged as the principal originator of the algorithm in its modern context.

We assume that the system is represented by the N-vector discrete model

$$\mathbf{x}(k+1) = \boldsymbol{\phi}[\mathbf{x}(k), \mathbf{a}, k] + \mathbf{w}(k) \tag{6.4-1}$$

where $\mathbf{a}$ is an unknown constant parameter m-vector and $\mathbf{w}(k)$ is plant noise. The elements of the vector-valued function $\phi[x(k), a, k]$ are assumed to be of the form

$$\{\boldsymbol{\phi}[\mathbf{x}(k), \mathbf{a}, k]\}_j = g_j^{\mathrm{T}}(\mathbf{a})\, h_j(\mathbf{x}, k) \tag{6.4-2}$$

where the dimension of the vectors $\mathbf{g}_j$ and $\mathbf{h}_j$ is finite. This restriction on the form of $\boldsymbol{\phi}$ is not serious but does rule out such functions as $\sin[ax(k)]$ or $\exp\{-ax(k)\}$. Any known input is contained in $\boldsymbol{\phi}$.

Unfortunately, it is also necessary to assume that the state vector $\mathbf{x}(k)$ is known for all values of k over some finite interval $[k_0, k_f]$. Later we will discuss how this requirement may be weakened but, for the moment, we will simply assume the existence of the required state record.

We wish to select the parameter vector $\mathbf{a}$ so that the following performance index is minimized:

$$J = \sum_{k=k_0}^{k_f-1} \| \mathbf{x}(k+1) - \boldsymbol{\phi}[\mathbf{x}(k), \mathbf{a}, k]\|^2_{\mathbf{Q}(k)} \tag{6.4-3}$$

where $\mathbf{Q}(k)$ is an arbitrary positive definite matrix. Often we will let $\mathbf{Q}(k) = \mathbf{V}_{\mathbf{w}}^{-1}(k)$ to obtain a maximum a posteriori type result. We can handle this optimization problem directly by setting the partial derivative of J with respect to $\mathbf{a}$ to zero to obtain for $\mathbf{Q} = \mathbf{I}$,

$$\sum_{k=0}^{k_f-1} \frac{\partial \boldsymbol{\phi}^T[\mathbf{x}(k), \hat{\mathbf{a}}, k]}{\partial \hat{\mathbf{a}}} \boldsymbol{\phi}[\mathbf{x}(k), \hat{\mathbf{a}}, k] = \sum_{k=k_0}^{k_f-1} \frac{\partial \boldsymbol{\phi}^T[\mathbf{x}(k), \hat{\mathbf{a}}, k]}{\partial \hat{\mathbf{a}}} \mathbf{x}(k+1) \tag{6.4-4}$$

Because of the assumed form for $\boldsymbol{\phi}[\mathbf{x}(k), \mathbf{a}, k]$ [Eq. (6.4-2)], Eq. (6.4-4) becomes a set of m simultaneous algebraic equations whose solution yields the desired plant parameter vector $\mathbf{a}$.

If $\boldsymbol{\phi}[\mathbf{x}(k), \mathbf{a}, k]$ is linear in $\mathbf{a}$, then Eq. (6.4-4) will produce linear algebraic equations which can be easily solved. In this case, $\boldsymbol{\phi}[\mathbf{x}(k), \mathbf{a}, k]$ takes the following form:

$$\boldsymbol{\phi}[\mathbf{x}(k), \mathbf{a}, k] = \mathbf{c}[\mathbf{x}(k), k] + \mathbf{F}[\mathbf{x}(k), k]\mathbf{a} \tag{6.4-5}$$

where $\mathbf{c}[\mathbf{x}(k), k]$ is an N-vector and $\mathbf{F}[\mathbf{x}(k), k]$ is a $N \times m$ matrix. If we substitute this expression for $\boldsymbol{\phi}[\mathbf{x}(k), \mathbf{a}, k]$ into Eq. (6.4-4), we obtain

$$\sum_{k=k_0}^{k_f-1} \mathbf{F}^T[\mathbf{x}(k), k]\{\mathbf{c}[\mathbf{x}(k), k] + \mathbf{F}[\mathbf{x}(k), k]\hat{\mathbf{a}}\} = \sum_{k=k_0}^{k_f-1} \mathbf{F}^T[\mathbf{x}(k), k]\, \mathbf{x}(k+1)$$

or

$$\left\{\sum_{k=k_0}^{k_f-1} \mathbf{F}^T[\mathbf{x}(k), k]\, \mathbf{F}[\mathbf{x}(k), k]\right\} \hat{\mathbf{a}} = \sum_{k=k_0}^{k_f-1} \mathbf{F}^T[\mathbf{x}(k), k]\{\mathbf{x}(k+1) - \mathbf{c}[\mathbf{x}(k), k]\} \tag{6.4-6}$$

Therefore we find that the optimum parameter vector is given by

$$\hat{\mathbf{a}} = \left\{ \sum_{k=k_0}^{k_f-1} \mathbf{F}^T[\mathbf{x}(k), k]\,\mathbf{F}[\mathbf{x}(k), k] \right\} \sum_{k=k_0}^{k_f-1} \mathbf{F}^T[\mathbf{x}(k), k]\{\mathbf{x}(k+1) - \mathbf{c}[\mathbf{x}(k), k]\} \tag{6.4-7}$$

Note that the system can be quite nonlinear in the state as long as it is linear in the parameter vector $\mathbf{a}$.

Example 6.4-1. In order to illustrate the method, let us use the differential approximation algorithm to determine the parameter α in the following scalar system:

$$x(k+1) = x(k) + \alpha x^2(k) + w(k) + u(k)$$

where $u(k)$ is known. This system is linear in α, and hence the result of Eq. (6.4-7) is applicable. In this case, we have $c[x(k), k] = x(k) + u(k)$ and $F[x(k), k] = x^2(k)$, so that the estimate of α becomes

$$\hat{\alpha} = \frac{\sum_{k=k_0}^{k_f-1} x^2(k)[x(k+1) - x(k) - u(k)]}{\sum_{k=k_0}^{k_f-1} x^4(k)}$$

The major difficulty with the use of the above algorithm is the need for a complete knowledge of the system state over a finite time interval. One possible approach for circumventing this difficulty is to use a point estimation technique to make a rough estimate of the state from the available measurements. Then one could use the difference approximation algorithm to obtain parameter estimates. These estimates might not be too accurate if only noisy measurements were initially known. However, the parameter and state estimate would often provide a sufficiently accurate initial trial trajectory for the use of gradient or quasilinearization algorithms.

The difference algorithm can be applied to continuous systems. In this case, the system model has the following form:

$$\dot{\mathbf{x}}(t) = \mathbf{f}[\mathbf{x}(t), \mathbf{a}, t] + \mathbf{w}(t) \tag{6.4-8}$$

where $\mathbf{f}[\mathbf{x}(t), \mathbf{a}, t]$ is assumed to have the same structure as before, i.e., the elements of $\mathbf{f}$ satisfy Eq. (6.4-2). The performance index is given by

$$J = \int_{t_0}^{t_f} \| \dot{\mathbf{x}}(t) - \mathbf{f}[\mathbf{x}(t), \mathbf{a}, t] \|^2_{\mathbf{Q}(t)}\, dt \tag{6.4-9}$$

where $\mathbf{Q}(t)$ is an arbitrary positive definite matrix as before. Here $\mathbf{x}(t)$ and $\dot{\mathbf{x}}(t)$ are assumed to be known over the time interval $[t_0, t_f]$.

If we set the partial derivative of J with respect to $\mathbf{a}$ to zero, we obtain, for $\mathbf{Q} = \mathbf{I}$, the following algebraic equations for $\mathbf{a}$:

$$\int_{t_0}^{t_f} \frac{\partial \mathbf{f}^T[\mathbf{x}(t), \mathbf{a}, t]}{\partial \mathbf{a}} \mathbf{f}[\mathbf{x}(t), \mathbf{a}, t] = \int_{t_0}^{t_f} \frac{\partial \mathbf{f}^T[\mathbf{x}(t), \mathbf{a}, t]}{\partial \mathbf{a}} \dot{\mathbf{x}}(t) \quad (6.4\text{-}10)$$

The continuous algorithm has the added disadvantage, as compared to the discrete algorithm, of a need to know the derivative $\dot{\mathbf{x}}(t)$. One approach is to use a finite difference to approximate the derivative in which case one essentially returns to the discrete algorithm.

6.5. Summary

The method of quasilinearization developed in this chapter may be used to overcome the computational difficulty of two-point boundary value problems associated with certain formulations of system identification problems. Although quasilinearization is primarily used as an indirect computational method, it can also be directly applied to some classes of system identification problems to develop simple and efficient algorithms. Both the discrete and continuous forms of the quasilinearization algorithm were presented.

The technique of difference of differential approximation was also examined. Although these algorithms have some serious limitations, there are problems for which they are applicable or can be used as initialization techniques for iterative approaches such as quasi-linearization or gradient methods.

7

INVARIANT IMBEDDING AND SEQUENTIAL IDENTIFICATION

7.1. Introduction

This chapter is concerned with an approach to system identification which differs from most of the methods previously discussed. Here we will develop sequential or on-line algorithms for system identification. This approach is to be contrasted with the off-line or nonsequential nature of the algorithms discussed in the preceding chapters. In many situations (as, for example, in many control applications), it is desirable to have a "running time" or sequential estimate of system parameters as the observation data is received.

In essence, the sequential estimation of system parameters is nothing more or less than a problem of nonlinear filtering. Voluminous amounts of literature are available on linear and nonlinear filtering (Sage and Melsa, 1971; Jazwinski, 1970, and their references); the interested reader is directed to these sources for extensive discussions of nonlinear filtering. Only one approach to nonlinear filtering is developed here. The method of invariant imbedding is developed and applied to the two-point boundary value problems of system identification developed in Chap. 3. This method was selected because it is conceptually simple and very flexible. The continuous case is developed first because it is easier than the discrete case.

7.2. Continuous Systems

In this section, we will consider the continuous formulation of the invariant imbedding algorithm. In order to simplify the notation in the development and to maximize the generality of the results, we will base the derivation on a general two-point boundary value problem. We wish to find the solution of the TPBVP represented by

$$\dot{\mathbf{x}}(t) = \boldsymbol{\gamma}[\mathbf{x}(t), \boldsymbol{\lambda}(t), t] \tag{7.2-1}$$

$$\dot{\boldsymbol{\lambda}}(t) = \boldsymbol{\beta}[\mathbf{x}(t), \boldsymbol{\lambda}(t), t] \tag{7.2-2}$$

with the split boundary conditions

$$\boldsymbol{\lambda}(t_0) = \mathbf{A}\mathbf{x}(t_0) + \mathbf{b}, \qquad \boldsymbol{\lambda}(t_f) = \mathbf{0} \tag{7.2-3}$$

This formulation encompasses all of the two-point boundary value problems developed in Chap. 3.

The basic concept of invariant imbedding is to change a specific problem into a more general problem. If one can solve the general problem, then the specific problem is automatically solved. The amazing fact is that it is often easier to solve the general problem than the specific problem.

Here we will generalize (that is, invariantly imbed) the TPBVP of Eqs. (7.2-1)–(7.2-3) by letting the terminal boundary condition on $\boldsymbol{\lambda}(t)$ take a general value $\mathbf{c}$ rather $\mathbf{0}$. In other words, the boundary condition $\boldsymbol{\lambda}(t_f) = \mathbf{0}$ will be replaced by $\boldsymbol{\lambda}(t_f) = \mathbf{c}$. In addition, we will assume that both the boundary value $\mathbf{c}$ and the terminal time t_f are variable; in particular, we wish to consider two neighboring trajectories, one which satisfies the boundary condition $\boldsymbol{\lambda}(t_f) = \mathbf{c}$ and one which satisfies $\boldsymbol{\lambda}(t_f + \epsilon) = \mathbf{c} + \Delta\mathbf{c}$.

For the trajectory which satisfies $\boldsymbol{\lambda}(t_f) = \mathbf{c}$, we assume that the terminal value for $\mathbf{x}(t)$ is given by

$$\mathbf{x}(t_f) = \mathbf{r}(\mathbf{c}, t_f) \tag{7.2-4}$$

In other words, the function $\mathbf{r}(\mathbf{c}, t_f)$ represents the relation between the boundary condition $\boldsymbol{\lambda}(t_f) = \mathbf{c}$ and the terminal value of $\mathbf{x}(t)$. If this function were known, then the TPBVP could be trivially solved by integrating $\mathbf{x}$ and $\boldsymbol{\lambda}$ backward in time from the boundary conditions $\boldsymbol{\lambda}(t_f) = \mathbf{c}$ and $\mathbf{x}(t_f) = \mathbf{r}(\mathbf{c}, t_f)$. But we do not know $\mathbf{r}(\mathbf{c}, t_f)$, and hence we wish to develop a technique for determining it.

Let us suppose that we have a trajectory which ends at t_f with $\boldsymbol{\lambda}(t_f) = \mathbf{c}$, so that $\mathbf{x}(t_f)$ is $\mathbf{r}(\mathbf{c}, t_f)$. Now let us extend the terminal time slightly so that it becomes $t_f + \epsilon$ and

$$\boldsymbol{\lambda}(t_f + \epsilon) = \mathbf{c} + \Delta\mathbf{c} \tag{7.2-5}$$

and correspondingly

$$\mathbf{x}(t_f + \epsilon) = \mathbf{x}(t_f) + \Delta\mathbf{x} = \mathbf{r}(\mathbf{c}, t_f) + \Delta\mathbf{x} \tag{7.2-6}$$

where $\Delta\mathbf{c}$ and $\Delta\mathbf{x}$ are $O(\epsilon)$.[1] But $\mathbf{x}(t_f + \epsilon)$ is also given by

$$\mathbf{x}(t_f + \epsilon) = \mathbf{r}(\mathbf{c} + \Delta\mathbf{c}, t_f + \epsilon) \tag{7.2-7}$$

Now let us equate Eqs. (7.2-6) and (7.2-7) to obtain

$$\mathbf{r}(\mathbf{c}, t_f) + \Delta\mathbf{x} = \mathbf{r}(\mathbf{c} + \Delta\mathbf{c}, t_f + \epsilon) \tag{7.2-8}$$

If we make a Taylor series expansion of the right-hand side of Eq. (7.2-8) about $\mathbf{c}$ and t_f, we obtain

$$\Delta\mathbf{x} \doteq \frac{\partial\mathbf{r}(\mathbf{c}, t_f)}{\partial\mathbf{c}}\Delta\mathbf{c} + \frac{\partial\mathbf{r}(\mathbf{c}, t_f)}{\partial t_f}\epsilon + O(\epsilon^2) \tag{7.2-9}$$

By the use of Eqs. (7.2-1) and (7.2-2), we see that

$$\Delta\mathbf{x} = \boldsymbol{\gamma}[\mathbf{r}(\mathbf{c}, t_f), \mathbf{c}, t_f]\epsilon + O(\epsilon^2) \tag{7.2-10}$$

and

$$\Delta\mathbf{c} = \boldsymbol{\beta}[\mathbf{r}(\mathbf{c}, t_f), \mathbf{c}, t_f]\epsilon + O(\epsilon^2) \tag{7.2-11}$$

so that Eq. (7.2-9) becomes

$$\boldsymbol{\gamma}[\mathbf{r}(\mathbf{c}, t_f), \mathbf{c}, t_f]\epsilon = \frac{\partial\mathbf{r}(\mathbf{c}, t_f)}{\partial\mathbf{c}}\boldsymbol{\beta}[\mathbf{r}(\mathbf{c}, t_f), \mathbf{c}, t_f]\epsilon + \frac{\partial\mathbf{r}(\mathbf{c}, t_f)}{\partial t_f}\epsilon + O(\epsilon^2) \tag{7.2-12}$$

If we divide by ϵ and let ϵ approach zero, we obtain the following partial differential equation which $\mathbf{r}(\mathbf{c}, t_f)$ must satisfy:

$$\boldsymbol{\gamma}(\mathbf{r}, \mathbf{c}, t_f) = \frac{\partial\mathbf{r}(\mathbf{c}, t_f)}{\partial\mathbf{c}}\boldsymbol{\beta}(\mathbf{r}, \mathbf{c}, t_f) + \frac{\partial\mathbf{r}(\mathbf{c}, t_f)}{\partial t_f} \tag{7.2-13}$$

[1] The notation $O(\epsilon^j)$ is used to indicate that $\lim_{\epsilon\to 0} O(\epsilon^j)/\epsilon^{j-1} = \mathbf{0}$.

Unfortunately, this partial differential equation has no known general solution. However, we can often approximate the solution by a linear function given by

$$\mathbf{r}(\mathbf{c}, t_f) = \hat{\mathbf{x}}(t_f) + \mathbf{P}(t_f)\mathbf{c} \tag{7.2-14}$$

Here $\hat{\mathbf{x}}(t_f)$ is the solution if $\mathbf{c} = \mathbf{0}$, which is the correct solution of the original TPBVP with $\boldsymbol{\lambda}(t_f) = \mathbf{0}$. In other words, we are assuming that $\mathbf{r}(\mathbf{c}, t_f)$ is the optimal value for $\mathbf{x}(t_f)$ if $\mathbf{c} = \mathbf{0}$ (as yet unknown) plus a linear weighting of the deviation of $\boldsymbol{\lambda}(t_f)$ from zero given by $\mathbf{c}$. This should be a good approximation if $\mathbf{c} \sim \mathbf{0}$, that is, if we are close to the true optimal solution. We will assume that $\mathbf{c}$ is small in the following development and ignore terms of $O(\| \mathbf{c} \|^2)$ or higher.

If we substitute the assumed form of $\mathbf{r}(\mathbf{c}, t_f)$ as given by Eq. (7.2-14) into Eq. (7.2-13), we obtain

$$\boldsymbol{\gamma}[\hat{\mathbf{x}} + \mathbf{P}\mathbf{c}, \mathbf{c}, t_f] = \mathbf{P}\boldsymbol{\beta}(\hat{\mathbf{x}} + \mathbf{P}\mathbf{c}, \mathbf{c}, t_f) + \dot{\hat{\mathbf{x}}} + \dot{\mathbf{P}}\mathbf{c} \tag{7.2-15}$$

Next we expand $\boldsymbol{\gamma}$ and $\boldsymbol{\beta}$ in a Taylor series about $\mathbf{x}$, $\mathbf{c}$, and t_f, retaining only first-order terms in $\mathbf{c}$. Equation (7.2-15) becomes

$$\boldsymbol{\gamma}(\hat{\mathbf{x}}, \mathbf{c}, t_f) + \frac{\partial \boldsymbol{\gamma}(\hat{\mathbf{x}}, \mathbf{c}, t_f)}{\partial \hat{\mathbf{x}}} \mathbf{P}\mathbf{c} = \mathbf{P}\boldsymbol{\beta}(\hat{\mathbf{x}}, \mathbf{c}, t_f) + \mathbf{P} \frac{\partial \boldsymbol{\beta}(\hat{\mathbf{x}}, \mathbf{c}, t_f)}{\partial \hat{\mathbf{x}}} \mathbf{P}\mathbf{c} + \dot{\hat{\mathbf{x}}} + \dot{\mathbf{P}}\mathbf{c} \tag{7.2-16}$$

In order to continue the development, it is necessary to substitute the actual functions for $\boldsymbol{\gamma}$ and $\boldsymbol{\beta}$ associated with the TPBVP of system identification. For simplicity, we will use the TPBVP of Eqs. (3.2-37)–(3.2-39); the extension to the more general TPBVP of Eqs. (3.2-63)–(3.2-65) is direct. From Eqs. (3.2-37) and (3.2-38), the functions $\boldsymbol{\gamma}$ and $\boldsymbol{\beta}$ are given by

$$\boldsymbol{\gamma}(\hat{\mathbf{x}}, \mathbf{c}, t_f) = \mathbf{f}(\hat{\mathbf{x}}, t_f) - \mathbf{G}(\hat{\mathbf{x}}, t_f)\, \boldsymbol{\Psi}_w(t_f)\, \mathbf{G}^T(\hat{\mathbf{x}}, t_f)\mathbf{c} \tag{7.2-17}$$

$$\boldsymbol{\beta}(\hat{\mathbf{x}}, \mathbf{c}, t_f) = \frac{\partial \mathbf{h}^T(\hat{\mathbf{x}}, t_f)}{\partial \hat{\mathbf{x}}} \boldsymbol{\Psi}_v^{-1}[\mathbf{z} - \mathbf{h}(\hat{\mathbf{x}}, t_f)] - \frac{\partial \mathbf{f}^T(\hat{\mathbf{x}}, t_f)}{\partial \hat{\mathbf{x}}} \mathbf{c} + \frac{\partial [\mathbf{c}^T \mathbf{G}(\hat{\mathbf{x}}, t_f)\, \boldsymbol{\Psi}_w(t_f)\, \mathbf{G}^T(\hat{\mathbf{x}}, t_f)]}{\partial \hat{\mathbf{x}}} \mathbf{c} \tag{7.2-18}$$

We note that we can delete the last term in the expression for $\boldsymbol{\beta}$ because it is quadratic in $\mathbf{c}$ and we are retaining only linear terms. Now if we substitute Eqs. (7.2-17) and (7.2-18) into Eq. (7.2-16), we obtain the following expression where no terms which are of higher order than the first $\mathbf{c}$ have been retained:

$$\mathbf{f}(\hat{\mathbf{x}}, t_f) - \mathbf{G}(\hat{\mathbf{x}}, t_f)\,\boldsymbol{\Psi}_w(t_f)\,\mathbf{G}^T(\hat{\mathbf{x}}, t_f)\mathbf{c} + \frac{\partial \mathbf{f}(\hat{\mathbf{x}}, t_f)}{\partial \hat{\mathbf{x}}}\mathbf{P}\mathbf{c}$$

$$= \mathbf{P}\frac{\partial \mathbf{h}^T(\hat{\mathbf{x}}, t_f)}{\partial \hat{\mathbf{x}}}\boldsymbol{\Psi}_v^{-1}[\mathbf{z} - \mathbf{h}(\hat{\mathbf{x}}, t_f)] - \mathbf{P}\frac{\partial \mathbf{f}^T(\hat{\mathbf{x}}, t_f)}{\partial \hat{\mathbf{x}}}\mathbf{c}$$

$$+ \mathbf{P}\frac{\partial}{\partial \hat{\mathbf{x}}}\left\{\frac{\partial \mathbf{h}^T(\hat{\mathbf{x}}, t_f)}{\partial \hat{\mathbf{x}}}\boldsymbol{\Psi}_v^{-1}[\mathbf{z} - \mathbf{h}(\hat{\mathbf{x}}, t_f)]\right\}\mathbf{P}\mathbf{c} + \dot{\hat{\mathbf{x}}} + \dot{\mathbf{P}}\mathbf{c} \tag{7.2-19}$$

Since the expression is to be true for all $\mathbf{c}$ (which are sufficiently small), we may separately equate the coefficients of the zero and first power of $\mathbf{c}$ to obtain

$$\dot{\hat{\mathbf{x}}} = \mathbf{f}(\hat{\mathbf{x}}, t_f) - \mathbf{P}\frac{\partial \mathbf{h}^T(\hat{\mathbf{x}}, t_f)}{\partial \hat{\mathbf{x}}}\boldsymbol{\Psi}_v^{-1}(t_f)[\mathbf{z} - \mathbf{h}(\hat{\mathbf{x}}, t_f)] \tag{7.2-20}$$

$$\dot{\mathbf{P}} = -\mathbf{G}(\hat{\mathbf{x}}, t_f)\,\boldsymbol{\Psi}_w(t_f)\,\mathbf{G}^T(\hat{\mathbf{x}}, t_f) + \mathbf{P}\frac{\partial \mathbf{f}^T(\hat{\mathbf{x}}, t_f)}{\partial \hat{\mathbf{x}}} + \frac{\partial \mathbf{f}(\hat{\mathbf{x}}, t_f)}{\partial \hat{\mathbf{x}}}\mathbf{P}$$

$$- \mathbf{P}\frac{\partial}{\partial \mathbf{x}}\left\{\frac{\partial \mathbf{h}^T(\hat{\mathbf{x}}, t_f)}{\partial \hat{\mathbf{x}}}\boldsymbol{\Psi}_v^{-1}(t_f)[\mathbf{z} - \mathbf{h}(\hat{\mathbf{x}}, t_f)]\right\}\mathbf{P} \tag{7.2-21}$$

We may put this result into a slightly more familiar form by substituting $\mathbf{P} = -\mathbf{P}$. At the same time, we recognize that the terminal time is variable and may be written as just t. With these changes, Eqs. (7.2-20) and (7.2-21) become

$$\dot{\hat{\mathbf{x}}}(t) = \mathbf{f}[\hat{\mathbf{x}}(t), t] + \mathbf{P}(t)\frac{\partial \mathbf{h}^T[\hat{\mathbf{x}}(t), t]}{\partial \hat{\mathbf{x}}(t)}\boldsymbol{\Psi}_v^{-1}(t)\{\mathbf{z}(t) - \mathbf{h}[\hat{\mathbf{x}}(t), t]\} \tag{7.2-22}$$

$$\dot{\mathbf{P}}(t) = \mathbf{G}[\hat{\mathbf{x}}(t), t]\,\boldsymbol{\Psi}_w(t)\,\mathbf{G}^T[\hat{\mathbf{x}}(t), t] + \mathbf{P}(t)\frac{\partial \mathbf{f}^T[\hat{\mathbf{x}}(t), t]}{\partial \hat{\mathbf{x}}(t)}$$

$$+ \frac{\partial \mathbf{f}[\hat{\mathbf{x}}(t), t]}{\partial \hat{\mathbf{x}}(t)}\mathbf{P}(t) - \mathbf{P}(t)\frac{\partial}{\partial \hat{\mathbf{x}}(t)}\left[\frac{\partial \mathbf{h}^T[\hat{\mathbf{x}}(t), t]}{\partial \hat{\mathbf{x}}(t)}\boldsymbol{\Psi}_v^{-1}(t)\right.$$

$$\left.\times \{\mathbf{z}(t) - \mathbf{h}[\hat{\mathbf{x}}(t), t]\}\right]\mathbf{P}(t) \tag{7.2-23}$$

Initial conditions for Eqs. (7.2-22) and (7.2-23) may be obtained from the boundary condition of Eq. (3.2-39) which is

$$\boldsymbol{\lambda}(t_0) = -\mathbf{V}_{x0}^{-1}[\hat{\mathbf{x}}(t_0) - \boldsymbol{\mu}_{x0}] \tag{7.2-24}$$

If we solve this expression for $\hat{\mathbf{x}}(t_0)$, we obtain

$$\hat{\mathbf{x}}(t_0) = \boldsymbol{\mu}_{x0} - \mathbf{V}_{x0}\boldsymbol{\lambda}(t_0) \tag{7.2-25}$$

A simple comparison of Eqs. (7.2-14) and (7.2-25) reveals that we should use the initial conditions

$$\hat{\mathbf{x}}(t_0) = \boldsymbol{\mu}_{x0} \tag{7.2-26}$$

$$\mathbf{P}(t_0) = \mathbf{V}_{x0} \tag{7.2-27}$$

for Eqs. (7.2-22) and (7.2-23). The reader should remember that we made the substitution $\mathbf{P} = -\mathbf{P}$ in writing Eqs. (7.2-22) and (7.2-23) from Eqs. (7.2-20) and (7.2-21), which explains the lack of a minus sign in Eq. (7.2-27).

Now we may integrate Eqs. (7.2-22) and (7.2-23) forward in time from the initial conditions of Eqs. (7.2-26) and (7.2-27) to obtain the sequential estimate of the "state" $\mathbf{x}(t)$. In the case of the constant parameters, which are the items of major interest in system identification, the value obtained at $t = t_f$ is both the filtered and smoothed estimate. Hence the terminal values of the parameter estimates are identical to the values which would be obtained by solving the TPBVP in a nonsequential fashion. The algorithm is summarized in Table 7.2-1 for easy reference.

Equation (7.2-23) looks very similar to the error variance algorithm for Kalman filtering. If $\mathbf{f}$, $\mathbf{g}$, and $\mathbf{h}$ are linear functions of $\mathbf{x}$, it is identical to the error variance in filtering. It is not a difficult task to show (Sage and Melsa, 1971) that Eq. (7.2-23) is a first-order approximation to the conditional error variance in estimation

$$\mathbf{V}_{\tilde{\mathbf{x}}}(t) = \text{var}\{\mathbf{x}(t) \mid Z(t)\}$$

if the initial condition $\mathbf{x}(t_0)$ and the noise inputs $\mathbf{w}(t)$ and $\mathbf{v}(t)$ are Gaussian, as assumed in Chap. 3.

Example 7.2-1. In order to illustrate the use of the above algorithm for system identification, let us apply the technique to the problem of identifying a time-varying parameter in a second-order system (Detchmendy and Sridhar, 1965). The system model is

$$\dot{x}_1 = x_2 + w_1(t)$$

$$\dot{x}_2 = -2x_1 - a(t)\, x_1^3 - 3x_2 + 5 \sin t + w_2(t)$$

TABLE 7.2-1

CONTINUOUS INVARIANT IMBEDDING ALGORITHM

System model	$\dot{\mathbf{x}}(t) = \mathbf{f}[\mathbf{x}(t), t] + \mathbf{G}[\mathbf{x}(t), t]\mathbf{w}(t)$	(3.2-5)
Observation model	$\mathbf{z}(t) = \mathbf{h}[\mathbf{x}(t), t] + \mathbf{v}(t)$	(3.2-6)
Statistical parameters	$\mathscr{E}\{\mathbf{x}(t_0)\} = \boldsymbol{\mu}_{\mathbf{x}0}, \quad \operatorname{var}\{\mathbf{x}(t_0)\} = \mathbf{V}_{\mathbf{x}0}$ $\mathscr{E}\{\mathbf{w}(t)\} = \mathscr{E}\{\mathbf{v}(t)\} = \mathbf{0}$ $\operatorname{cov}\{\mathbf{w}(t), \mathbf{w}(\tau)\} = \boldsymbol{\Psi}_{\mathbf{w}}(t)\delta_{\mathrm{D}}(t-\tau)$ $\operatorname{cov}\{\mathbf{v}(t), \mathbf{v}(\tau)\} = \boldsymbol{\Psi}_{\mathbf{v}}(t)\delta_{\mathrm{D}}(t-\tau)$	
Filter algorithm	$\dot{\hat{\mathbf{x}}}(t) = \mathbf{f}[\hat{\mathbf{x}}(t), t] + \mathbf{P}(t)\dfrac{\partial \mathbf{h}^T[\hat{\mathbf{x}}(t), t]}{\partial \hat{\mathbf{x}}(t)}\boldsymbol{\Psi}_{\mathbf{v}}^{-1}(t)\{\mathbf{z}(t) - \mathbf{h}[\hat{\mathbf{x}}(t), t]\}$	(7.2-22)
Error variance algorithm	$\dot{\mathbf{P}}(t) = \mathbf{G}[\hat{\mathbf{x}}(t), t]\boldsymbol{\Psi}_{\mathbf{w}}(t)\mathbf{G}^{\mathrm{T}}[\hat{\mathbf{x}}(t), t] + \mathbf{P}(t)\dfrac{\partial \mathbf{f}^{\mathrm{T}}[\hat{\mathbf{x}}(t), t]}{\partial \hat{\mathbf{x}}(t)} + \dfrac{\partial \mathbf{f}[\hat{\mathbf{x}}(t), t]}{\partial \hat{\mathbf{x}}(t)}\mathbf{P}(t) - \mathbf{P}(t)\dfrac{\partial}{\partial \hat{\mathbf{x}}(t)} \times \left[\dfrac{\partial \mathbf{h}^{\mathrm{T}}[\hat{\mathbf{x}}(t), t]}{\partial \hat{\mathbf{x}}(t)}\boldsymbol{\Psi}_{\mathbf{v}}^{-1}(t)\{\mathbf{z}(t) - \mathbf{h}[\hat{\mathbf{x}}(t), t]\}\right]\mathbf{P}(t)$	(7.2-23)
Initial conditions	$\hat{\mathbf{x}}(t_0) = \boldsymbol{\mu}_{\mathbf{x}0}, \quad \mathbf{P}(t_0) = \mathbf{V}_{\mathbf{x}0}$	

where

$$a(t) = 2e^{-0.1t}$$

and the observation model is

$$z(t) = x_1(t) + v(t)$$

Suppose that the form of $a(t)$ is known, but that its time constant and initial value are unknown. We may model $a(t)$ as

$$\dot{a}(t) = \dot{x}_3(t) = -x_4(t)\,x_3(t) + w_3(t)$$

$$\dot{x}_4(t) = -w_4(t)$$

where the initial conditions on $x_3(t)$ and $x_4(t)$ are unknown. We assume that $\mu_{\mathbf{x}0} = \mathbf{0}$ and

$$\mathbf{P}(0) = \frac{1}{2}\begin{bmatrix} 3 & 1 & 1 & 1 \\ 1 & 3 & 1 & 1 \\ 1 & 1 & 3 & 1 \\ 1 & 1 & 1 & 3 \end{bmatrix}$$

and $\Psi_v = 1$ and $\mathbf{\Psi}_\mathbf{w} = \mathbf{I}$. The estimator equation for this problem are given by

$$\dot{\hat{x}}_1(t) = \hat{x}_2(t) + P_{11}(t)[z(t) - \hat{x}_1(t)]$$

$$\dot{\hat{x}}_2(t) = -2\hat{x}_1(t) - \hat{x}_3(t)\,\hat{x}_1^3(t) - 3\hat{x}_2(t) + 5\sin t + P_{11}(t)[z(t) - \hat{x}_1(t)]$$

$$\dot{\hat{x}}_3(t) = -\hat{x}_4(t)\,\hat{x}_3(t) + P_{31}(t)[z(t) - \hat{x}_1(t)]$$

$$\dot{\hat{x}}_4(t) = P_{41}(t)[z(t) - \hat{x}_1(t)]$$

and

$$\dot{\mathbf{P}} = -\mathbf{P}\mathbf{H}\mathbf{H}^T\mathbf{P} + \frac{\partial \mathbf{f}[\hat{\mathbf{x}}, t]}{\partial \hat{\mathbf{x}}}\mathbf{P} + \mathbf{P}\frac{\partial \mathbf{f}^T[\hat{\mathbf{x}}, t]}{\partial \hat{\mathbf{x}}} + \mathbf{I}$$

where

$$\mathbf{H}^T = \begin{bmatrix} 1 \\ 0 \\ 0 \\ 0 \end{bmatrix}, \qquad \mathbf{f}[\mathbf{x}, t] = \begin{bmatrix} x_2 \\ -2x_1 - x_3 x_1^3 - 3x_2 \\ -x_4 x_3 \\ 0 \end{bmatrix}$$

Figure 7.2-1 shows the estimates of $\mathbf{x}(t)$ for this problem. Note that the estimates of $x_1(t)$ and $x_2(t)$ "track" much sooner than $x_3(t) = a(t)$ and $x_4(t) = b$.

Example 7.2-2. As another example of the use of the invariant imbedding technique to solve system identification problems, let us consider the determination of bias errors in linear sequential estimation (Sage, 1970; Sage and Lin, 1971). The message model is

$$\dot{\mathbf{x}}(t) = \mathbf{F}(t)\,\mathbf{x}(t) + \mathbf{G}(t)\,\mathbf{w}(t) \tag{1}$$

while the observation model is

$$\mathbf{z}(t) = \mathbf{H}(t)\,\mathbf{x}(t) + \mathbf{v}(t) \tag{2}$$

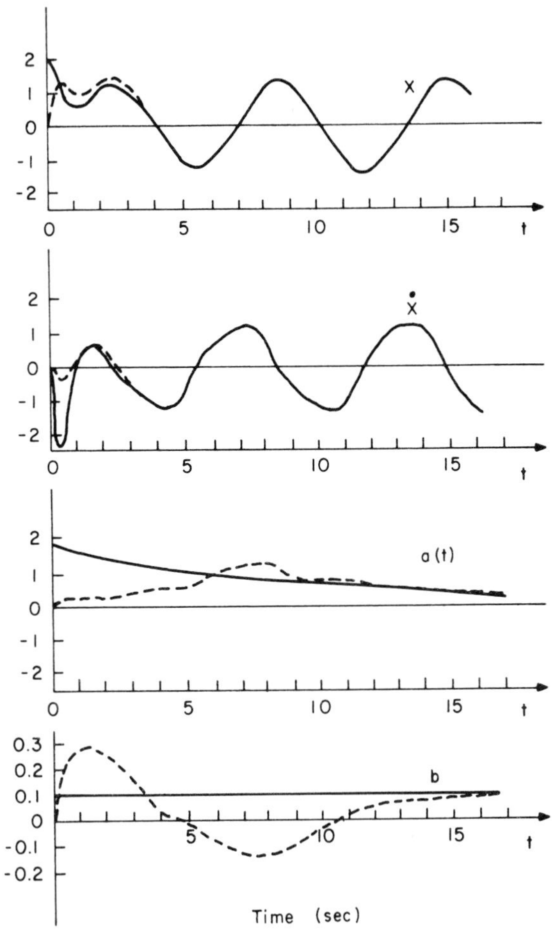

FIG. 7.2-1. State and parameter estimates for Example 7.2-1. ——— true value, - - - estimated value.

The sequential filtering problem in which one desires to estimate $\mathbf{x}(t)$ based on the observation sequence $Z(t) = \{\mathbf{z}(\tau), 0 \leqslant \tau \leqslant t\}$ has been solved by numerous authors (Sage and Melsa, 1971). Usually the assumption is made that the means of the plant and measurement noise, $\mathbf{w}(t)$ and $\mathbf{v}(t)$, are known. If this is not the case, serious error and possibly filter divergence may result. Let us consider the use of the method of maximum likelihood estimation in order to determine adaptive sequential estimation algorithms for unknown plant and measurement noise means. It is known that maximizing the likelihood function is equivalent to solving a deterministic optimization problem. This leads to a two-point boundary value problem which is resolved by the method of invariant imbedding so as to obtain a sequential estimator for plant and measurement noise mean values.

It is a relatively straightforward task to show that maximizing the likelihood function $p[Z(t_f) \mid \boldsymbol{\mu}_w, \boldsymbol{\mu}t]$ for the message and observation models of (1) and (2), where $\mathbf{x}(0)$, $\mathbf{w}(t)$, and $\mathbf{v}(t)$ are uncorrelated Gaussian random variables with

$$\begin{aligned} \boldsymbol{\mu}_w &\triangleq \mathscr{E}\{w(t)\}, & \operatorname{cov}\{\mathbf{w}(t), \mathbf{w}(\tau)\} &= \boldsymbol{\Psi}_w(t)\,\delta(t-\tau) \\ \boldsymbol{\mu}_v &\triangleq \mathscr{E}\{v(t)\}, & \operatorname{cov}\{\mathbf{v}(t), \mathbf{v}(\tau)\} &= \boldsymbol{\Psi}_v(t)\,\delta(t-\tau) \\ \mathscr{E}\{\mathbf{x}(0)\} &\triangleq \boldsymbol{\mu}_{x0}, & \operatorname{var}\{\mathbf{x}(0)\} &= \mathbf{V}_{x0} \end{aligned} \tag{3}$$

is equivalent to minimization of the cost function

$$J = \frac{1}{2}\int_0^{t_f} \| \mathbf{z}(t) - \boldsymbol{\mu}_v - \mathbf{H}(t)\,\mathbf{x}(t)\|^2_{\boldsymbol{\Psi}_v^{-1}(t)}\, dt \tag{4}$$

where $\mathbf{x}(t)$ is the minimum variance estimate given by

$$\dot{\hat{\mathbf{x}}} = \mathbf{F}\hat{\mathbf{x}} + \mathbf{G}\boldsymbol{\mu}_w + \mathbf{K}(\mathbf{z} - \boldsymbol{\mu}_v - \mathbf{H}\mathbf{x}) \tag{5}$$

$$\mathbf{K} = \mathbf{V}\mathbf{H}^{\mathrm{T}}\boldsymbol{\Psi}_v^{-1} \tag{6}$$

$$\dot{\mathbf{V}} = \mathbf{F}\mathbf{V} + \mathbf{V}\mathbf{F}^{\mathrm{T}} + \mathbf{G}\boldsymbol{\Psi}_w\mathbf{G} - \mathbf{V}\mathbf{H}^{\mathrm{T}}\boldsymbol{\Psi}_v^{-1}\mathbf{H}\mathbf{V} \tag{7}$$

$$\hat{\mathbf{x}}(0) = \boldsymbol{\mu}_{x0} \tag{8}$$

$$\mathbf{V}(0) = \mathbf{V}_{x0} \tag{9}$$

The plant and measurement noise means are assumed to be constant in time such that

$$\dot{\boldsymbol{\mu}}_w = \mathbf{0} \tag{10}$$

$$\dot{\boldsymbol{\mu}}_v = \mathbf{0} \tag{11}$$

are necessary equations which describe the constant means.

Thus the problem of maximum likelihood estimation of the bias in sequential filtering has been reduced to an optimal control problem. It is desired to minimize the cost function of (4) subject to the equality constraints of (5), (8), (10), and (11).

The optimization problem developed in the previous section is a standard one to which the techniques of optimum systems control may readily be applied. There results the two-point boundary value problem of (5), (10), and (11) with the adjoint equations

$$\dot{\boldsymbol{\lambda}} = \mathbf{H}^T \boldsymbol{\Psi}_v^{-1}(\mathbf{z} - \boldsymbol{\mu}_v - \mathbf{H}\hat{\mathbf{x}}) - (\mathbf{F}^T - \mathbf{H}^T\mathbf{K}^T)\boldsymbol{\lambda} \tag{12}$$

$$\dot{\boldsymbol{\omega}} = -\mathbf{G}^T\boldsymbol{\lambda} \tag{13}$$

$$\dot{\boldsymbol{\nu}} = \boldsymbol{\Psi}_v^{-1}(\mathbf{z} - \boldsymbol{\mu}_v - \mathbf{H}\mathbf{x}) + \mathbf{K}^T\boldsymbol{\lambda} \tag{14}$$

The two-point boundary conditions for the foregoing are

$$\hat{\mathbf{x}}(0) = \boldsymbol{\mu}_{x0}, \qquad \boldsymbol{\lambda}(t_f) = \mathbf{0}$$
$$\boldsymbol{\omega}(0) = \mathbf{0}, \qquad \boldsymbol{\omega}(t_f) = \mathbf{0}$$
$$\boldsymbol{\nu}(0) = \mathbf{0}, \qquad \boldsymbol{\nu}(t_f) = \mathbf{0}$$

If this TPBVP is solved for $t \in [0, t_f]$, fixed interval smoothing solutions for $\boldsymbol{\mu}_w$ and $\boldsymbol{\mu}_v$ are obtained.

The method of invariant imbedding may now be used to obtain a sequential form for the maximum likelihood estimates of $\boldsymbol{\mu}_v$ and $\boldsymbol{\mu}_w$. This approach leads to the following algorithm:

$$\boldsymbol{\mu}_w = (\mathbf{P}_{12}^T\mathbf{H}^T + \mathbf{P}_{23})\, \boldsymbol{\Psi}_v^{-1}(\mathbf{z} - \hat{\boldsymbol{\mu}}_v - \mathbf{H}\hat{\mathbf{x}}) \tag{15}$$

$$\hat{\boldsymbol{\mu}}_v = (\mathbf{P}_{13}^T\mathbf{H}^T + \mathbf{P}_{33})\, \boldsymbol{\Psi}_v^{-1}(\mathbf{z} - \hat{\boldsymbol{\mu}}_v - \mathbf{H}\hat{\mathbf{x}}) \tag{16}$$

$$\dot{\hat{\mathbf{x}}} = \mathbf{F}\hat{\mathbf{x}} + \mathbf{G}\hat{\boldsymbol{\mu}}_w + (\mathbf{K} + \mathbf{P}_{11}\mathbf{H}^T\boldsymbol{\Psi}_v^{-1} + \mathbf{P}_{13}\boldsymbol{\Psi}_v^{-1})(\mathbf{z} - \hat{\boldsymbol{\mu}}_v - \mathbf{H}\hat{\mathbf{x}}) \tag{17}$$

$$\dot{\mathbf{P}}_{11} = (\mathbf{F} - \mathbf{K}\mathbf{H})\,\mathbf{P}_{11} + \mathbf{P}_{11}(\mathbf{F} - \mathbf{K}\mathbf{H})^T + \mathbf{G}\mathbf{P}_{12}^T + \mathbf{P}_{12}\mathbf{G}^T - \mathbf{K}\mathbf{P}_{13}^T - \mathbf{P}_{13}\mathbf{K}^T - (\mathbf{P}_{11}\mathbf{H}^T + \mathbf{P}_{13})\, \boldsymbol{\Psi}_v^{-1}(\mathbf{H}\mathbf{P}_{11} + \mathbf{P}_{13}^T) \tag{18}$$

$$\dot{\mathbf{P}}_{12} = (\mathbf{F} - \mathbf{K}\mathbf{H})\,\mathbf{P}_{12} + \mathbf{G}\mathbf{P}_{22} - \mathbf{K}\mathbf{P}_{23}^T - (\mathbf{P}_{11}\mathbf{H}^T + \mathbf{P}_{13})\, \boldsymbol{\Psi}_v^{-1}(\mathbf{H}\mathbf{P}_{12} + \mathbf{P}_{23}^T) \tag{19}$$

$$\dot{\mathbf{P}}_{13} = (\mathbf{F} - \mathbf{K}\mathbf{H})\,\mathbf{P}_{13} + \mathbf{G}\mathbf{P}_{23} - \mathbf{K}\mathbf{P}_{33} - (\mathbf{P}_{11}\mathbf{H}^T + \mathbf{P}_{13})\, \boldsymbol{\Psi}_v^{-1}(\mathbf{H}\mathbf{P}_{13} + \mathbf{P}_{33}) \tag{20}$$

$$\dot{\mathbf{P}}_{22} = -(\mathbf{P}_{12}^T\mathbf{H}^T + \mathbf{P}_{23})\, \boldsymbol{\Psi}_v^{-1}(\mathbf{H}\mathbf{P}_{12} + \mathbf{P}_{23}^T) \tag{21}$$

$$\dot{\mathbf{P}}_{23} = -(\mathbf{P}_{12}^T\mathbf{H}^T + \mathbf{P}_{23})\, \boldsymbol{\Psi}_v^{-1}(\mathbf{H}\mathbf{P}_{12} + \mathbf{P}_{33}) \tag{22}$$

$$\dot{\mathbf{P}}_{33} = -(\mathbf{P}_{13}^T\mathbf{H}^T + \mathbf{P}_{33})\, \boldsymbol{\Psi}_v^{-1}(\mathbf{H}\mathbf{P}_{13} + \mathbf{P}_{33}) \tag{23}$$

Since this is maximum likelihood estimation, it is meaningless to attempt to determine optimum initial conditions for other than $\hat{\mathbf{x}}(0)$, which should be set at the prior mean $\mu_x(0)$. "Reasonable" values for other initial conditions may be used. Alternatively, the TPBVP may be solved for a short interval of time to provide appropriate initial conditions.

It should be noted that the $\hat{\mathbf{x}}(t)$ solution of (17) is not the Kalman filter estimate of $\mathbf{x}(t)$ of (5), although for sufficiently large time, (17) converges to (5). In fact, it is not necessary even to solve (5) in order to determine the maximum likelihood estimate of μ_w and μ_v. It is necessary, however, to solve (6) and (7), as they are an integral part of the maximum likelihood estimation algorithms. There is considerable computational merit inherent in this approach. It is possible to show that the $\mathbf{K}$ term in (17) predominates the terms $(\mathbf{P}_{11}\mathbf{H}^T+\mathbf{P}_{13})\Psi_v^{-1}$, and thus computational errors in determining these terms will not greatly affect the estimate $\hat{\mathbf{x}}$, in (17). Computational errors in determining the $(\mathbf{P}_{11}\mathbf{H}^T + \mathbf{P}_{13})\Psi_v^{-1}$ terms are quite likely to be large compared with computational errors in determining $\mathbf{K}$ or $\mathbf{V}$, since the order of the matrix Riccati equation for $\mathbf{P}$ may be considerably higher than the order of the Riccati equation for $\mathbf{V}$.

The adaptive filter problem with unknown bias (unknown plant and measurement noise mean) may also be solved by adjoining the relations $\dot{\mu}_w = \mathbf{0}$, $\dot{\mu}_v = \mathbf{0}$ to the message model of (1). In this method, we are attempting a maximum a posteriori or minimum variance estimate of $\mathbf{x}$, μ_w, and μ_v. Since prior statistics for μ_w and μ_v are unknown, it is not possible to obtain initial conditions for the final estimation algorithms. The initial algorithms which are obtained by this approach are unfortunate in that the algorithms obtained are

$$\dot{\hat{\mathbf{x}}} = \mathbf{F}\hat{\mathbf{x}} + \mathbf{G}\hat{\mu}_w + (\bar{\mathbf{P}}_{11}\mathbf{H}^T + \bar{\mathbf{P}}_{13})\Psi_v^{-1}(\mathbf{z} - \mathbf{H}\hat{\mathbf{x}} - \hat{\mu}_v) \tag{24}$$

$$\dot{\hat{\mu}}_w = (\bar{\mathbf{P}}_{21}\mathbf{H}^T + \bar{\mathbf{P}}_{23})\Psi_v^{-1}(\mathbf{z} - \mathbf{H}\hat{\mathbf{x}} - \hat{\mu}_v) \tag{25}$$

$$\dot{\hat{\mu}}_v = (\bar{\mathbf{P}}_{31}\mathbf{H}^T + \bar{\mathbf{P}}_{33})\Psi_v^{-1}(\mathbf{z} - \mathbf{H}\hat{\mathbf{x}} - \hat{\mu}_v) \tag{26}$$

where

$$\begin{gathered}\dot{\Xi} = \bar{\mathbf{F}}\Xi + \Xi\bar{\mathbf{F}}^T - \Xi\bar{\mathbf{H}}^T\Psi_v^{-1}\bar{\mathbf{H}}\Xi + \bar{\Psi}_w \\ \Xi = \begin{bmatrix} \bar{\mathbf{P}}_{11} & \bar{\mathbf{P}}_{12} & \bar{\mathbf{P}}_{13} \\ \bar{\mathbf{P}}_{12}^T & \bar{\mathbf{P}}_{22} & \bar{\mathbf{P}}_{23} \\ \bar{\mathbf{P}}_{13}^T & \bar{\mathbf{P}}_{23}^T & \bar{\mathbf{P}}_{33} \end{bmatrix}, \qquad \bar{\mathbf{F}} = \begin{bmatrix} \mathbf{F} & \mathbf{G} & 0 \\ 0 & 0 & 0 \\ 0 & 0 & 0 \end{bmatrix} \\ \bar{\mathbf{H}} = [\mathbf{H} \quad 0 \quad \mathbf{I}], \qquad \bar{\Psi}_w = \begin{bmatrix} \Psi_w & 0 & 0 \\ 0 & 0 & 0 \\ 0 & 0 & 0 \end{bmatrix}\end{gathered} \tag{27}$$

While these algorithms may appear simpler, the fact that the $\bar{\mathbf{P}}_{11}$ equation is not coupled with the other matrix Riccati equations will greatly increase the problem of computational accuracy. Friedland (1969) points out that a transformation of variables may be used to decouple the **V** equation of (7) from the higher order equations of (27). The resulting algorithms from this transformation are similar to those of (15)–(23). The approach taken here, using maximum likelihood, optimization theory, and invariant imbedding, is advantageous from a computational sensitivity standpoint in that the original error variance Riccati equation is incorporated into the optimization equations as an equality constraint. Also, the optimization theory–invariant imbedding approach is applicable to nonlinear systems.

7.3. Discrete Systems

The invariant imbedding technique may also be used to develop sequential estimation algorithms for system identification in discrete time systems. The purpose of this section is to illustrate this fact. To a great extent, the development of this section will follow closely the derivation in the preceding section; for this reason, some of the explanation is eliminated here since it would be almost identical to that in the previous section.

Once again we consider a general TPBVP, in discrete time however, given by[2]

$$\mathbf{x}(k+1) = \boldsymbol{\alpha}[\mathbf{x}(k), \boldsymbol{\lambda}(k), k] \tag{7.3-1}$$

$$\boldsymbol{\lambda}(k+1) = \boldsymbol{\eta}[\mathbf{x}(k), \boldsymbol{\lambda}(k), k] \tag{7.3-2}$$

with the split boundary conditions

$$\boldsymbol{\lambda}(k_0) = \mathbf{A}\mathbf{x}(k_0) + \mathbf{b}, \qquad \boldsymbol{\lambda}(k_f) = \mathbf{0} \tag{7.3-3}$$

We replace the terminal boundary $\boldsymbol{\lambda}(k_f) = \mathbf{0}$ by the more general expression $\boldsymbol{\lambda}(k_f) = \mathbf{c}$, and let both k_f and $\mathbf{c}$ be variable. The related terminal value of $\mathbf{x}$ will be given by

$$\mathbf{x}(k_f) = \mathbf{r}[\mathbf{c}, k_f] \tag{7.3-4}$$

[2] For notational simplicity, the sample interval T will be omitted when it appears in combination with an index as kT or $(k+1)T$.

In other words, if we obtain a solution of the TPBVP with

$$\boldsymbol{\lambda}(k_0) = \mathbf{A}\mathbf{x}(k_0) + \mathbf{b}, \qquad \boldsymbol{\lambda}(k_f) = \mathbf{c}$$

then $\mathbf{x}(k_f) = \mathbf{r}(\mathbf{c}, k_f)$ which will depend on both the terminal time k_f and the terminal value of $\boldsymbol{\lambda}$.

We assume that the solution of the above TPBVP is extended for one sample interval, so that k_f becomes $k_f + 1$ and $\mathbf{c}$ becomes $\mathbf{c} + \Delta\mathbf{c}$. The corresponding new terminal value of $\mathbf{x}$ becomes

$$\mathbf{x}(k_f + 1) = \mathbf{x}(k_f) + \Delta\mathbf{x} = \mathbf{r}(\mathbf{c} + \Delta\mathbf{c}, k_f + 1) \tag{7.3-5}$$

But $\mathbf{x}(k_f) = \mathbf{r}(\mathbf{c}, k_f)$, so that we have

$$\mathbf{r}(\mathbf{c}, k_f) + \Delta\mathbf{x} = \mathbf{r}(\mathbf{c} + \Delta\mathbf{c}, k_f + 1) \tag{7.3-6}$$

It is possible to write $\mathbf{r}(\mathbf{c} + \Delta c, k_f + 1)$ in the following form:

$$\mathbf{r}(\mathbf{c} + \Delta\mathbf{c}, k_f + 1) = \mathbf{r}(\mathbf{c}, k_f) + \frac{\delta\mathbf{r}(\mathbf{c}, k_f)}{\delta\mathbf{c}} \Delta\mathbf{c} + \frac{\delta\mathbf{r}(\mathbf{c}, k_f)}{\delta k_f} T + \frac{\delta^2\mathbf{r}(\mathbf{c}, k_f)}{\delta\mathbf{c}\,\delta k_f} T\Delta\mathbf{c} \tag{7.3-7}$$

which is easily verified by direct substitution, where $\delta\mathbf{r}/\delta\mathbf{c}$ is the first partial difference defined by

$$\left[\frac{\delta\mathbf{r}(\mathbf{c}, k_f)}{\delta\mathbf{c}}\right]_{ij} = \frac{r_i(\mathbf{c} + \Delta\mathbf{c}, k_f) - r_i(\mathbf{c}, k_f)}{\Delta c_j} \tag{7.3-8}$$

and

$$\left[\frac{\delta\mathbf{r}(\mathbf{c}, k_f)}{\delta k_f}\right]_{i} = \frac{r_i(\mathbf{c}, k_f + 1) - r_i(\mathbf{c}, k_f)}{T} \tag{7.3-9}$$

It should be noted that Eq. (7.3-7) is an exact expression and is not a Taylor series expansion, although the form of Eq. (7.3-7) is quite similar to that of a Taylor series. If we substitute Eq. (7.3-7) into Eq. (7.3-6), we obtain

$$\Delta\mathbf{x} = \left[\frac{\delta\mathbf{r}(\mathbf{c}, k_f)}{\delta\mathbf{c}} + \frac{\delta^2\mathbf{r}(\mathbf{c}, k_f)}{\delta\mathbf{c}\,\delta k_f} T\right] \Delta\mathbf{c} + \frac{\delta\mathbf{r}(\mathbf{c}, k_f)}{\delta k_f} T \tag{7.3-10}$$

We can obtain $\Delta\mathbf{x}$ and $\Delta\mathbf{c}$ from Eqs. (7.3-1) and (7.3-2) as

$$\Delta\mathbf{x} = \mathbf{x}(k_f + 1) - \mathbf{x}(k_f) = \boldsymbol{\alpha}[\mathbf{r}(\mathbf{c}, k_f), \mathbf{c}, k_f] - \mathbf{r}(\mathbf{c}, k_f) \tag{7.3-11}$$

and

$$\Delta \mathbf{c} = \boldsymbol{\lambda}(k_{\mathrm{f}} + 1) - \boldsymbol{\lambda}(k_{\mathrm{f}}) = \boldsymbol{\eta}[\mathbf{r}(\mathbf{c}, k_{\mathrm{f}}), \mathbf{c}, k_{\mathrm{f}}] - \mathbf{c} \tag{7.3-12}$$

so that Eq. (7.3-10) becomes

$$\boldsymbol{\alpha}[\mathbf{r}(\mathbf{c}, k_{\mathrm{f}}), \mathbf{c}, k_{\mathrm{f}}] - \mathbf{r}(\mathbf{c}, k_{\mathrm{f}}) = \left[\frac{\delta \mathbf{r}(\mathbf{c}, k_{\mathrm{f}})}{\delta \mathbf{c}} + \frac{\delta^2 \mathbf{r}(\mathbf{c}, k_{\mathrm{f}})}{\delta \mathbf{c}\, \delta k_{\mathrm{f}}} T\right] \times \{\boldsymbol{\eta}[\mathbf{r}(\mathbf{c}, k_{\mathrm{f}}), \mathbf{c}, k_{\mathrm{f}}] - \mathbf{c}\} + \frac{\delta \mathbf{r}(\mathbf{c}, k_{\mathrm{f}})}{\delta k_{\mathrm{f}}} T \tag{7.3-13}$$

If this partial difference equation could be solved for $\mathbf{r}(\mathbf{c}, k_{\mathrm{f}})$, then the TPBVP would be completely solved. However, just as in the continuous case, there is no general analytic solution to Eq. (7.3-13), and hence one normally resorts to the development of an approximate solution. Before continuing this chain of reasoning, it is instructive to consider the relationship between the discrete and continuous invariant imbedding equations, that is, Eqs. (7.2-13) and (7.3-13).

A possible discrete form for the continuous equations (7.2-1) and (7.2-2) is

$$\mathbf{x}(k + 1) = T\boldsymbol{\gamma}[\mathbf{x}(k), \boldsymbol{\lambda}(k), k] + \mathbf{x}(k)$$

$$\boldsymbol{\lambda}(k + 1) = T\boldsymbol{\beta}[\mathbf{x}(k), \boldsymbol{\lambda}(k), k] + \boldsymbol{\lambda}(k)$$

so that $\boldsymbol{\alpha}$ and $\boldsymbol{\eta}$ of Eqs. (7.3-1) and (7.3-2) are obviously given by

$$\boldsymbol{\alpha}[\mathbf{x}(k), \boldsymbol{\lambda}(k), k] = T\boldsymbol{\gamma}[\mathbf{x}(k), \boldsymbol{\lambda}(k), k] + \mathbf{x}(k) \tag{7.3-14}$$

$$\boldsymbol{\eta}[\mathbf{x}(k), \boldsymbol{\lambda}(k), k] = T\boldsymbol{\beta}[\mathbf{x}(k), \boldsymbol{\lambda}(k), k] + \boldsymbol{\lambda}(k) \tag{7.3-15}$$

Here T is assumed to be small, so that the continuous derivative may be approximated by the first difference. The substitution of Eqs. (7.3-14) and (7.3-15) into Eq. (7.3-13) yields

$$T\boldsymbol{\gamma}(\mathbf{r}, \mathbf{c}, k_{\mathrm{f}}) = \left[\frac{\delta \mathbf{r}(\mathbf{c}, k_{\mathrm{f}})}{\delta \mathbf{c}} + \frac{\delta^2 \mathbf{r}(\mathbf{c}, k_{\mathrm{f}})}{\delta \mathbf{c}\, \delta k_{\mathrm{f}}} T\right] T\boldsymbol{\beta}(\mathbf{r}, \mathbf{c}, k_{\mathrm{f}}) + \frac{\delta \mathbf{r}(\mathbf{c}, k_{\mathrm{f}})}{\delta k_{\mathrm{f}}} T \tag{7.3-16}$$

Now if we divide by T and let T approach zero while we let $k_{\mathrm{f}} T = t_{\mathrm{f}}$, then Eq. (7.3-16) becomes

$$\boldsymbol{\gamma}(\mathbf{r}, \mathbf{c}, t_{\mathrm{f}}) = \frac{\partial \mathbf{r}(\mathbf{c}, t_{\mathrm{f}})}{\partial \mathbf{c}} \boldsymbol{\beta}(\mathbf{r}, \mathbf{c}, t_{\mathrm{f}}) + \frac{\partial \mathbf{r}(\mathbf{c}, t_{\mathrm{f}})}{\partial t_{\mathrm{f}}} \tag{7.3-17}$$

since the partial differences become partial derivatives when $T \to 0$. Note that the second derivative term vanishes at $T \to 0$, since there are two T's in that term and there is no second partial derivative in the continuous case. In this sense, the discrete equation is more general than the continuous version; it would be impossible to generate the second partial difference by discretizing the continuous equations. In cases where the sample interval is not small, it is possible to show (Sage, 1968) that the second partial difference term is of significance.

Let us return now to the basic question of obtaining a solution for our TPBVP. Since there is no known general solution for Eq. (7.3-13), we will assume that $\mathbf{r}(\mathbf{c}, k_f)$ is once again by the linear form

$$\mathbf{r}(\mathbf{c}, k_f) = \mathbf{x}(k_f) - \mathbf{P}(k_f)\mathbf{c} \tag{7.3-18}$$

Using this assumed form for $\mathbf{r}(\mathbf{c}, k_f)$, the partial differences needed in Eq. (7.3-13) are found to be

$$\frac{\delta \mathbf{r}(\mathbf{c}, k_f)}{\delta \mathbf{c}} = -\mathbf{P}(k_f) \tag{7.3-19}$$

$$\frac{\delta^2 \mathbf{r}(\mathbf{c}, k_f)}{\delta \mathbf{c}\, \delta k_f} = -[\mathbf{P}(k_f + 1) - \mathbf{P}(k_f)]/T \tag{7.3-20}$$

and

$$\frac{\delta \mathbf{r}(\mathbf{c}, k_f)}{\delta k_f} = [\hat{\mathbf{x}}(k_f + 1) - \hat{\mathbf{x}}(k_f) - \mathbf{P}(k_f + 1)\mathbf{c} + \mathbf{P}(k_f)\mathbf{c}]/T \tag{7.3-21}$$

If we substitute these expressions into Eq. (7.3-13), we obtain

$$\boldsymbol{\alpha}[\hat{\mathbf{x}}(k_f) - \mathbf{P}(k_f)\mathbf{c}, \mathbf{c}, k_f] = -\mathbf{P}(k_f + 1)\, \boldsymbol{\eta}[\hat{\mathbf{x}}(k_f) - \mathbf{P}(k_f)\mathbf{c}, \mathbf{c}, k_f] + \hat{\mathbf{x}}(k_f + 1) \tag{7.3-22}$$

By making Taylor series expansions of $\boldsymbol{\alpha}$ and $\boldsymbol{\eta}$ about $\hat{\mathbf{x}}(k_f)$, $\mathbf{0}$, and k_f, and ignoring high-order terms, we may write Eq. (7.3-22) as

$$\begin{aligned}
&\boldsymbol{\alpha}[\hat{\mathbf{x}}(k_f), \mathbf{0}, k_f] + \left.\frac{\partial \boldsymbol{\alpha}[\hat{\mathbf{x}}(k_f) - \mathbf{P}(k_f)\mathbf{c}, \mathbf{c}, k_f]}{\partial \mathbf{c}}\right|_{\mathbf{c}=\mathbf{0}} \mathbf{c} \\
&\quad = -\mathbf{P}(k_f + 1) \left\{ \boldsymbol{\eta}[\hat{\mathbf{x}}(k_f), \mathbf{0}, k_f] + \left.\frac{\partial \boldsymbol{\eta}[\hat{\mathbf{x}}(k_f) - \mathbf{P}(k_f)\mathbf{c}, \mathbf{c}, k_f]}{\partial \mathbf{c}}\right|_{\mathbf{c}=\mathbf{0}} \mathbf{c} \right\} \\
&\qquad + \hat{\mathbf{x}}(k_f + 1)
\end{aligned} \tag{7.3-23}$$

This expression must be valid for all values of $\mathbf{c}$ as long as $\mathbf{c}$ is sufficiently small, so that we may equate the coefficients of the zero and first power of $\mathbf{c}$ to obtain

$$\hat{\mathbf{x}}(k_f+1) = \boldsymbol{\alpha}[\hat{\mathbf{x}}(k_f), \mathbf{0}, k_f] + \mathbf{P}(k_f+1)\, \boldsymbol{\eta}[\hat{\mathbf{x}}(k_f), \mathbf{0}, k_f] \tag{7.3-24}$$

$$\mathbf{P}(k_f+1)\left\{\frac{\partial\boldsymbol{\eta}[\hat{\mathbf{x}}(k_f) - \mathbf{P}(k_f)\mathbf{c}, \mathbf{c}, k_f]}{\partial\mathbf{c}}\bigg|_{\mathbf{c}=\mathbf{0}}\right\} = -\frac{\partial\boldsymbol{\alpha}[\hat{\mathbf{x}}(k_f) - \mathbf{P}(k_f)\mathbf{c}, \mathbf{c}, k_f]}{\partial\mathbf{c}}\bigg|_{\mathbf{c}=\mathbf{0}} \tag{7.3-25}$$

In order to continue the derivation, it is necessary to substitute the actual expressions for $\boldsymbol{\alpha}$ and $\boldsymbol{\eta}$. We will use the discrete TPBVP given by Eqs. (3.2-30)–(3.2-32) and (3.2-34), which if we neglect the terms that are quadratic in $\boldsymbol{\lambda}(k_f) = \mathbf{c}$, becomes

$$\hat{\mathbf{x}}(k_f+1) = \boldsymbol{\phi}[\hat{\mathbf{x}}(k_f), k_f] - \boldsymbol{\Gamma}[\hat{\mathbf{x}}(k_f), k_f]\, \mathbf{V}_w(k_f)\, \boldsymbol{\Gamma}^T[\hat{\mathbf{x}}(k_f), k_f]\, \frac{\partial\boldsymbol{\phi}^{-T}[\hat{\mathbf{x}}(k_f), k_f]}{\partial\hat{\mathbf{x}}(k_f)}\, \mathbf{c} \tag{7.3-26}$$

$$\boldsymbol{\lambda}(k_f+1) = \frac{\partial\boldsymbol{\phi}^{-T}[\hat{\mathbf{x}}(k_f), k_f]}{\partial\hat{\mathbf{x}}(k_f)}\, \mathbf{c} + \frac{\partial\mathbf{h}^T[\hat{\mathbf{x}}(k_f+1), k_f+1]}{\partial\hat{\mathbf{x}}(k_f+1)} \times \mathbf{V}_v^{-1}(k_f+1)\{\mathbf{z}(k_f+1) - \mathbf{h}[\hat{\mathbf{x}}(k_f+1), k_f+1]\} \tag{7.3-27}$$

$$\boldsymbol{\lambda}(k_0) = -\mathbf{V}_{x0}^{-1}[\hat{\mathbf{x}}(k_0) - \boldsymbol{\mu}_{x0}] \tag{7.3-28}$$

$$\boldsymbol{\lambda}(k_f) = \mathbf{c} = \mathbf{0} \tag{7.3-29}$$

Note that these expressions are not quite in the form of Eqs. (7.3-1) and (7.3-2), because $\hat{\mathbf{x}}(k_f+1)$ appears on the right-hand side of Eq. (7.3-27). We may remove this difficulty by substituting Eq. (7.3-26) for $\hat{\mathbf{x}}(k_f+1)$; then if we expand about $\boldsymbol{\phi}[\hat{\mathbf{x}}(k_f), k_f]$, we obtain the following expression for $\boldsymbol{\alpha}$ and $\boldsymbol{\eta}$:

$$\boldsymbol{\alpha}[\hat{\mathbf{x}}(k_f), \mathbf{c}, k_f] = \boldsymbol{\phi}[\hat{\mathbf{x}}(k_f), k_f] - \boldsymbol{\Gamma}[\hat{\mathbf{x}}(k_f), k_f]\, \mathbf{V}_w(k_f)\, \boldsymbol{\Gamma}^T[\hat{\mathbf{x}}(k_f), k_f] \times \frac{\partial\boldsymbol{\phi}^{-T}[\hat{\mathbf{x}}(k_f), k_f]}{\partial\hat{\mathbf{x}}(k_f)}\, \mathbf{c} \tag{7.3-30}$$

and

$$\begin{aligned}\boldsymbol{\eta}[\hat{\mathbf{x}}(k_f), \mathbf{c}, k_f] = {} & \frac{\partial\boldsymbol{\phi}^{-T}[\hat{\mathbf{x}}(k_f), k_f]}{\partial\hat{\mathbf{x}}(k_f)}\, \mathbf{c} + \frac{\partial\mathbf{h}^T[\hat{\mathbf{x}}(k_f+1 \mid k_f), k_f+1]}{\partial\hat{\mathbf{x}}(k_f+1 \mid k_f)}\, \mathbf{V}_v^{-1}(k_f+1) \\ & \times \{\mathbf{z}(k_f+1) - \mathbf{h}[\hat{\mathbf{x}}(k_f+1 \mid k_f), k_f+1]\} - \frac{\partial}{\partial\hat{\mathbf{x}}(k_f+1 \mid k_f)} \\ & \times \left\{\frac{\partial\mathbf{h}^T[\hat{\mathbf{x}}(k_f+1 \mid k_f), k_f+1]}{\partial\hat{\mathbf{x}}(k_f+1 \mid k_f)}\, \mathbf{V}_v^{-1}(k_f+1) \right. \\ & \left. \times \{\mathbf{z}(k_f+1) - \mathbf{h}[\hat{\mathbf{x}}(k_f+1 \mid k_f), k_f+1]\}\right\} \\ & \times \boldsymbol{\Gamma}[\hat{\mathbf{x}}(k_f), k_f]\, \mathbf{V}_w(k_f)\, \boldsymbol{\Gamma}^T[\hat{\mathbf{x}}(k_f), k_f]\, \frac{\partial\boldsymbol{\phi}^{-T}[\hat{\mathbf{x}}(k_f), k_f]}{\partial\hat{\mathbf{x}}(k_f)}\, \mathbf{c}\end{aligned} \tag{7.3-31}$$

where

$$\hat{\mathbf{x}}(k_f + 1 \mid k_f) \triangleq \boldsymbol{\phi}[\hat{\mathbf{x}}(k_f), k_f] \tag{7.3-32}$$

To simplify the development, we will drop some of the arguments in the expressions of Eqs. (7.3-31) and (7.3-32). From Eqs. (7.3-31) and (7.3-32), we can easily determine the expressions which we need for Eqs. (7.3-24) and (7.3-25) as

$$\boldsymbol{\alpha}[\hat{\mathbf{x}}(k_f), \mathbf{0}, k_f] = \boldsymbol{\phi}[\hat{\mathbf{x}}(k_f), k_f] = \hat{\mathbf{x}}(k_f + 1 \mid k_f) \tag{7.3-33}$$

$$\left.\frac{\partial \boldsymbol{\alpha}[\hat{\mathbf{x}} - \mathbf{Pc}, \mathbf{c}, k_f]}{\partial \mathbf{c}}\right|_{\mathbf{c}=\mathbf{0}} = -\frac{\partial \boldsymbol{\phi}[\hat{\mathbf{x}}, k_f]}{\partial \hat{\mathbf{x}}} \mathbf{P}(k_f) - \boldsymbol{\Gamma}\mathbf{V}_w\boldsymbol{\Gamma}^T \frac{\partial \boldsymbol{\phi}^{-T}[\hat{\mathbf{x}}, k_f]}{\partial \hat{\mathbf{x}}} \tag{7.3-34}$$

$$\boldsymbol{\eta}[\hat{\mathbf{x}}(k_f), \mathbf{0}, k_f] = \mathbf{M}[\hat{\mathbf{x}}(k_f + 1 \mid k_f), k_f + 1] \tag{7.3-35}$$

and

$$\begin{aligned}\left.\frac{\partial \boldsymbol{\eta}[\hat{\mathbf{x}} - \mathbf{Pc}, \mathbf{c}, k_f]}{\partial \mathbf{c}}\right|_{\mathbf{c}=\mathbf{0}} &= \frac{\partial \boldsymbol{\phi}^{-T}[\hat{\mathbf{x}}, k_f]}{\partial \hat{\mathbf{x}}} \mathbf{P}(k_f) \\ &\quad + \frac{\partial \mathbf{M}[\hat{\mathbf{x}}(k_f + 1 \mid k_f), k_f + 1]}{\partial \hat{\mathbf{x}}(k_f + 1 \mid k_f)} \frac{\partial \boldsymbol{\phi}[\hat{\mathbf{x}}, k_f]}{\partial \hat{\mathbf{x}}} \mathbf{P}(k_f) \\ &\quad - \frac{\partial \mathbf{M}[\hat{\mathbf{x}}(k_f + 1 \mid k_f), k_f + 1)}{\partial \hat{\mathbf{x}}(k_f + 1 \mid k_f)} \boldsymbol{\Gamma}\mathbf{V}_w\boldsymbol{\Gamma}^T \frac{\partial \boldsymbol{\phi}^{-T}[\hat{\mathbf{x}}, k_f]}{\partial \hat{\mathbf{x}}}\end{aligned} \tag{7.3-36}$$

where the matrix $\mathbf{M}[\hat{\mathbf{x}}(k_f + 1 \mid k_f), k_f + 1]$ is defined as

$$\begin{aligned}\mathbf{M}[\hat{\mathbf{x}}(k_f + 1 \mid k_f), k_f + 1] &\triangleq \frac{\partial \mathbf{h}^T[\hat{\mathbf{x}}(k_f + 1 \mid k_f), k_f + 1]}{\partial \hat{\mathbf{x}}(k_f + 1 \mid k_f]} \mathbf{V}_v^{-1}(k_f + 1) \\ &\quad \times \{\mathbf{z}(k_f + 1) - \mathbf{h}[\hat{\mathbf{x}}(k_f + 1 \mid k_f]\end{aligned} \tag{7.3-37}$$

Now if we substitute these results in Eqs. (7.3-24) and (7.3-25), we obtain

$$\begin{aligned}\hat{\mathbf{x}}(k_f + 1) &= \boldsymbol{\phi}[\hat{\mathbf{x}}(k_f), k_f] + \mathbf{P}(k_f + 1) \frac{\partial \mathbf{h}^T[\hat{\mathbf{x}}(k_f + 1 \mid k_f), k_f + 1]}{\partial \hat{\mathbf{x}}(k_f + 1 \mid k_f)} \mathbf{V}_v^{-1}(k_f + 1) \\ &\quad \times \{\mathbf{z}(k_f + 1) - \mathbf{h}[\hat{\mathbf{x}}(k_f + 1 \mid k_f), k_f + 1]\}\end{aligned} \tag{7.3-38}$$

$$\begin{aligned}&\mathbf{P}(k_f + 1) \left\{ \frac{\partial \boldsymbol{\phi}^{-T}}{\partial \hat{\mathbf{x}}} - \frac{\partial \mathbf{M}[\hat{\mathbf{x}}(k_f + 1 \mid k_f), k_f + 1]}{\partial \hat{\mathbf{x}}(k_f + 1 \mid k_f)} \right. \\ &\qquad \left. \times \left[\boldsymbol{\Gamma}\mathbf{V}_w\boldsymbol{\Gamma}^T \frac{\partial \boldsymbol{\phi}^{-T}[\hat{\mathbf{x}}, k_f]}{\partial \hat{\mathbf{x}}} + \frac{\partial \boldsymbol{\phi}[\hat{\mathbf{x}}, k_f]}{\partial \hat{\mathbf{x}}} \mathbf{P}(k_f) \right] \right\} \\ &\qquad = \boldsymbol{\Gamma}\mathbf{V}_w\boldsymbol{\Gamma}^T \frac{\partial \boldsymbol{\phi}^{-T}}{\partial \hat{\mathbf{x}}} + \frac{\partial \boldsymbol{\phi}}{\partial \hat{\mathbf{x}}} \mathbf{P}(k_f)\end{aligned} \tag{7.3-39}$$

Now if we postmultiply Eq. (7.3-39) by $\partial\boldsymbol{\phi}^{\mathrm{T}}/\partial\hat{\mathbf{x}}$, and define $\mathbf{P}(k_{\mathrm{f}}+1 \mid k_{\mathrm{f}})$ as

$$\mathbf{P}(k_{\mathrm{f}}+1 \mid k_{\mathrm{f}}) = \boldsymbol{\Gamma}[\hat{\mathbf{x}}(k_{\mathrm{f}}), k_{\mathrm{f}}]\, \mathbf{V}_{\mathrm{w}}(k_{\mathrm{f}})\, \boldsymbol{\Gamma}^{\mathrm{T}}[\hat{\mathbf{x}}(k_{\mathrm{f}}), k_{\mathrm{f}}] + \frac{\partial\boldsymbol{\phi}[\hat{\mathbf{x}}(k_{\mathrm{f}}), k_{\mathrm{f}}]}{\partial\hat{\mathbf{x}}(k_{\mathrm{f}})}\, \mathbf{P}(k_{\mathrm{f}})\, \frac{\partial\boldsymbol{\phi}^{\mathrm{T}}[\hat{\mathbf{x}}(k_{\mathrm{f}}), k_{\mathrm{f}}]}{\partial\hat{\mathbf{x}}(k_{\mathrm{f}})} \tag{7.3-40}$$

then Eq. (7.3-39) becomes

$$\mathbf{P}(k_{\mathrm{f}}+1) = \left\{\mathbf{I} - \frac{\partial\mathbf{M}[\hat{\mathbf{x}}(k_{\mathrm{f}}+1 \mid k_{\mathrm{f}}), k_{\mathrm{f}}+1]}{\partial\hat{\mathbf{x}}(k_{\mathrm{f}}+1 \mid k_{\mathrm{f}})}\, \mathbf{P}(k_{\mathrm{f}}+1 \mid k_{\mathrm{f}})\right\}^{-1} \mathbf{P}(k_{\mathrm{f}}+1 \mid k_{\mathrm{f}}) \tag{7.3-41}$$

The combination of Eqs. (7.3-32), (7.3-37), (7.3-38), (7.3-40), and (7.3-41) generates a complete sequential algorithm for the estimation of the state $\mathbf{x}(k)$, if k_{f} is treated as the running current time variable k. The initial conditions for this algorithm are determined from Eqs. (7.3-28) and (7.3-29) as

$$\mathbf{P}(k_0) = \mathbf{V}_{\mathrm{x0}} \tag{7.3-42}$$

$$\hat{\mathbf{x}}(k_0) = \boldsymbol{\mu}_{\mathrm{x0}} \tag{7.3-43}$$

The above algorithm can also be placed in a form which is often more convenient. If we factor the symmetric matrix $\mathbf{M}[\hat{\mathbf{x}}(k+1 \mid k), k+1]$ defined by Eq. (7.3-37) into the form

$$\frac{\partial\mathbf{M}[\hat{\mathbf{x}}(k+1 \mid k), k+1]}{\partial\hat{\mathbf{x}}(k+1 \mid k)} = \mathbf{H}^{\mathrm{T}}(k+1)\, \mathbf{V}_{\mathrm{v}}^{-1}(k+1)\, \mathbf{H}(k+1) \tag{7.3-44}$$

where, in general, $\mathbf{H}(k+1)$ will depend on $\hat{\mathbf{x}}(k+1 \mid k)$ and $\mathbf{z}(k+1)$, then we may rewrite Eq. (7.3-41) in the following form by the use of the matrix inversion lemma:

$$\mathbf{P}(k+1) = \mathbf{P}(k+1 \mid k) - \mathbf{P}(k+1 \mid k)\, \mathbf{H}^{\mathrm{T}}(k+1)[\mathbf{H}(k+1)\, \mathbf{P}(k+1 \mid k) \times \mathbf{H}^{\mathrm{T}}(k+1) + \mathbf{V}_{\mathrm{v}}(k+1)]^{-1}\, \mathbf{H}(k+1)\, \mathbf{P}(k+1 \mid k) \tag{7.3-45}$$

The advantage of this form is that the required matrix inversion is generally of lower order, since the observation is generally of lower order than the state. The major disadvantage of this form is that one must accomplish the factorization specified by Eq. (7.3-44), which may not be a trivial feat. Of course, if the observation is linear so that

$$\mathbf{z}(k) = \mathbf{H}(k)\, \mathbf{x}(k) + \mathbf{v}(k)$$

then the form of Eq. (7.3-44) arises naturally.

The complete discrete invariant imbedding-MAP algorithm for system identification and state estimation is summarized for easy reference in Table 7.3-1. The use of this algorithm is illustrated by the following example.

TABLE 7.3-1

DISCRETE INVARIANT IMBEDDING ALGORITHM

System model	$\mathbf{x}(k+1) = \boldsymbol{\phi}[\mathbf{x}(k), k] + \boldsymbol{\Gamma}[\mathbf{x}(k), k]\mathbf{w}(k)$	(3.2-1)
Observation model	$\mathbf{z}(k) = \mathbf{h}[\mathbf{x}(k), k] + \mathbf{v}(k)$	(3.2-2)
Statistical parameters	$\mathscr{E}\{\mathbf{x}(k_0)\} = \boldsymbol{\mu}_{\mathbf{x}0}, \quad \operatorname{var}\{\mathbf{x}(k_0)\} = \mathbf{V}_{\mathbf{x}0}$ $\mathscr{E}\{\mathbf{w}(k)\} = \mathscr{E}\{\mathbf{v}(k)\} = \mathbf{0}$ $\operatorname{cov}\{\mathbf{w}(k), \mathbf{w}(j)\} = \mathbf{V}_{\mathbf{w}}(k)\delta_{\mathrm{K}}(k-j)$ $\operatorname{cov}\{\mathbf{v}(k), \mathbf{v}(j)\} = \mathbf{V}_{\mathbf{v}}(k)\delta_{\mathrm{K}}(k-j)$	
One-stage predictor	$\hat{\mathbf{x}}(k+1 \mid k) = \boldsymbol{\phi}[\hat{\mathbf{x}}(k), k]$	(7.3-32)
Filter algorithm	$\hat{\mathbf{x}}(k+1) = \hat{\mathbf{x}}(k+1 \mid k) + \mathbf{P}(k+1) \dfrac{\partial \mathbf{h}^{\mathrm{T}}[\hat{\mathbf{x}}(k+1 \mid k), k+1]}{\partial \hat{\mathbf{x}}(k+1 \mid k)} \times \mathbf{V}_{\mathbf{v}}^{-1}(k+1)\{\mathbf{z}(k+1) - \mathbf{h}[\hat{\mathbf{x}}(k+1 \mid k), k+1)]\}$	(7.3-38)
Prior variance algorithm	$\mathbf{P}(k+1 \mid k) = \boldsymbol{\Gamma}[\hat{\mathbf{x}}(k), k]\mathbf{V}_{\mathbf{w}}(k)\boldsymbol{\Gamma}^{\mathrm{T}}[\hat{\mathbf{x}}(k), k] + \dfrac{\partial \boldsymbol{\phi}[\hat{\mathbf{x}}(k), k]}{\partial \hat{\mathbf{x}}(k)} \mathbf{P}(k) \dfrac{\partial \boldsymbol{\phi}^{\mathrm{T}}[\hat{\mathbf{x}}(k), k]}{\partial \hat{\mathbf{x}}(k)}$	(7.3-40)
Error variance algorithm	$\mathbf{P}(k+1) = \left\{\mathbf{I} - \dfrac{\partial \mathbf{M}[\hat{\mathbf{x}}(k+1 \mid k), k+1]}{\partial \hat{\mathbf{x}}(k+1 \mid k)} \mathbf{P}(k+1 \mid k)\right\}^{-1} \times \mathbf{P}(k+1 \mid k)$	(7.3-41)
	$\mathbf{M}[\hat{\mathbf{x}}(k+1 \mid k), k+1] = \dfrac{\partial \mathbf{h}^{\mathrm{T}}[\hat{\mathbf{x}}(k+1 \mid k), k+1]}{\partial \hat{\mathbf{x}}(k+1 \mid k)} \times \mathbf{V}_{\mathbf{v}}^{-1}(k+1)\{\mathbf{z}(k+1) - \mathbf{h}[\hat{\mathbf{x}}(k+1 \mid k), k+1$	
Alternate error variance algorithms	$\mathbf{P}(k+1) = \mathbf{P}(k+1 \mid k) - \mathbf{P}(k+1 \mid k)\mathbf{H}^{\mathrm{T}}(k+1) \times [\mathbf{H}(k+1)\mathbf{P}(k+1 \mid k)\mathbf{H}^{\mathrm{T}}(k+1) + \mathbf{V}_{\mathbf{v}}(k+1)]^{-1} \times \mathbf{H}(k+1)\mathbf{P}(k+1 \mid k)$	(7.3-45)
	$\dfrac{\partial \mathbf{M}[\hat{\mathbf{x}}(k+1 \mid k), k+1]}{\partial \hat{\mathbf{x}}(k+1 \mid k)} = \mathbf{H}^{\mathrm{T}}(k+1)\mathbf{V}_{\mathbf{v}}^{-1}(k+1)\mathbf{H}(k+1)$	
Initial condition	$\hat{\mathbf{x}}(k_0) = \boldsymbol{\mu}_{\mathbf{x}0}, \quad \mathbf{P}(k_0) = \mathbf{V}_{\mathbf{x}0}$	

Example 7.3-1. The system equations to be considered are the discrete first-order approximations for the state equations of a linear, underdamped, second-order system with displacement $x(t)$, rate $r(t)$, a damping ratio of d, an undamped natural frequency of 5 rad/sec, and a constant driving term of 12.5 with additive input noise $w(t)$. The corresponding continuous dynamic equation is:

$$\ddot{x} + 10d\dot{x} + 25x = 12.5 + w(t)$$

The discrete state equations are third order and nonlinear. The system (model) equations are:

$$\begin{aligned}
x(k+1) &= x(k) + Tr(k) \\
r(k+1) &= -25Tx(k) + [1 - 10Td(k)]\, r(k) + 12.5T + Tw(k) \\
d(k+1) &= d(k) \\
z(k) &= x(k) + v(k)
\end{aligned}$$

Employing Table 7.3-1 yields the sequential estimation equations:

$$\begin{aligned}
\hat{x}(k+1) &= \hat{x}(k) + T\hat{r}(k) + 2TV_v^{-1}V_{11}(k+1)[z(k+1) - \hat{x}(k) - T\hat{r}(k)] \\
\hat{r}(k+1) &= 25T\hat{x}(k) + \hat{r}(k) - 10T\hat{d}(k)\,\hat{r}(k) \\
&\quad + 12.5T + 2TV_v^{-1}V_{21}(k+1)[z(k+1) - \hat{x}(k) - T\hat{r}(k)] \\
\hat{d}(k+1) &= \hat{d}(k) + 2TV_v^{-1}V_{31}(k+1)[z(k+1) - \hat{x}(k) - T\hat{r}(k)]
\end{aligned}$$

$$\begin{aligned}
P_{11}(k+1 \mid k) &= P_{11}(k) + 2TP_{21}(k) + T^2P_{22}(k) \\
P_{21}(k+1 \mid k) &= -25TP_{11}(k) + [1 - 10T\hat{d}(k) - 25T^2]\, P_{21}(k) \\
&\quad - 10T\hat{r}(k)\, P_{31}(k) + [1 - 10T\hat{d}(k)]\, TP_{22}(k) \\
&\quad - 10T^2\hat{r}(k)\, P_{32}(k) \\
P_{31}(k+1 \mid k) &= P_{31}(k) + TP_{32}(k) \\
P_{22}(k+1 \mid k) &= 625T^2P_{11}(k) - 50T[1 - 10T\hat{d}(k)]\, P_{21}(k) \\
&\quad + 500T^2\hat{r}(k)\, P_{31}(k) + [1 - 10T\hat{d}(k)]^2\, P_{22}(k) \\
&\quad - 20T\hat{r}(k)[1 - 10T\hat{d}(k)]\, P_{32}(k) \\
&\quad + 100T^2\hat{d}(k)\,\hat{r}(k)\, P_{33}(k) + V_w(k) \\
P_{32}(k+1 \mid k) &= -25TP_{31}(k) + [1 - 10T\hat{d}(k)]\, P_{32}(k) \\
&\quad - 10T\hat{r}(k)\, P_{33}(k) \\
P_{33}(k+1 \mid k) &= P_{33}(k)
\end{aligned}$$

$$P_{11}(k+1) = \frac{P_{11}(k+1 \mid k)\, V_v(k+1)}{P_{11}(k+1 \mid k) + V_v(k+1)}$$

$$P_{21}(k+1) = \frac{P_{21}(k+1 \mid k)\, V_v(k+1)}{P_{11}(k+1 \mid k) + V_v(k+1)}$$

$$P_{31}(k+1) = \frac{P_{31}(k+1 \mid k)\, V_v(k+1)}{P_{11}(k+1 \mid k) + V_v(k+1)}$$

$$P_{22}(k+1) = P_{22}(k+1 \mid k) - \frac{P_{21}^2(k+1 \mid k)}{P_{11}(k+1 \mid k) + V_v(k+1)}$$

$$P_{32}(k+1) = P_{32}(k+1 \mid k) - \frac{P_{31}(k+1 \mid k)\, P_{21}(k+1 \mid k)}{P_{11}(k+1 \mid k) + V_v(k+1)}$$

$$P_{33}(k+1) = P_{33}(k+1 \mid k) - \frac{P_{31}^2(k+1 \mid k)}{P_{11}(k+1 \mid k) + V_v(k+1)}$$

Note that we have taken advantage of the symmetry of error variance matrix to eliminate three of the nine equations for the matrix elements.

Computational results using these algorithms are shown in Fig. 7.3–1 with

$$\mathbf{V}_{\mathbf{x}0} = \begin{bmatrix} 2 & 1 & 1 \\ 1 & 2 & 1 \\ 1 & 1 & 2 \end{bmatrix}, \qquad \boldsymbol{\mu}_{\mathbf{x}0} = \begin{bmatrix} 0 \\ 0 \\ 0 \end{bmatrix}$$

and $V_v = V_w = 20$, $T = 0.002$. The estimates of both state variables and the parameter $d(k)$ are seen to converge to a close approximation of their true values within a fraction of the system natural period.

In many system identification problems, one finds that the plant noise $\mathbf{w}$ and the observation noise $\mathbf{v}$ are correlated. This often happens, not because the origin plant and measurement noises are correlated, but because the way in which the problem is modeled leads to such correlation. This is the case when noisy observations of the plant input are available, or when the observations contain colored noise.

The most direct way to handle the problem of correlated plant and measurement noises is to redefine the problem in such a way as to uncorrelate the noises. Let us consider the message model

$$\mathbf{x}(k+1) = \boldsymbol{\phi}[\mathbf{x}(k), k] + \boldsymbol{\Gamma}(k)\, \mathbf{w}(k) \qquad (7.3\text{-}46)$$

and the observation model

$$\mathbf{z}(k) = \mathbf{h}[\mathbf{x}(k), k] + \mathbf{v}(k) \qquad (7.3\text{-}47)$$

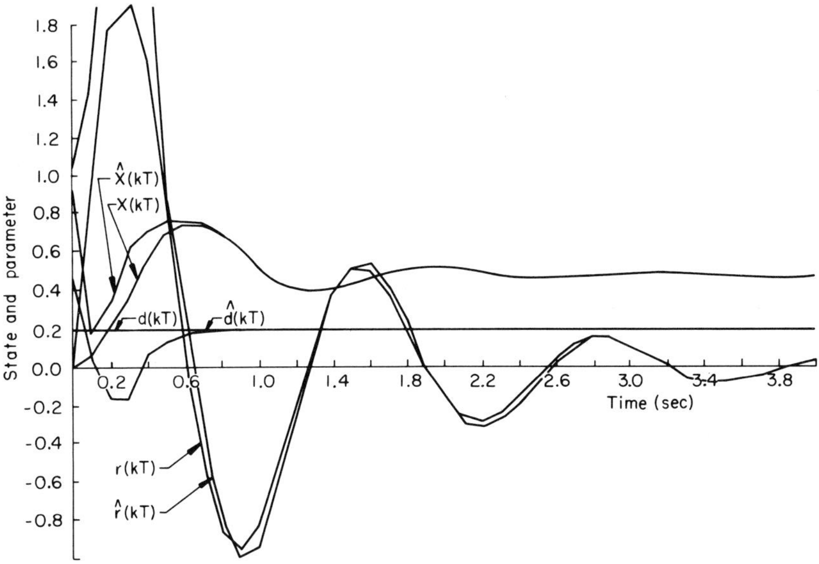

FIG. 7.3-1. Combined state and parameter estimation in a noisy second-order system.

Here, for simplicity, we have assumed that $\mathbf{\Gamma}$ is *not* a function of $\mathbf{x}(k)$. As usual, $\mathbf{w}(k)$ and $\mathbf{v}(k)$ are assumed to be zero mean, discrete white sequences with covariance matrices $\mathbf{V}_w(k)$ and $\mathbf{V}_v(k)$. Now, however, it is no longer assumed that $\mathbf{v}(k)$ and $\mathbf{w}(k)$ are independent, but rather that

$$\operatorname{cov}\{\mathbf{w}(k), \mathbf{v}(j)\} = \mathbf{V}_{wv}(k)\, \delta_K(k-j) \tag{7.3-48}$$

In order to "uncorrelate" the input and observation noises, let us rewrite the message of Eq. (7.3-46) in the following form:

$$\mathbf{x}(k+1) = \boldsymbol{\phi}[\mathbf{x}(k), k] + \mathbf{\Gamma}(k)\,\mathbf{w}(k) + \mathbf{K}_p(k)[\mathbf{z}(k) - \mathbf{h}[\mathbf{x}(k), k] - \mathbf{v}(k)] \tag{7.3-49}$$

Clearly, the model is equivalent to Eq. (7.3-46) as "nothing" has been added. Now we express Eq. (7.3-49) as

$$\mathbf{x}(k+1) = \boldsymbol{\phi}^*[\mathbf{x}(k), k] + \mathbf{w}^*(k) + \mathbf{K}_p(k)\,\mathbf{z}(k) \tag{7.3-50}$$

where

$$\boldsymbol{\phi}^*[\mathbf{x}(k), k] = \boldsymbol{\phi}[\mathbf{x}(k), k] - \mathbf{K}_p(k)\,\mathbf{h}[\mathbf{x}(k), k] \tag{7.3-51}$$

and

$$\mathbf{w}^*(k) = \mathbf{\Gamma}(k)\,\mathbf{w}(k) - \mathbf{K}_p(k)\,\mathbf{v}(k) \tag{7.3-52}$$

The process $\mathbf{w}^*(k)$ is still zero mean and white with covariance

$$\operatorname{cov}\{\mathbf{w}^*(k), \mathbf{w}^*(j)\} = [\mathbf{\Gamma}(k)\,\mathbf{V}_{\mathrm{w}}(k)\,\mathbf{\Gamma}^{\mathrm{T}}(k) - \mathbf{\Gamma}(k)\,\mathbf{V}_{\mathrm{wv}}(k)\,\mathbf{K}_p{}^{\mathrm{T}}(k)$$
$$- \mathbf{K}_p(k)\,\mathbf{V}_{\mathrm{wv}}^{\mathrm{T}}\mathbf{\Gamma}^{\mathrm{T}}(k) + \mathbf{K}_p(k)\,\mathbf{V}_{\mathrm{v}}(k)\,\mathbf{K}_p{}^{\mathrm{T}}(k)]\,\delta_{\mathrm{K}}(k-j) \tag{7.3-53}$$

and is, in general, still correlated with $\mathbf{v}(k)$. However by proper selection of $\mathbf{K}_p(k)$ we may "uncorrelate" $\mathbf{w}^*(k)$ and $\mathbf{v}(k)$. In order to see this, we need only consider the covariance of $\mathbf{w}^*(k)$ and $\mathbf{v}(k)$ given by

$$\operatorname{cov}\{\mathbf{w}^*(k), \mathbf{v}(k)\} = \mathbf{\Gamma}(k)\,\mathbf{V}_{\mathrm{wv}}(k) - \mathbf{K}_p(k)\,\mathbf{V}_{\mathrm{v}}(k)$$

If we let

$$\mathbf{K}_p(k) = \mathbf{\Gamma}(k)\,\mathbf{V}_{\mathrm{wv}}(k)\,\mathbf{V}_{\mathrm{v}}^{-1}(k) \tag{7.3-54}$$

then we see that $\operatorname{cov}\{\mathbf{w}^*(k), \mathbf{v}(k)\} = \mathbf{0}$, and we have an equivalent problem with uncorrelated plant and measurement noises; hence the algorithm of Table 7.3-1 can be applied. The resulting algorithm for correlated plant and measurements noises is summarized in Table 7.3-2. Note that because of the known input $\mathbf{z}(k)$ in Eq. (7.3-50), the one-stage prediction equation takes the form

$$\hat{\mathbf{x}}(k+1 \mid k) = \boldsymbol{\phi}^*[\mathbf{x}(k), k] + \mathbf{K}_p(k)\,\mathbf{z}(k) \tag{7.3-55}$$

The above result can be extended to the adaptive situation in which the means or variances (Sage and Husa, 1969; Sage, 1970; Sage and Wakefield, 1970) are unknown. The combinations and permutations of these algorithms and their associated assumed models are almost limitless. We consider a simple example to illustrate the use of these concepts in system identification.

Example 7.3-2. Let us consider the problem of identification with noise corrupted observation of the plant input. The message model is still

$$\mathbf{x}(k+1) = \boldsymbol{\phi}[\mathbf{x}(k), k] + \mathbf{\Gamma}(k)\,\mathbf{w}(k)$$

and we have the observation

$$\mathbf{z}_1(k) = \mathbf{h}[\mathbf{x}(k), k] + \mathbf{v}_1(k)$$

TABLE 7.3-2

DISCRETE INVARIANT IMBEDDING ALGORITHM FOR CORRELATED PLANT AND MEASUREMENT NOISES

Message model	$\mathbf{x}(k+1) = \boldsymbol{\phi}[\mathbf{x}(k), k] + \boldsymbol{\Gamma}\mathbf{w}(k)$ (7.3-46)
Observation model	$\mathbf{z}(k) = \mathbf{h}[\mathbf{x}(k), k] + \mathbf{v}(k)$ (7.3-47)
Prior statistics	$\boldsymbol{\mu}_{\mathbf{x}}(k_0) = \boldsymbol{\mu}_{\mathbf{x}_0}$, $\mathbf{V}_{\mathbf{x}}(k_0) = \mathbf{V}_{\mathbf{x}_0}$ $\boldsymbol{\mu}_{\mathbf{w}}(k) = \mathbf{0}$, $\operatorname{cov}\{\mathbf{w}(k), \mathbf{w}(j)\} = \mathbf{V}_{\mathbf{w}}(k)\delta_K(k-j)$ $\boldsymbol{\mu}_{\mathbf{v}}(k) = \mathbf{0}$, $\operatorname{cov}\{\mathbf{v}(k), \mathbf{v}(j)\} = \mathbf{V}_{\mathbf{v}}(k)\delta_K(k-j)$ $\operatorname{cov}\{\mathbf{w}(k), \mathbf{x}(k_0)\} = \mathbf{0}$ $\operatorname{cov}\{\mathbf{v}(k), \mathbf{w}(j)\} = \mathbf{V}_{\mathbf{vw}}(k)\delta_K(k-j)$
Filter algorithm	$\hat{\mathbf{x}}(k+1) = \hat{\mathbf{x}}(k+1 \mid k) + \mathbf{K}(k+1)\{\mathbf{z}(k+1) - \mathbf{h}[\hat{\mathbf{x}}(k+1 \mid k), k+1]\}$
One-stage prediction algorithm	$\hat{\mathbf{x}}(k+1 \mid k) = \boldsymbol{\phi}[\hat{\mathbf{x}}(k), k] + \boldsymbol{\Gamma}\mathbf{V}_{\mathbf{wv}}(k)\mathbf{V}_{\mathbf{v}}^{-1}(k) \times \{\mathbf{z}(k) - \mathbf{h}[\hat{\mathbf{x}}(k), k]\}$
Gain algorithm	$\mathbf{K}(k+1) = \mathbf{P}(k+1)\dfrac{\partial \mathbf{h}^T[\hat{\mathbf{x}}(k+1 \mid k), k+1]}{\partial \hat{\mathbf{x}}(k+1 \mid k)}\mathbf{V}_{\mathbf{v}}^{-1}(k+1 \mid k)$
Prior error variance algorithm	$\mathbf{P}(k+1 \mid k) = \boldsymbol{\phi}^*(k)\mathbf{P}(k)\boldsymbol{\phi}^{*T}(k)\boldsymbol{\Gamma} + \boldsymbol{\Gamma}\mathbf{V}_{\mathbf{w}}(k)\boldsymbol{\Gamma}^T - \mathbf{V}_{\mathbf{wv}}(k)\mathbf{V}_{\mathbf{v}}^{-1}(k)\mathbf{V}_{\mathbf{vw}}(k)\boldsymbol{\Gamma}^T$
Error variance algorithm	$\mathbf{P}(k+1) = \mathbf{P}(k+1 \mid k) - \mathbf{P}(k+1 \mid k)\mathbf{H}^T(k+1) \times [\mathbf{H}(k+1)\mathbf{P}(k+1 \mid k)\mathbf{H}^T(k+1) + \mathbf{V}_{\mathbf{v}}(k+1 \mid k)]^{-1} \times \mathbf{H}(k+1)\mathbf{P}(k+1 \mid k)$
Transition and observation matrix algorithms	$\boldsymbol{\phi}^*(k) = \dfrac{\partial \boldsymbol{\phi}[\hat{\mathbf{x}}(k), k]}{\partial \hat{\mathbf{x}}(k)} - \boldsymbol{\Gamma}\mathbf{V}_{\mathbf{wv}}(k)\mathbf{V}_{\mathbf{v}}^{-1}(k \mid k-1)\dfrac{\partial \mathbf{h}[\hat{\mathbf{x}}(k), k]}{\partial \hat{\mathbf{x}}(k)}$ $\mathbf{H}^T(k+1)\mathbf{V}_{\mathbf{v}}^{-1}(k+1 \mid k)\mathbf{H}(k+1) = -\dfrac{\partial}{\partial \hat{\mathbf{x}}(k+1 \mid k)}\left[\dfrac{\partial \mathbf{h}^T[\hat{\mathbf{x}}(k+1 \mid k), k+1]}{\partial \hat{\mathbf{x}}(k+1 \mid k)} \times \mathbf{V}_{\mathbf{v}}^{-1}(k+1 \mid k)\{\mathbf{z}(k+1) - \mathbf{h}[\hat{\mathbf{x}}(k+1 \mid k), k+1]\}\right]$
Initial conditions	$\hat{\mathbf{x}}(0) = \boldsymbol{\mu}_{\mathbf{x}_0}$, $\mathbf{V}_{\tilde{\mathbf{x}}}(0) = \mathbf{V}_{\mathbf{x}_0}$

However, in addition, we also have the observation

$$\mathbf{z}_2(k) = \mathbf{H}(k)\,\mathbf{w}(k) + \mathbf{v}_2(k)$$

Here, we assume that $\mathbf{w}(k)$, $\mathbf{v}_1(k)$, and $\mathbf{v}_2(k)$ are all independent zero-mean, discrete white sequences with variance matrices $\mathbf{V}_\mathbf{w}(k)$, $\mathbf{V}_{\mathbf{v}_1}(k)$, and $\mathbf{V}_{\mathbf{v}_2}(k)$, respectively. The two observations can be modeled as the single observation

$$\mathbf{z}(k) = \mathbf{h}^*[\mathbf{x}(k), k] + \mathbf{v}^*(k)$$

where

$$\mathbf{z}(k) = \begin{bmatrix} \mathbf{z}_1(k) \\ \hline \mathbf{z}_2(k) \end{bmatrix}, \quad \mathbf{h}^*[\mathbf{x}(k), k] = \begin{bmatrix} \mathbf{h}[\mathbf{x}(k), k] \\ \hline \mathbf{0} \end{bmatrix}, \quad \mathbf{v}^*(k) = \begin{bmatrix} \mathbf{v}_1(k) \\ \hline \mathbf{H}(k)\,\mathbf{w}(k) + \mathbf{v}_2(k) \end{bmatrix}$$

Note that now the input and observation noises are correlated. An application of the algorithm of Table 7.3-2 leads to the following equations for the estimates:

$$\hat{\mathbf{x}}(k+1) = \hat{\mathbf{x}}(k+1 \mid k) + \mathbf{K}_1(k+1)\{\mathbf{z}_1(k+1) - \mathbf{h}[\hat{\mathbf{x}}(k), k]\}$$

$$\hat{\mathbf{x}}(k+1 \mid k) = \boldsymbol{\phi}[\hat{\mathbf{x}}(k), k] + \mathbf{K}_2(k)\,\mathbf{z}_2(k)$$

where

$$\mathbf{K}_1(k+1) = \mathbf{P}(k+1)\,\frac{\partial \mathbf{h}^T[\hat{\mathbf{x}}(k+1 \mid k), k+1]}{\partial \hat{\mathbf{x}}(k+1 \mid k)}\,\mathbf{V}_{\mathbf{v}_1}^{-1}(k+1)$$

$$\mathbf{K}_2(k) = \boldsymbol{\Gamma}(k)\,\mathbf{V}_\mathbf{w}(k)\,\mathbf{H}^T(k)[\mathbf{V}_{\mathbf{v}_2}(k) + \boldsymbol{\Gamma}(k)\,\mathbf{V}_\mathbf{w}(k)\,\boldsymbol{\Gamma}^T(k)]^{-1}$$

and

$$\begin{aligned} \mathbf{P}(k+1 \mid k) = {} & \frac{\partial \boldsymbol{\phi}[\hat{\mathbf{x}}(k), k]}{\partial \hat{\mathbf{x}}(k)}\,\mathbf{P}(k)\,\frac{\partial \boldsymbol{\phi}^T[\hat{\mathbf{x}}(k), k]}{\partial \hat{\mathbf{x}}(k)} + \boldsymbol{\Gamma}(k)\,\mathbf{V}_\mathbf{w}(k)\,\boldsymbol{\Gamma}^T(k) \\ & - \boldsymbol{\Gamma}(k)\,\mathbf{V}_\mathbf{w}(k)\,\mathbf{H}^T(k)[\mathbf{H}(k)\,\mathbf{V}_\mathbf{w}(k)\,\mathbf{H}^T(k) + \mathbf{V}_{\mathbf{v}_2}(k)]^{-1} \\ & \times \mathbf{H}(k)\,\mathbf{V}_\mathbf{w}(k)\,\boldsymbol{\Gamma}^T(k) \end{aligned}$$

$$\begin{aligned} \mathbf{P}(k+1) = {} & \mathbf{P}(k+1 \mid k) - \mathbf{P}(k+1 \mid k)\,\mathbf{R}^T(k+1)[\mathbf{R}(k+1)\,\mathbf{V}\text{-}(k+1 \mid k) \\ & \times \mathbf{R}^T(k+1) + \mathbf{V}_{\mathbf{v}_1}(k+1)]^{-1}\,\mathbf{R}(k+1)\,\mathbf{P}(k+1 \mid k) \end{aligned}$$

while

$$\mathbf{R}^{\mathrm{T}}(k+1)\,\mathbf{V}_{\mathbf{v}_1}^{-1}(k+1)\,\mathbf{R}(k+1)$$

$$= \frac{-\partial}{\partial\hat{\mathbf{x}}(k+1 \mid k)}\left[\frac{\partial\mathbf{h}^{\mathrm{T}}[\hat{\mathbf{x}}(k+1 \mid k), k+1]}{\partial\hat{\mathbf{x}}(k+1 \mid k)}\right.$$

$$\left.\times\,\mathbf{V}_{\mathbf{v}_1}^{-1}(k+1)\{\mathbf{z}_1(k+1) - \mathbf{h}[\hat{\mathbf{x}}(k+1 \mid k), k+1]\}\right]$$

These identification algorithms are applied in order to identify parameters a and c in the stochastic linear system

$$x(k+1) = (1 + 0.01a)\,x(k) + 0.015w(k)$$

$$z_1(k) = cx(k) + v_1(k)$$

$$z_2(k) = w(k) + v_2(k)$$

The true values of a and c are -0.5 and 0.8. Fixed values are

$$V_{v_1} = 25, \qquad V_w = 100 \qquad \mathrm{var}\begin{Bmatrix} x(0) \\ a(0) \\ c(0) \end{Bmatrix} = \begin{bmatrix} 1 & 0 & 0 \\ 0 & 1 & 0 \\ 0 & 0 & 1 \end{bmatrix}$$

$$\hat{a}(0) = 1, \qquad \hat{c}(0) = 5$$

The algorithms of Table 7.3-2 were processed for three different values of var$\{v_2\}$: 16, 49, and 400. Figures 7.3–2 and 7.3–3 illustrate the generally excellent results. As expected, the identification is "better" for lower values of V_{v_2}.

Example 7.3-3. One can pose the problem of identification of the constant scalar plant noise variance in a linear system as the problem of identifying a for the message and observation model

$$\begin{aligned} \mathbf{x}(k+1) &= \boldsymbol{\Phi}\mathbf{x}(k) + \boldsymbol{\Gamma} a^{1/2}(k) w(k) \\ \mathbf{z}(k) &= \mathbf{H}\mathbf{x}(k) + \mathbf{v}(k) \\ a(k+1) &= a(k) \end{aligned}$$

where the mean and variance of $w(k)$ is assumed to be 0 and 1. The quantity $a(k)$ represents the unknown variance of $w(k)$ and the relation $a(k+1) = a(k)$ is used to constrain a to be constant.

Use of Table 7.3-1 leads to results that are unfortunate in that one would normally set $\mathbf{P}_{a\mathbf{x}}(0) = \mathbf{0}$ and this constrains $\mathbf{P}_{a\mathbf{x}}(k)$ to be $\mathbf{0}$ for all k. The expression for the a estimator is

$$\hat{a}(k+1) = \hat{a}(k) + \mathbf{P}_{a\mathbf{x}}(k+1)\mathbf{H}^{\mathrm{T}}[\mathbf{z}(k+1) - \mathbf{H}\hat{\mathbf{x}}(k+1 \mid k)]$$

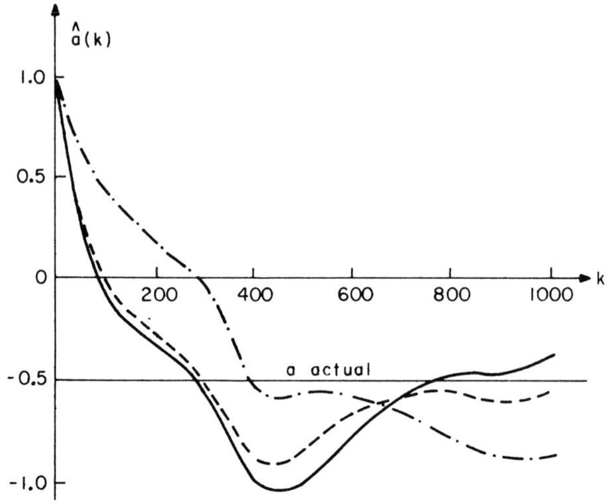

FIG. 7.3-2. Identification of *c*. ——— $V_{v2} = 16$; – – – $V_{v2} = 49$; – · – · – $V_{v2} = 400$.

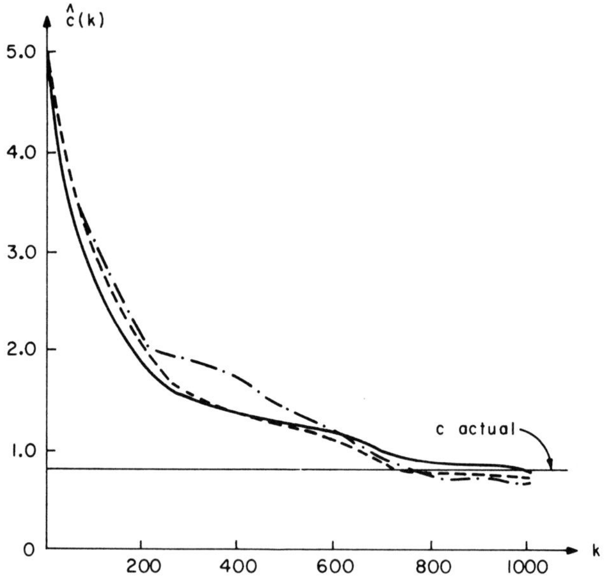

FIG. 7.3-3. Identification of *a*. ——— $V_{v2} = 16$; – – – $V_{v2} = 49$; – · – · – $V_{v2} = 400$.

and consequently the algorithm does not converge to the correct value of a. However, if one models the problem by letting $a^{-1/2}(k)\mathbf{x}(k) = \mathbf{y}(k)$ then the message and observation model becomes

$$\begin{aligned} \mathbf{y}(k+1) &= \mathbf{\Phi}\mathbf{y}(k) + \mathbf{\Gamma}w(k) \\ \mathbf{z}(k) &= \mathbf{H}a^{1/2}(k)\mathbf{y}(k) + \mathbf{v}(k) \\ a(k+1) &= a(k) \end{aligned}$$

and the algorithms for this model perform nicely. Thus we reach the important conclusion that the method of formulation of the model for a system identification problem is of considerable importance, especially if approximations are used to develop the identification algorithms.

7.4. Summary

In this chapter, the method of invariant imbedding has been developed and applied to some of the two-point boundary value problems of system identification from Chap. 3. The result is a set of sequential algorithms for state and parameter estimation. Several examples have been considered to illustrate the theory and indicate the many extensions of the material which is possible.

There are many methods of system identifications, both off line and on line, which have not been considered in this brief monograph. Emphasis has been placed on the methods with which the authors have had actual computational experience and/or on methods which have gained wide acceptance in practice. The authors apologize to the numerous authors whose works have not been discussed here; many, but by no means all, of the articles on system identification are listed in the bibliography. It is hoped that the reader who is interested in examining techniques other than those presented here will find that the bibliography is a helpful source of information.

The field of system identification is a rapidly expanding one with many special and general theoretical papers as well as numerous applications being published, literally daily. It is entirely possible, and in fact likely, that the key to the system identification problem resides in one of the works not discussed here or in some work as yet unpublished. However, it is the authors' feelings that the methods developed in this monograph represent those with the most general promise for the solution of system identification problems at this time.

BIBLIOGRAPHY

A complete bibliography would require a text at least the size of this one in order to detail it. We have chosen to present a list of what we believe are the more pertinent papers related to a study of the system identification topics presented in this text. In order to relate the entries in the bibliography to our foregoing discussions, we have keyed the bibliography to the chapter or section in this text to which it most closely corresponds. General references are keyed to Chapter 1, references concerned with classical methods of system identification are keyed to the various sections of Chapter 2, and so forth. We apologize to the authors of many fine system identification papers whose papers are not included here.

Acker, W. F., and Sage, A. P. (1967). Identification and control of systems with striction and coulomb friction, ISA Preprint P16-1-PHYMMID-67, September 1967. [1]

Albert, A. E., and Gardner, L. A. (1967). "Stochastic Approximation and Nonlinear Regression." M.I.T. Press, Cambridge, Massachusetts. [5]

Aoki, M., and Staley, R. M. (1970). On input signal synthesis in parameter identification, *Proc. 4th IFAC Congr., Warsaw, 1969*; *Automatica* **6**, 431–440. [2]

Aoki, M., and Yue, P. C. (1970). On certain convergence questions in system identification, *SIAM J. Control* **8**, 239–250. [2]

Åström, K. J., and Bohlin, T. (1966). Numerical identification of linear dynamic systems from normal operating records, *in* "Theory of Self-Adaptive Control Systems" (Hammond, P. H., ed.). Plenum Press, New York. [4]

Åström, M., and Eykhoff, P. (1971). System identification, *Automatica* **7**, March 1971. [1]

Balakrishnan, A. V. (1968). A new computing technique in system identification, *J. Comput. System Sci.* **2**, 102–116. [1]

Balakrishnan, A. V. (1968a). Stochastic system identification techniques, *Proc. Advanced Seminar Stochastic Optimization and Control, 2–4 October, 1967, Madison, Wisconsin.* Wiley, New York. [1, 5]

Balakrishnan, A. V., and Lions, J. L. (1967). State estimation for infinite-dimensional systems, *J. Comput. System Sci.* **1**, 391–403. [1]

Balakrishnan, A. V., and Neustadt, L. W., eds. (1964). "Computing Methods in Optimization Problems." Academic Press, New York. [1, 4, 6, 7]

Balakrishnan, A. V., and Peterka, V. (1969). Identification in automatic control systems, *Automatica* **5**, 817–829. [1]

Bekey, G. A., and Karplus, W. J. (1968). "Hybrid Computation." Wiley, New York. [1, 2, 4]

Bellman, R. E. (1961). "Adaptive Control Processes, A Guided Tour." Princeton Univ. Press, Princeton, New Jersey. [1]

Bellman, R. E. (1965). Dynamic programming, system identification, and suboptimization, RAND Corp., RM-4593-PR, June 1965. [1]

Bellman, R. E., Kagiwada, H. H., Kalaba, R. E., and Sridhar, R. (1964). Invariant imbedding and nonlinear filtering theory, RAND Corp., RM-4374-PR, December 1964. [7.3]

Bellman, R. E., and Kalaba, R. E. (1965). "Quasilinearization and Nonlinear Boundary Value Problems." Elsevier, Amsterdam. [6.2]

Bellman, R. E., Kagiwada, H. H., and Kalaba, R. E. (1965). Identification of linear systems via numerical inversion of Laplace transforms, *IEEE Trans. Automatic Control* **AT-10**, 111–112. [6.2]

Blandhol, E., and Balchen, J. G. (1963). Determination of system dynamics by use of adjustable models, *Proc. 2nd IFAC Cong., Basil, Switzerland, 1963.* [2.5]

Blum, J. (1954). Multidimensional stochastic approximation procedures, *Ann. Math. Statist.* **25**, 737–744. [5]

Boxer, R. (1957). A note on numerical transform calculus, *Proc. IRE* **45**, 1401–1406. [2.2]

Boxer, R., and Thaler, S. (1956). A simplified method of solving linear and nonlinear systems, *Proc. IRE* **44**, 89–101. [2.2]

Breakwell, J. V., Speyer, J. L., and Bryson, A. E. (1963). Optimization and control of nonlinear systems using the second variation, *J. SIAM Control Ser. A* **1**, 193–223. [4.3]

Bryson, A. E., and Denham, W. F. (1962). A steepest ascent method for solving optimum programming problems, *J. Appl. Mech.* **29**, 247–257. [4.3]

Bryson, A. E., and Ho, Y. C. (1969). "Applied Optimal Control." Ginn (Blaisdell), Boston. [1, 4, 6, 7]

Bullock, T. E., and Franklin, G. F. (1967). A Second-order feedback method for optimal control computations, *IEEE Trans. Automatic Control* **AC-12**, 666–673. [4.3]

Comer, J. P. (1964). Some stochastic approximation procedures for use in process control, *Ann. Math. Statist.* **35**, 1136–1146. [5]

Cox, H. (1964). On the estimation of state variables and parameters for nonlinear dynamic systems, *IEEE Trans. Automatic Control* **AC-9**, 5–12. [7.4]

Cuenod, M., and Sage, A. P. (1968). Comparison of some methods used for process identification, *Automatica* **4**, 235–269. [1]

Denham, W. F., and Bryson, A. E. (1964). Optimal programming problems with inequality constraints II: Solution by steepest-ascent, *AIAA J.* **2**, 25–34. [4.2]

Detchmendy, D. M., and Sridhar, R. (1965). On the experimental determination of the dynamical characteristics of physical systems, *Proc. Nat. Electron. Conf.* **21**, 575–580. [6.2]

Detchmendy, D. M., and Sridhar, R. (1966). Sequential estimation of state and parameters in noisy non-linear dynamical systems, *J. Basic Eng. Ser. D* **88**, 362–366. [7.3]

Dupac, V. (1965). A dynamic stochastic approximation method, *Ann. Math. Statist.* **36**, 1695–1702. [5]

Dvoretzky, A. (1956). On stochastic approximation, *Proc. 3rd Berkeley Symp. Math. Statist. and Prob.* (J. Neyman, ed.), pp. 39-55. Univ. of California Press, Berkeley, California. [5]

Elliott, D. F., and Sworder, D. D. (1969). Applications of a simplified multidimensional stochastic approximation algorithm, *Proc. Joint Automatic Control Conf.*, 148–154. [5]

Ellis, T. W., and Sage, A. P. (1968). Application of a method for on-line combined estimation and control, *Proc. Southwest IEEE Conf., April 1968.* [7.3]

Eveleigh, V. W. (1967). "Adaptive Control and Optimization Techniques." McGraw-Hill, New York. [2]

Eykhoff, P. (1963). Some fundamental aspects of process-parameter estimation, *IEEE Trans. Automatic Control* **AC-8**, 347–357. [3]

Eykhoff, P. (1968). Process parameter and state estimation, *Automatica* **4**, 205–233. [1]

Eykhoff, P. (1971). "System Parameter and State Estimation." Wiley, New York. [1]

Friedland, B. (1969). Treatment of bias in recursive filtering, *IEEE Trans. Automatic Control* **AC-14**, 359–367. [7.2]

Fletcher, R., and Powell, M. J. D. (1963). A rapid descent method for minimization, *Comput. J.* **6**, 163–168. [4.2]

Giese, C., and McGhee, G. (1965). Estimation of nonlinear system states and parameters by regression methods, *Proc. Joint Automatic Control Conf., Troy, New York*, 46–53.

Goodman, T. P., and Reswick, J. B. (1956). Determination of system characteristics from normal operating records, *Trans. ASME* **78**, 259–271. [1]

Graupe, D., Swanick, B. H., and Cassir, G. R. (1968). Reduction and identification of multivariable processes using regression analysis, *IEEE Trans. Autommatic Control* **AC-13**, 564–567. [1]

Gray, K. B. (1964). The application of stochastic approximation for the optimization of random circuits, *Proc. Symp. Appl. Math.* **16**, 178–192. [4]

Henrici, P. (1962). "Discrete Variable Methods in Ordinary Differential Equations." Wiley, New York. [6]

Hilborn, C. G., and Lainiotis, D. G. (1969). Optimal estimation in the presence of unknown parameters, *IEEE Trans. Systems Sci. Cybernetics* **SSC-5**, 38–43. [1]

Ho, Y. C. (1962). Stochastic approximation method and optimal filter theory, *J. Math. Anal. Appl.* **6**, 152–154. [5.2]

Ho, Y. C., and Agrawala, A. K. (1968). On pattern classification algorithms: Introduction and survey, *Proc. IEEE* **56**, 2101–2176. [1]

Ho, Y. C., and Bryson, A. E. (1969). "Applied Optimal Control." Ginn (Blaisdell), Boston. [1]

Ho. Y. C., and Lee, R. C. K. (1965). Identification of linear dynamic systems, *Proc. 1965 Nat. Electron. Conf.*, pp. 647–651. [5.3]

Ho, Y. C., and Lee, R. C. K. (1965a). Identification of linear dynamic systems, *Inform. Control* **8**, 93–110. [5.3]

Ho, Y. C., and Newbold, P. M. (1967). A descent algorithm for constrained stochastic extrema, *J. Optimization Theory Appl.* **1**, 215–231. [5.3]

Ho, Y. C., and Whalen, B. H. (1963). An approach to the identification and control of linear dynamic systems with unknown parameters, *IEEE Trans. Automatic Control* **AC-8**, 255–256. [1]

Holmes, J. K. (1968). System identification from noise corrupted measurements, *J. Optimization Theory Appl.* **2**, 102–116. [5]

Holmes, J. K. (1969). Two stochastic approximation procedures for identifying linear systems, *IEEE Trans. Automatic Control* **AC-14**, 292–295. [5]

Hsia, T. C., and Landgrebe, D. A. (1967). On a method for estimating power spectra, *IEEE Trans. Instr. Meas.* **IM-16**, 255–257. [3.4]

Hsia, T. C., and Vimolvanich, V. (1969). An on line technique for system identification, *IEEE Trans. Automatic Control*, **AC-14**, 92–96. [2]

Identification in automatic control systems, (1967). *Preprints of the IFAC Symposiums*, Parts I and II, Prague, Czechoslovakia, June, 12–17, 1967; *also* Prague, Czechoslovakia, June, 1970. [1]

Jazwinski, A. H. (1969). Adaptive filtering, *Automatica* **5**, 475–485. [7]

Jazwinski, A. H. (1970). "Stochastic Processes and Filtering Theory." Academic Press, New York. [1,7]

Joseph, P., Lewis, L., and Tou, J. (1961). Plant identification in the presence of disturbances and application to digital adaptive systems, *Trans. AIEE* **80**, Part II *(Appl. and Ind.)*, 18–24. [1]

Kalaba, R. E. (1959). On nonlinear differential equations, the maximum operation, and monotone convergence, *J. Math. Mech.* **8**, 519–574. [6.2]

Kalaba, R. E. (1963). Some aspects of quasilinearization, *in* "Nonlinear Differential Equations and Nonlinear Mechanics" (J. P. LaSalle and S. Lefschetz, eds.). Academic Press, New York. [6.2]

Kalman, R. E., and Bucy, R. (1961). New results in linear filtering and prediction theory, *J. Basic Eng.* **83**, 95–108. [1]

Kashyap, R. L. (1970). Maximum likelihood identification of stochastic linear systems, *IEEE Trans. Automatic Control* **AC-15**, 25–34. [3]

Kashyap, R. L. (1970a). A new method of recursive estimation in discrete linear systems, *IEEE Trans. Automatic Control* **AC-15**, 34–43. [2]

Kelley, H. J. (1960). Gradient theory of optimal flight paths, *ARS J.* **30**, 947–954. [4.3]

Kelley, H. J. (1962). Guidance theory and extremal fields, *IRE Trans. Automatic Control* **AC-7**, 75–81. [4.3]

Kelley, H. J. (1962a). Method of gradients, *in* "Optimization Techniques" (G. Leitman, ed.), Chapter 6. Academic Press, New York. [4.3]

Kelley, H. J., Kopp, R. E., and Moyer, H. G. (1964). A trajectory optimization technique based upon the theory of the second variation, *Progr. Astronaut. Aeronaut.* **14**, 559–582. [4.3]

Kenneth, P., and McGill, R. (1966). Two point boundary value problem techniques, *Advan. Contr. Syst.* **3**, 69–110. [4,6]

Kenneth, P., and Taylor, G. E. (1966). Solution of variational problems with bounded control variables by means of the generalized Newton–Raphson method, *in* "Recent Advances in Optimization Techniques" (A. Lavi, ed.). Wiley, New York. [6.2]

Kesten, H. (1958). Accelerated stochastic approximation, *Ann. Math. Statist.* **29**, 111. [5]

Kiefer, J., and Wolfowitz, J. (1952). Statistical estimation of the maximum of a regression function, *Ann. Math. Statist.* **23**, 462–466. [5]

Kirvaitis, K., and Fu, K. S. (1966). Identification of nonlinear systems by stochastic approximation, *Proc. Joint Automatic Control Conf., Seattle, June, 1966.* [5.2]

Kopp, R. E., and McGill, R. (1964). Several trajectory optimization techniques; Part I: Discussion, *in* "Computing Methods in Optimization Problems" (L. W. Neustadt and A. V. Balakrishnan, eds.), pp. 65–89. Academic Press, New York. [6.2, 7.3]

Kopp, R. E., and Orford, R. J. (1963). Linear regression applied to system identification for adaptive control systems, *AIAA J.* **5**, 2300–2306. [7.4]

Kroy, W. H., and Stubberud, A. R. (1967). Identification via nonlinear filtering, *Int. J. Contr.* **6**, 499–522, 1967; *Proc. Joint Automatic Control Conf. Ann Arbor, Michigan, June 1968*, pp. 394–412. [7.4]

Kumar, K. S. P., and Sridhar, R. (1964). On the identification of control systems by the quasilinearization method, *IEEE Trans. Automatic Control* **AC-9**, 151–154. [6.2]

Kumar, K. S. P., and Sridhar, R. (1964a). On the identification of linear systems, *1964 JACC Preprints*, Stanford Univ. Press, Stanford, California, pp. 361–365, June 1964. [1]

Kushner, H. (1963). A simple iterative procedure for the identification of the unknown parameters of a linear time-varying discrete system, *J. Basic Eng. Ser. D*, **85**, 227–235. [5.3]

Kushner, H. (1963a). Hill climbing methods for the optimization of multiparameter noise disturbed systems, *J. Basic Eng. Ser. D*, **85**, 157–164. [5.3]

Kushner, H. (1965). On stochastic extremum problems: Calculus, *J. Math. Anal. Appl.* **10**, 354–367. [5]

Lasdon, L. S., Mitter, S. K., and Waren, A. D. (1967). The conjugate gradient method for optimal control problems, *IEEE Trans. Automatic Control* **AC-12**, 132–138. [4.3]

Lavi, A., and Strauss, J. C. (1965). Parameter identification in continuous dynamic systems, *IEEE Int. Conv. Rec. Symp. Automatic Control; Systems Sci., Cybernetics; Human Factors, Part VI*, pp. 49–61. [6.2]

Lee, R. C. K. (1964). "Optimal Estimation Identification and Control." M.I.T. Press, Cambridge, Massachusetts. [1,5.3]

Lee, R. C. K., and Ho, Y. C. (1965). Identification of linear dynamic systems, *Inform. Control* **8**, 93–110. [5.3]

Lee, Y. W., and Schetzen, M. (1965). Some aspects of the Wiener theory of nonlinear systems, *Proc. Nat. Electron. Conf.* **21**, 759–764. [2.5]

Levin, M. J. (1964). Optimum estimation of impulse response in the presence of noise, *IRE Trans. Circuit Theory* **CT-7**, 50–56. [2.2]

Lichtenberger, W. W. (1961). A technique of linear system identification using correlating filters, *IRE Trans. Automatic Control* **AC-6**, 183–199. [2.4]

Lindenlaub, J. C., and Cooper, G. R. (1963). Noise limitations of system identification techniques, *IEEE Trans. Automatic Control* **AC-8**, 43–48. [2.4]

Loginov, N. V. (1966). Methods of stochastic approximation, *Automat. Remote Contr.* (*USSR*) **27**, 706–728. [5]

Margolis, M., and Leondes, C. T. (1959). A parameter tracking series for adaptive control systems, *IRE Trans. Automatic Control* **AC-4**, 100–111. [2.5]

McLendon, J. T., and Sage, A. P. (1967). Computational algorithms for discrete detection and likelihood ratio computation, *Inform. Sci.* **2**, 273–298. [3]

McReynolds, S. R. (1967). The successive sweep method and dynamic programming, *J. Math. Anal. Appl.* **19**, 565–598. [4.3]

McReynolds, S. R., and Bryson, A. E. (1965). A successive sweep method for solving optimal programming problems, *Proc. Joint Automatic Control Conf., Troy, New York, August 1965*, pp. 551–555. [4.3]

Mehra, R. K. (1969). On the identification of variances and adaptive Kalman filtering, *Proc. Joint Automatic Control Conf., August 1969*, pp. 494–505. [3.4,7]

Mehra, R. K. (1969a). Identification of stochastic linear dynamic systems, *Proc. 8th Symp. Adaptive Processes*, pp. 6-*f*-1–6-*f*-5, November 1969. [4.3]

Meisel, W. S. (1968). Least-square methods in abstract pattern recognition, *Inform. Sci.* **1**, 43–54. [1]

Meissinger, H. F., and Bekey, G. A. (1966). An analysis of continuous parameter identification methods, *Simulations* **6**, 95–102. [2]

Melsa, J. L., and Schultz, D. G. (1969). "Linear Control Systems." McGraw-Hill, New York. [2,4]

Mendel, J. M. (1968). Gradient, error-correction identification algorithms, *Inform. Sci.* **1**, 23–42. [4.2]

Merriam, C. W. (1964). "Optimization Theory and the Design of Feedback Control Systems." McGraw-Hill, New York. [1]

Mishkin, E., and Braun, L. (1961). "Adaptive Control Systems." McGraw-Hill, New York. [1,2]

Moyer, H. G., and Pinkham, G. (1964). Several trajectory optimization techniques; Part II: Application, *in* "Computing Methods in Optimization Problems" (A. V. Balakrishnan and L. W. Neustadt, eds.), pp. 91–105. Academic Press, New York. [4.3, 6.2]

Oza, K. G., and Juy, E. I. (1970). Adaptive algorithms for identification problem, *Automatica* **6**, 795–820. [1]

Panuska, V. (1968). A stochastic approximation method for identification of linear systems using adaptive filtering, *Proc. Joint Automatic Control Conf., Ann Arbor, Michigan, June 1968*, pp. 1014–1021. [5.3]

Powell, M. J. D. (1964). An efficient method for finding the minimum of a function of several variables without calculating derivatives, *Comput. J.* **7**, 155–162. [4.2]

Rauch, H. E., Tung, F., and Striebel, C. T. (1965). Maximum likelihood estimates of linear dynamic systems, *AIAA J.* **3**, 1445–1450. [1,3.3]

Ricker, D. W., and Saridis, G. N. (1968). Analog methods for on-line system identification using noisy measurements, *Simulation* **8**, 241–248. [1]

Robbins, H., and Monro, S. (1957). A stochastic approximation method, *Ann. Math. Statist.* **22**, 400–407. [4]

Rogers, A. E., and Steiglitz, K. (1967). Maximum likelihood estimation of rational transfer function parameters, *IEEE Trans. Automatic Control* **AC-14**, 594–597. [3.4,3.2]

Rosen, J. B. (1960). The gradient projection method for nonlinear programming, Part I, Linear constraints, *J. Soc. Indust. Appl. Math.* **8**, 181–217. [4.2]

Rosen, J. B. (1961). The gradient projection method for nonlinear programming, Part II, Nonlinear constraints, *J. Soc. Indust. Appl. Math.* **9**, 514–532. [4.2]

Sage, A. P. (1968). "Optimum Systems Control." Prentice-Hall, Englewood Cliffs, New Jersey. [1,4,6,7]

Sage, A. P. (1970). Maximum likelihood estimation of bias in sequential estimation, *Proc. Univ. Missouri at Rolla, Mervin J. Kelly Communications Conference, October 1970.* [3,7]

Sage, A. P., and Burt, R. W. (1965). Optimum design and error analysis of digital integrators for discrete system simulation, *A.F.I.P.S. Proc. FJCC* **27**, *Las Vegas, Nevada*, pp. 903–914.

Sage, A. P., and Choate, W. C. (1965). Minimum time identification of non-stationary dynamic processes, *Proc. Nat. Electron. Conf.* **21**, 587–592. [2.4]

Sage, A. P., and Eisenberg, B. R., (1965). Experiments in nonlinear and nonstationary system identification via quasilinearization and differential approximation, *Proc. Joint Automatic Control Conf.*, pp. 522–530. [6.2]

Sage, A. P., and Ellis, T. W. (1966). Sequential suboptimal adaptive control of nonlinear systems, *Proc. Nat. Electron. Conf.* **22**, 692–697. [7.3]

Sage, A. P., and Husa, G. W. (1969). Adaptive filtering with unknown prior statistics, *Proc. 1969 Joint Automatic Control Conf.* [3.3]

Sage, A. P., and Husa, G. W. (1969a). Algorithms for sequential adaptive estimation of prior statistics, *Proc. 1969 Symp. Adaptive Processes.* [3]

Sage, A. P., and Lin, J. L. (1971). Algorithms for continuous sequential maximum likelihood bias estimation and associated error analysis, *Inform. Sci.* **3** (to appear). [7]

Sage, A. P., and Masters, G. W. (1966). On-line estimation of states and parameters for discrete non-linear dynamic systems, *Proc. Nat. Electron. Conf.* **22**, 677–682. [7.2]

Sage, A. P., and Masters, G. W. (1967). Identification and modeling of states and parameters of nuclear reactor systems, *IEEE Trans. Nucl. Sci.*, pp. 279–285, February 1967. [7.2]

Sage, A. P., and Melsa, J. L. (1971). "Estimation Theory with Application to Communications and Control." McGraw-Hill, New York. [1,3,7]

Sage, A. P., and Smith, S. L. (1966). Real-time digital simulation for systems control, *Proc. IEEE* **54**, 1802–1812. [6.3]

Sage, A. P., and Wakefield, C. D. (1970). System identification with noise corrupted input observation. *SWIEEECO Record Technical Papers*, pp. 333–337, April 22–24, 1970, Dallas, Texas. [7.3]

Sakrison, D. J. (1966). Stochastic Approximation, *Advan. Commun. Syst.* **2**, 51–106.

Sakrison, D. J. (1967). The use of stochastic approximation to solve the system identification problem, *IEEE Trans. Automatic Control* **AC-12**, 563–567. [5.3]

Saridis, G. N., and Stein, G. (1968). Stochastic approximation algorithms for linear discrete-time system identification, *IEEE Trans. Automatic Control* **AC-13**, 515–523. [5.3]

Saridis, G. N., Nikolic, Z. J., and Fu, K. S. (1969). Stochastic approximation algorithms for system identification, estimation and decomposition of mixtures, *IEEE Trans. Syst. Sci. Cybernetics* **SSC-5**, 8–15. [5.3]

Schley, C. H., Jr., and Lee, I. (1967). Optimal control computation by the Newton–Raphson method and the Riccati transformation, *IEEE Trans. Automatic Control* **AC-12**, 139–144. [6.2]

Schultz, E. R. (1968). Estimation of pulse transfer function parameters by quasilinearization, *IEEE Trans. Automatic Control* **AC-13**, 424–426. [6.3]

Shellenbarger, J. C. (1960). Estimation of covariance parameters for an adaptive Kalman filter, *Proc. Nat. Electron. Conf.*, pp. 698–702. [3.4]

Spang, H. A. (1962]. A review of minimization techniques for nonlinear functions, *SIAM Rev.* **4**, 343–365. [1]

Stancil, R. T. (1964). A new approach to steepest ascent trajectory optimization, *AIAA J.* **2**, 1365–1370. [4.3]

Steiglitz, K., and McBride, L. E. (1965). A technique for the identification of linear systems, *IEEE Trans. Automatic Control* **10**, 461–464. [4.2]

Sworder, D. D. (1966). A study of the relationship between identification and optimization in adaptive control problems, *J. Franklin Inst.* **281**, 198–213. [1]

Sylvester, R. J., and Meyer, F. (1965). Two point boundary value problems by quasilinearization, *J. Soc. Indus. Appl. Math.* **13**, 586–602. [6.2]

Tretter, S. A., and Steiglitz, K. (1967). Power spectrum identification in terms of rational models, *IEEE Trans. Automatic Control* **AC-12**, 185–188. [2.3]

Turin, G. L. (1957). On the estimation in the presence of noise of the impulse response of a random linear filter, *IRE Trans. Inform. Theory* **IT-3**, 5–10. [2.3]

Venter, J. H. (1967). An extension of the Robbins–Munro procedure, *Ann. Math. Statist.* **38**, 181–190. [5]

Widrow, B., Mantey, P. E., Griffiths, L. J., and Goode, B. B. (1967). Adaptive antenna systems, *Proc. IEEE* **55**, 2143–2159. [1]

Wilde, D. J. (1964). "Optimum Seeking Methods." Prentice-Hall, Englewood Cliffs, New Jersey. [1]

Wong, K. Y., and Polak, E. (1967). Identification of linear discrete time systems using the instrumental variable method, *IEEE Trans. Automatic Control* **AC-12**, 707–718. [2]

Yore, E. E., and Takahashi, Y. (1967). Identification of dynamic systems by digital computer modeling in state space, *J. Basic Eng.* **89**, 295–297. [1]

Zadeh, L. A. (1950). Frequency analysis of variable networks, *Proc. IRE* **38**, 291–299. [2.4]

Ziv, J. (1969). Some lower bounds on signal parameter estimation, *IEEE Trans. Inform. Theory* **IT-15**, 386–391. [1]

INDEX

Mathematics in Science and Engineering

A Series of Monographs and Textbooks

Edited by RICHARD BELLMAN, *University of Southern California*

1. T. Y. Thomas. Concepts from Tensor Analysis and Differential Geometry. Second Edition. 1965

2. T. Y. Thomas. Plastic Flow and Fracture in Solids. 1961

3. R. Aris. The Optimal Design of Chemical Reactors: A Study in Dynamic Programming. 1961

4. J. LaSalle and S. Lefschetz. Stability by by Liapunov's Direct Method with Applications. 1961

5. G. Leitmann (ed.). Optimization Techniques: With Applications to Aerospace Systems. 1962

6. R. Bellman and K. L. Cooke. Differential-Difference Equations. 1963

7. F. A. Haight. Mathematical Theories of Traffic Flow. 1963

8. F. V. Atkinson. Discrete and Continuous Boundary Problems. 1964

9. A. Jeffrey and T. Taniuti. Non-Linear Wave Propagation: With Applications to Physics and Magnetohydrodynamics. 1964

10. J. T. Tou. Optimum Design of Digital Control Systems. 1963.

11. H. Flanders. Differential Forms: With Applications to the Physical Sciences. 1963

12. S. M. Roberts. Dynamic Programming in Chemical Engineering and Process Control. 1964

13. S. Lefschetz. Stability of Nonlinear Control Systems. 1965

14. D. N. Chorafas. Systems and Simulation. 1965

15. A. A. Pervozvanskii. Random Processes in Nonlinear Control Systems. 1965

16. M. C. Pease, III. Methods of Matrix Algebra. 1965

17. V. E. Benes. Mathematical Theory of Connecting Networks and Telephone Traffic. 1965

18. W. F. Ames. Nonlinear Partial Differential Equations in Engineering. 1965

19. J. Aczel. Lectures on Functional Equations and Their Applications. 1966

20. R. E. Murphy. Adaptive Processes in Economic Systems. 1965

21. S. E. Dreyfus. Dynamic Programming and the Calculus of Variations. 1965

22. A. A. Fel'dbaum. Optimal Control Systems. 1965

23. A. Halanay. Differential Equations: Stability, Oscillations, Time Lags. 1966

24. M. N. Oguztoreli. Time-Lag Control Systems. 1966

25. D. Sworder. Optimal Adaptive Control Systems. 1966

26. M. Ash. Optimal Shutdown Control of Nuclear Reactors. 1966

27. D. N. Chorafas. Control System Functions and Programming Approaches (In Two Volumes). 1966

28. N. P. Erugin. Linear Systems of Ordinary Differential Equations. 1966

29. S. Marcus. Algebraic Linguistics; Analytical Models. 1967

30. A. M. Liapunov. Stability of Motion. 1966

31. G. Leitmann (ed.). Topics in Optimization. 1967

32. M. Aoki. Optimization of Stochastic Systems. 1967

33. H. J. Kushner. Stochastic Stability and control. 1967

34. M. Urabe. Nonlinear Autonomous Oscillations. 1967

35. F. Calogero. Variable Phase Approach to Potential Scattering. 1967

36. A. Kaufmann. Graphs, Dynamic Programming, and Finite Games. 1967

37. A. Kaufmann and R. Cruon. Dynamic Programming: Sequential Scientific Management. 1967

38. J. H. Ahlberg, E. N. Nilson, and J. L. Walsh. The Theory of Splines and Their Applications. 1967

39. Y. Sawaragi, Y. Sunahara, and T. Nakamizo. Statistical Decision Theory in Adaptive Control Systems. 1967

40. R. Bellman. Introduction to the Mathematical Theory of Control Processes, Volume I. 1967; Volume II. 1971 (Volume III in preparation)

41. E. S. Lee. Quasilinearization and Invariant Imbedding. 1968

42. W. Ames. Nonlinear Ordinary Differential Equations in Transport Processes. 1968

43. W. Miller, Jr. Lie Theory and Special Functions. 1968

44. P. B. Bailey, L. F. Shampine, and P. E. Waltman. Nonlinear Two Point Boundary Value Problems. 1968

45. Iu. P. Petrov. Variational Methods in Optimum Control Theory. 1968

46. O. A. Ladyzhenskaya and N. N. Ural'tseva. Linear and Quasilinear Elliptic Equations. 1968

47. A. Kaufmann and R. Faure. Introduction to Operations Research. 1968

48. C. A. Swanson. Comparison and Oscillation Theory of Linear Differential Equations. 1968

49. R. Hermann. Differential Geometry and the Calculus of Variations. 1968

50. N. K. Jaiswal. Priority Queues. 1968

51. H. Nikaido. Convex Structures and Economic Theory. 1968

52. K. S. Fu. Sequential Methods in Pattern Recognition and Machine Learning. 1968

53. Y. L. Luke. The Special Functions and Their Approximations (In Two Volumes). 1969

54. R. P. Gilbert. Function Theoretic Methods in Partial Differential Equations. 1969

55. V. Lakshmikantham and S. Leela. Differential and Integral Inequalities (In Two Volumes). 1969

56. S. H. Hermes and J. P. LaSalle. Functional Analysis and Time Optimal Control. 1969

57. M. Iri. Network Flow, Transportation, and Scheduling: Theory and Algorithms. 1969

58. A. Blaquiere, F. Gerard, and G. Leitmann. Quantitative and Qualitative Games. 1969

59. P. L. Falb and J. L. de Jong. Successive Approximation Methods in Control and Oscillation Theory. 1969

60. G. Rosen. Formulations of Classical and Quantum Dynamical Theory. 1969

61. R. Bellman. Methods of Nonlinear Analysis, Volume I. 1970

62. R. Bellman, K. L. Cooke, and J. A. Lockett. Algorithms, Graphs, and Computers. 1970

63. E. J. Beltrami. An Algorithmic Approach to Nonlinear Analysis and Optimization. 1970

64. A. H. Jazwinski. Stochastic Processes and Filtering Theory. 1970

65. P. Dyer and S. R. McReynolds. The Computation and Theory of Optimal Control. 1970

66. J. M. Mendel and K. S. Fu (eds.). Adaptive, Learning, and Pattern Recognition Systems: Theory and Applications. 1970

67. C. Derman. Finite State Markovian Decision Processes. 1970

68. M. Mesarovic, D. Macko, and Y. Takahara. Theory of Hierarchial Multilevel Systems. 1970

69. H. H. Happ. Diakoptics and Networks. 1971

70. Karl Astrom. Introduction to Stochastic Control Theory. 1970

71. G. A. Baker, Jr. and J. L. Gammel (eds.). The Padé Approximant in Theoretical Physics. 1970

72. C. Berge. Principles of Combinatorics. 1971

73. Ya. Z. Tsypkin. Adaptation and Learning in Automatic Systems. 1971

74. Leon Lapidus and John H. Seinfeld. Numerical Solution of Ordinary Differential Equations. 1971

75. L. Mirsky. Transversal Theory, 1971

76. Harold Greenberg. Integer Programming, 1971

77. E. Polak. Computational Methods in Optimization: A Unified Approach, 1971

78. Thomas G. Windeknecht. General Dynamical Processes: A Mathematical Introduction, 1971

79. M. A. Aiserman, L. A. Gusev, L. I. Rozonoer, I. M. Smirnova, and A. A. Tal'. Logic, Automata, and Algorithms, 1971

80. Andrew P. Sage and James L. Melsa. System Identification

In preparation

R. Boudarel, J. Delmas, and P. Guichet. Dynamic Programming and Its Application to Optimal Control

Alexander Weinstein and William Stenger. Methods of Intermediate Problems for Eigenvalues Theory and Ramifications